Science-Mart

Science-Mart

Privatizing American Science

PHILIP MIROWSKI

Harvard University Press
Cambridge, Massachusetts, and London, England | 2011

Printed in the United States of America

Library of Congress Cataloging-in-Publication Data

Mirowski, Philip, 1951-
Science-mart : privatizing American science / Philip Mirowski.
p. cm.
Includes bibliographical references and index.
ISBN 978-0-674-04646-7 (alk. paper)
1. Science—Economic aspects—United States. 2. Research—Economic aspects—United States. 3. Privatization—United States. 4. Science—United States—History—20th century. 5. Science—United States—History—21st century. I. Title.
Q127.U6M576 2011
338.973'06—dc22 2010038495

Contents

Science-Mart

1

Viridiana Jones and the Temple of Mammon

Or, Adventures in Neoliberal Science Studies

Meet Viridiana Jones

It's not easy making a living in the knowledge biz these days. Lately our heroine, the intrepid academic researcher Viridiana Jones, feels strung out between the Scylla of Disneyfication of higher education and the Charybdis of Free EnronPrise in securing a patron, any patron, to support her inquiries in an era of impending financial doom. Viridiana finds herself sometimes wistfully wondering what life might have been like if she had gone and gotten that law degree instead. She considers herself someone who keeps up with current events, but the news about her university these days just brings on a headache. Every visit to the department mailbox has turned into another occasion for heartburn. Viridiana used to enjoy reading the *Chronicle of Higher Education*, but now she just tosses it out. She sees a colleague's copy of the *Wall Street Journal* with the headline "Basic Research Loses Some Allure" (Clark and Rhoads 2009): too right, mate, she groans. The other day, she received a glossy flyer that said, "As a University of Phoenix instructor, you could be sharing your knowledge and skills with motivated adult students via the Internet. Because our web-based format is asynchronous, you can teach class at times and places that fit your schedule—without interrupting your full-time career."[1] Just prior to that, she got a missive from a publishing company threatening her with prosecution if she didn't remove one of her own journal articles from her local university Web page (Corbyn 2009). The memos from the administration at her home campus have hardly been more edifying. For instance, she cannot believe that someone would voluntarily want the Ken Lay Chair for the Study of Markets or accept Madoff grants for health research (Bernstein 2009), but in her sober moments she knows she can't revel in her own moral superiority. As for all those juvenile pipe dreams of serving mankind and speaking truth to power—well, the less said about them, the better.

Of late, Viridiana feels like a character trapped in a George Saunders short story. It's hard not to notice the theme park character of the modern university campus: intellectual crowd control at the intake gates, carny barkers flogging customized "majors," t-shirt and souvenir vendors, internship thrill rides promising accelerated plunges into employment, long queues for the most popular *son et lumière* entertainers, sports extravaganzas, science lite served up in postmodern special effects pavilions pitched toward crowds jaded by video games, package tours through the Tunnel of Love mislabeled as study semesters abroad, dorm/hotel package deals, binge drinking, outsourced functions to low-wage contractors, and academic convocations as choreographed as any performance in Tomorrowland. Sometimes Viridiana wonders for whom or for what her university really exists. But then a thought brings her up short: Is she beginning to sound like some creepy curmudgeon from another planet? Or maybe Marshall Sahlins (2009)? She looks at herself in the mirror and winces. So what if you have to coddle the customer a little? And anyway, who really cares about such wintertime discontents when there's an economic crisis brewing?

It has been commonplace in certain circles to bemoan the troubled relationship of science to the state, and by this, Viridiana doesn't mean the isolated hot-button issue of stem-cell research (not her field), but rather the demonstrated willingness by state organs to participate more directly in defining what would count as "high-quality research."[2] Of course, when the time comes to gather up the fruits of the projects they have funded, the patron of research has always enjoyed the option to take it or leave it; what seems different of late is that there exists a whole parallel universe of think tanks and shadowy "experts" having little to do with the kind of academic science Viridiana had been acclimatized to expect in her youth. The state has apparently become much more willing to dispense with internal peer quality controls, hastening to intervene in the early stages of dissemination of results, purchasing their preferred party line neatly packaged from some think tank, suppress or otherwise discourage that which is inconvenient or strays off-message, while cherry-picking whatever seems expedient to tout as proven knowledge. Whenever the truth is inconvenient, science patrons now seem inclined to shoot the messenger. Viridiana recalls reading a front-page article in the *New York Times* about NASA trying to silence one of its own on global warming; there are a thousand smaller acts of overt censorship that never make it into the newspapers, including one she has witnessed herself at her home institution.[3]

You could blame it all on the nanny state, but Viridiana has to suppress the darker misgiving that her own peers in the world of science aren't much better. She knows that money has always been needed to make science, but whoever anticipated that her colleagues would come to take it as axiomatic that science was just another way to make money? She picks up *American Scientist* off the department coffee table and flips to an article that suggests that the glitzy new way to fund science and garner public support is to have scientists float their research proposals on something like a stock exchange, with the affluent public placing bets on the kinds of theories they have gut instincts will pan out in the future (Schneider 2008). At first she suspects it must be a joke, but it is not April 1; then she sees Google was one of the sponsors pushing the idea. Who else? After all, isn't their mantra "Nobody is as smart as everybody?" (Shapin 2008b, 194).

Viridiana has always known in her bones that the pursuit of knowledge is wayward and easily deflected, potentially suffering all kinds of deformations and biases because of the way it is prosecuted, framed, generated, and conveyed. Those concerns have long been the province of academic disciplinary scrutiny, from philosophy to psychology to sociology of knowledge. Sometimes, over in the philosophy department, epistemology seems to have been discussed as if it were merely a matter of isolated solipsistic individuals hewing doggedly to the rules of deductive and inductive inference; yet a closer look always reveals that the "social" context has continually been situated at the core of many supposedly abstract epistemological disputes.[4] Indeed, the proper relationship of science to the state was sometimes deemed to be the most significant problem in coming to understand the conditions under which science could make progress. But Viridiana is loath to admit that she doesn't know or much care about all that, or indeed about politics in general. Of course she votes for Democrats in American elections and considers herself a liberal in the awkward American sense, calls herself a feminist, and used to ridicule George Bush the Lesser along with the rest of her colleagues, but the truth is she has never thought very long or hard about the implications of cultural, religious, or economic movements for her science, her university, or her future. She once heard a friend say that science should be more democratic, but she hasn't a clue what that would entail. Curiously for one so intelligent, she more or less subscribes to the sound bite that a conservative is a troglodyte who grunts: government bad, market good. Yet, it must be admitted, the *New York Times* has hardly been superior in that regard, in her experience.

At a dean's reception, Viridiana became acquainted with a couple of faculty members at her university who consider themselves representatives of a field called "science studies," which seemed to her like it ought to have some salient things to say about the climacteric she feels she is living through. Out of curiosity, she went to hear one or two of their more famous representatives at a conference held at the university—people like Bruno Latour, Steve Woolgar, Henry Etzkowitz, and Steve Shapin—but was distressed to find that when they weren't indulging in opaque jargon about "actants," "performativity," "constructivism versus essentialism," "triple helix," "Mode 1/Mode 2," and moral economies, they ended up sounding just like some of the more cynical deans she was forced to deal with in the course of her committee duties. Those savants seemed to suggest that shamelessly flogging yourself and your ideas was the pinnacle of strategic wisdom in science, and they tended to confuse "excellence" (whatever that was) with the crudest sorts of proxy measures for scientific output.[5] In her funk, Viridiana could barely tell them apart from some of her business school colleagues at the university who kept extolling the "marketplace of ideas" to the skies. Viridiana, by now inured to disappointment, chalked it all up to the pernicious effects of postmodernism.[6]

Consequently, Viridiana suspects that contemporary scholars have had little useful to say about all these changes that nag at her. But her policy of benign neglect doesn't stop there. She knows hardly anyone cares anymore about the trials and tribulations of college professors, disdained as a pampered and privileged bunch by hoi polloi. Maybe we are part of the problem, she mopes. Nonetheless, there is a needling suspicion that the changes perturbing Viridiana might also have consequences for the wider world: What happens to the average citizen when knowledge becomes reengineered, appropriated, and shrink-wrapped under the new regime? And to whom will they turn to when they want a dissenting analysis? (Okay—that possibility is pretty remote.) Another think-tank doppelganger wielding a cardboard "opposing" position in a staged faux debate? Viridiana is sick to death of pundits sneering at academic tenure as the last refuge of lazy charlatans; she notes no one ever says that about lawyers or accountants when they make partner in their firms, or about upper management with their golden parachutes, or about the bailed-out New York bankers or the occupants of sinecures at the American Enterprise Institute. But, conflicted, she simultaneously feels guilty about her own tenure. She knows her department has kept its costs down by using PhD temp labor to an ever-increasing degree; however, her colleagues tend to avoid broaching the subject, the same way they evade talking about their Salvadorian nannies.

Prophets of progress once exclaimed that we were living in a shiny new information society and that the Internet was going to democratize everything under the sun because information just wants to be free; the economic crisis of 2008 seems to have muffled a lot of that boosterism. Now Viridiana can hardly be bothered to glance at blogs she used to enjoy. How did something that once bore so much promise become so trivial and clogged with noise? Viridiana, a natural scientist, harbors a soft spot for technological determinism as a force for progress; but in her gut she feels that the "information economy" resembles yet another facet of the regime that has Disneyfied her university and made it harder to initiate and conduct serious long-term fundamental research. After the Hwang Woo-Suk debacle (involving fabricated research results), there was a brief tendency to question the quality and motives of some of the most august science journals; Viridiana herself knows of a case of ghost authorship, but has no idea how prevalent it is.[7] And what were those news reports concerning the American Chemical Society suppressing investigative reporting about the chemical industry and lobbying against open-source journals?[8] Every time she has had to fill out a ten-page materials transfer agreement form just to get a reagent from a friend at MIT, her faith in progress flags a little bit more. If the Internet has been a force for liberation, Viridiana has missed out on it. To her, it seems most people have become much more vulnerable to information manipulation in the last decade—witness the run-up to the Iraq war, the bipolar swings of opinion about medical research, or the notion bandied about in the presidential election of 2008 that the United States can drill its way out of an oil crisis. Knowledge may be power, but the Enlightenment conviction that knowledge is emancipatory, so crucial to her own upbringing, has begun to leave a bad taste. The worst part of all of this is that Viridiana can't let her students catch a glimmer of her doubts.

The Road to Microserfdom

Is Viridiana Jones to blame for her befuddlement concerning what has happened to her profession and her university over the course of her lifetime? By and large, I would be inclined to say no. Existing analyses of the commercialization of science and the transformation of the university on the part of economists, philosophers, sociologists, and science studies scholars have left much to be desired, to put it politely.[9] It is drudgery to try and cut through the contemporary fog surrounding the question of the health of contemporary science. One of the stranger recent developments has been the performance of a small number of econometric exercises by economists to *quantify* the extent of harm done to science by certain aspects of the modern commercial regime.[10] The barely concealed presumption that economists just naturally come equipped with a good feel for how science works, and that the marginal costs of the minor inconveniences of privatization could be captured by before-and-after citation analyses of patent-publication pairs, is a symptom of just how far the ability to think clearly about the issues has atrophied. The capacity to conceptualize solid bases of comparison between the two qualitatively different historical science regimes has been even rarer than a clear explanation of a collateralized debt obligation.

Various luminaries have pointed to this problem, including the current president of Harvard:

> As the world indulged in a bubble of false prosperity and excessive materialism, should universities have made greater efforts to expose the patterns of risk and denial? Should universities have presented a firmer counterweight to economic irresponsibility? Have universities become captive to the immediate and worldly purposes they serve? Has the market model become the fundamental and defining identity of higher education? (Faust 2009)

What is striking is that wherever such questions are broached, pointers toward the answers to these questions have been notable in their absence.

It is not as though there were some short pithy syllabus one could assign to Viridiana, which would force her to rethink her malaise from first principles. But beyond skewering scapegoats, there are some serious *conceptual* reasons why responses to the current crisis of knowledge production and dissemination have been so disappointing. Since 1980, we have lived through a period of profound transformation in the social practices, institutions, rules, and formats of the generation and conveyance of information, one that has slowly but inexorably transformed the very meaning of knowledge and the place it occupies in the modern polity. Viridiana's gut instinct that she is stranded in an alien landscape compared to that of the one she inhabited in her youth is basically correct; nonetheless, she needs a systematic survey of the new commercialized university landscape, not just a pat on the head and a couple of Valium.

Viridiana would undoubtedly wish to be told the *theory* of what is going on, but, indeed, there's the rub. In all likelihood, anyone poised conveniently ready and willing to supply an abstract theory (like the ones provided by some economists and legal theorists who are later cited) would almost certainly be misrepresenting the situation, because he or she would omit whole swaths of key recent events that, taken together, have brought us to the present impasse. The problem with providing a short précis of the modern predicament as a prelude to understanding the modern politics of knowledge is that the trends do not all uniformly point in the same direction. Depending on your standpoint, some developments might herald a new dawn of self-organized cooperative inquiry, the invisible college finally made manifest; others may portend a grim, brave new world of knowledge haves and have-nots, a road to microserfdom where every trope and concept comes indelibly attached with an electronic price tag, and every infobit is monitored from server to eyeball. Of course, there is the more immediate problem that most analysts can't or won't see beyond their own parochial concerns: Technogeeks only see the nifty technology, scientists only see the status of their own science, lawyers only see the law, economists only see market signals, philosophers only see epistemology, sociologists only see networks, NGOs only see globalization, technology transfer officers only see the color of money, and humanists only see the creeping demise of their own disciplines (Newfield 2003, 2008; Sahlins 2009). The Big Picture inevitably fades into the babel.

While the "correct" interpretation of events won't be settled in our lifetime, the service we might offer Viridiana is (1) to briefly enumerate the relevant range of economic and social phenomena that should factor into any assessment of the modern politics of knowledge and (2) to begin to describe the ways in which a particular modern theory of political economy—that is, the widely misunderstood doctrine of neoliberalism—has colored almost every discussion of the fate of the university and the "efficient organization of science" over the last three decades. This will serve to usher us into a series of empirical meditations on the state of contemporary science in the rest of this volume. Given the nature of the problems and their urgency, we can only hope that at the end of our survey Viridiana will be in a better position to make up her mind on the sources and implications of her disquiet.

One major lesson I hope to convey is the extent to which social and economic events turn out to be inseparable from the history of ideas put to use in order to make sense of them. In this book, I argue that much of the modern commercialization of science and commodification of the university has followed a script promulgated by neoliberal thinkers.[11] This was not due to omnipotent puppetmasters pulling the strings behind the proscenium, but rather because of a more subtle convergence of circumstances. The disdain of philosophers for the concrete, of the economist for the polysemous, the scientist for history, and the science policy maven for political theory will turn out to be a big part of the reason for the modern success of the neoliberal worldview. Neoliberalism, we shall discover, has managed to provide a grand integrative narrative, whereas all of the individual professions have seemingly absolved themselves of any responsibility to render the totality of academic life coherent. In this section, I will lay out a census of the six big trends that have shaped the modern regime of science management and funding in recent decades as it bears on the sciences. In the following section I provide a rudimentary primer in the theory of neoliberalism for those hesitant to plunge directly into the key texts. The eventual upshot will be to demonstrate that there can be no return to what many fondly imagine as the Golden Age of Science during the Cold War.

The Deindustrialization of the West, and the Putative Emergence of a "New Knowledge Economy"

Although it is difficult to point to any single phenomenon as paramount in our slate of late twentieth-century watersheds in science, it would be foolish to avoid the most significant economic development of the last quarter-century, namely, the loss of manufacturing base in most of the post–World War II self-identified industrial economies. This transformation has been comprehensively described by social scientists, and so I need not cover it in detail here. The relevance for our present analysis is that it was not simply some smooth shift from one indifferent economic "sector" to another, as in the adjustments to "comparative advantage" imagined by economists. An elaborate industrial base had previously defined many aspects of what it meant to live in a developed economy, in everything from the culture of consumption to the promotion of certain versions of science, so the erosion of the manufacturing base within these economies could not help but have far-reaching ripple effects, even for those who might have been proud never to have set foot on a shop floor. It was moreover a transnational phenomenon; indeed, current estimates for the older countries of the Organisation for Economic Co-operation and Development (OECD) suggest services and related industries now account for two-thirds of all economic value added.

The United States has been losing net manufacturing employment to production facilities overseas since 1989 (Burke, Epstein, and Choi 2004). Although the number of nonfarm jobs had been falling since the 1960s (due to employment growth in other sectors), the absolute number of workers employed in manufacturing has been collapsing since 2000, as can be observed in Figure 1.1. This decline had long preceded the Great Economic Contrac-

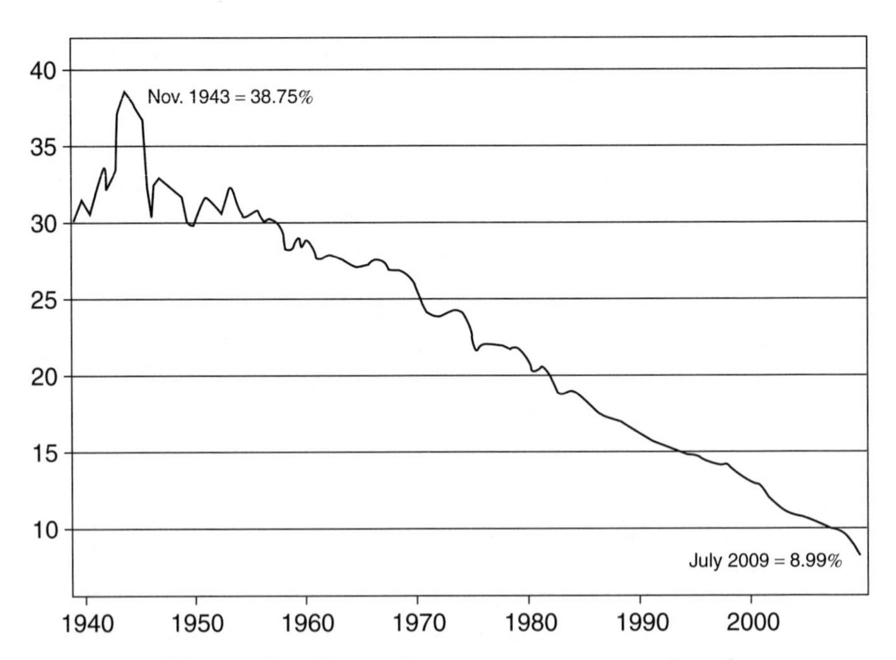

Figure 1.1. U.S. manufacturing employment as percentage total nonfarm employment. *Source:* http://www.bls.gov/iag/tgs/iag31-33.htm.

tion of 2008–2009, during which the percentage of manufacturing employment fell below 9 percent for the first time in recorded history. In 2002 the U.S. net trade balance in high-tech products was negative for the first time, and it has deteriorated since. Not only were entire geographical locales devastated but these prolonged losses also raised the issue of the aims and purposes of education under such circumstances. Had we done our economic rivals the favor of training their indigenous cadres in our universities, which were then responsible for their own industrialization? What were the causes and implications of U.S. citizens progressively giving the natural sciences a wide berth in higher education? Should we be concerned about waning public support for fields like science and engineering when many of the industries previously depending on them for new personnel were shifting not just production capacity but also supervisory functions overseas? Degrees awarded in engineering tend to mirror trends in the growth and decline of manufacturing, as one could observe fairly dramatically in world trends in engineering degrees awarded (see Figure 1.2).

Some pundits sought to paste a positive face on this phenomenon, by suggesting the advanced economies were becoming increasingly "weightless," embarking upon a third stage of capitalism consisting almost exclusively of the service sector, disengaged from gross physical production processes altogether.

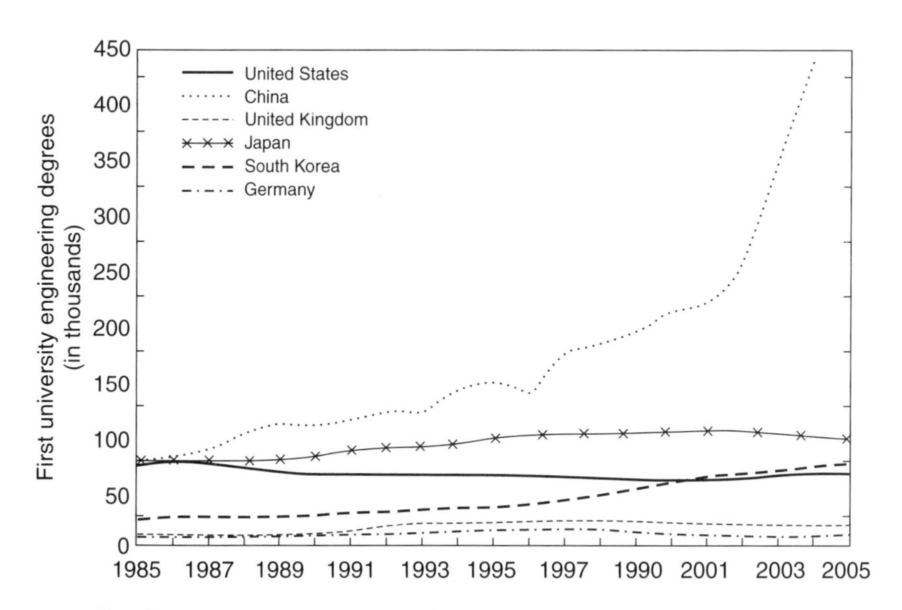

Figure 1.2. First university engineering degrees. *Source:* NSB, *Science and Engineering Indicators*, 2008, http://www.nsf.gov/statistics/seind08/.

Of course, most people recognized that such talk bordered on the delusional, but that didn't stop economists from reveling in locutions such as "the information society" or "the new knowledge economy."[12] It is striking the extent to which such notions have become entrenched as second nature among bureaucrats, businessmen, and the denizens of business schools. As Walter Powell put it, "If the knowledge economy is measured by the rise in knowledge management services among consulting firms or by the rapid growth in intellectual property as a legal specialty, then its growth has been considerable" (Powell and Snellman 2004, 199). The rise of "knowledge management" as its own professional specialization in business schools and management consultancies has had an underappreciated impact on the modern regime of science management.[13] Indeed, we can pinpoint the emergence of the information economy as a self-conscious statistical category with the revision of the Standard Industrial Classification (SIC) as the 1997 North American Industry Classification System (Malone and Elichirigoity 2003). This modification took various industrial activities that had been scattered by function throughout the previous SIC code and grouped them together for the first time as dealing with a product called "information," which itself could be rendered amenable to ownership and control.

Under these circumstances, the contemporary options for nervous students entering the job market after graduating from college must seem rather bipolar, with the future consisting either of slinging hamburgers for the minimum wage or else becoming a Knowledge Worker in some urban office building. Apparently, the current generation of American students has not been interested in apprenticeships to make things, as demonstrated by Figure 1.2. The temptation to reify these trends into some grand synthesis of a new Information Mode of Production has proven irresistible for a broad swath of sociologists, lawyers, and cultural theorists.[14] More to the point, this idea of knowledge management has had a profound influence on the very vocabulary that we use to discuss the economy. Neoclassical economics from the 1950s used to divide subject matter up along the lines of the physical world of commodities, thereby to contrast their solidity with the subjective world of mental impressions and preferences; growth was treated rather straightforwardly as more tangible stuff. Yet deindustrialization tended to threaten that complacent picture and to undermine the clear separation between mind and the world originally inscribed in the equations of general equilibrium.

All of this has put severe strain on neoclassical economists to provide an upbeat assessment of modern developments. Lest the reader think this merely the airy froth of the chattering classes, here is a quote from a recent Nobel economist:

> The scientific revolution of the past century has resulted in the systematization of change itself . . . Knowledge and information is being produced today as cars and steel were produced one hundred years ago . . . the standard theorems of welfare economics, which underlay the presumption of the efficiency of the market economy, *assume* that information and knowledge is unaffected by any action taken by any participant. Thus standard economic theory has little to say about the efficiency of the knowledge-based economy . . . We are slowly shedding the limitations of Matter to unleash the expansiveness of non-rivalrous Ideas. (Stiglitz 1999)

However much the siren song of Pure Transcendent Idealism did or did not manage to bewitch economists at the end of the millennium, or however much it was in the nature of the Capitalist Spirit actually to slip the surly bond ratings of earth, it seems fairly clear that neoclassical economists were coming under heavy pressure to analyze knowledge as well as they could analyze freight cars or steel. The fact that they had only sporadically attempted to do so in a desultory manner prior to the 1980s was not regarded as an insurmountable obstacle: Encomia to the "standard economics" of the knowledge economy flowed forth abundantly in the modern era. It was no accident that during this period one encountered repetitive claims about knowledge being a public good, science being the true "first cause" of all growth, and so forth. You couldn't attend a seminar without hearing about how knowledge was special, how it was nonrival and nonexclusive, though sometimes unfortunately "asymmetric," and how it gave rise to an infinite bounty of Good Things. Whether or not the newfound enthusiasm for the economics of information had been adequately logically grounded in the previous neoclassical theoretical tradition is something that needs not be decided here, since it rarely seemed to perturb anyone.[15] Viridiana needs to know that all these notions were symptoms of a repressed memory of a world where people actually made things for a living. And she also needs to be aware that, historically, every single country that ever managed to develop a growing vibrant science base had enjoyed a growing vibrant manufacturing sector.

The Spread of Computer Technology and the Rise of Decentralized Control via the Net

There are many fine histories of the postwar development of the computer and the construction of the Internet, such that it might appear almost superfluous to mention them as a prime determinant in the transformation of our understanding of the production and distribution of information in science.[16] Computers have, quite simply, changed the way we think about thinking. But have they altered the economics of science? Their joint impact on global communications and the control and storage of information is conventionally attributed to two principles said to be inscribed in the very technological trajectory of the development of computers: Moore's Law, which guaranteed that the per-unit cost of a calculation or memory unit was destined to fall exponentially over the horizon of its relevance; and the "end-to-end" principle in the packet-switching architecture of the Internet, where the coordination and control functions were decentralized to the edges of the network. Both of these technological characteristics were rapidly invested with deep social meanings, which tend to persist. In short, digital technologies were thought to *exemplify* what it meant to think and communicate as a human being, or as one of the gurus of this movement put it, "The digital world is closer to the world of ideas than the world of things" (Lessig 2001, 116).

Yet, curiously enough, the rise of computer technologies simultaneously made it *easier* to conflate information with a physical object, especially in formal economic models. In the case of Moore's Law, it was frequently asserted that a consequence would be that costs of storage and communication would have fallen so dramatically that the marginal costs of any piece of information being reproduced had to be very near zero, and hence information must be treated as "special" in comparison to other commodities. Because price could not be set to marginal cost (as in conventional neoclassical theory), some noncompetitive remedy was generally prescribed: Left-leaning commentators drew the inference that some system of state subsidies was required to guarantee sufficient knowledge generation. However, in a twist, corporate commentators drew the conclusion that monopoly controls and pricing would become inevitable under those circumstances (Boyle 2000). In the case of end-to-end packet-switching, Lawrence Lessig drew the inference that "architecture is politics" and that "end-to-end renders the Internet an information commons" (2001, 35, 40; Zittrain 2008). Popular pundits like John Perry Barlow, in a "Declaration of Independence of Cyberspace," converted the latter notions into even more dramatic claims concerning the democratic politics supposedly inherent in the Internet, such as "information just wants to be free" and "the Internet treats censorship as if it were a malfunction and routes around it." One might have been forgiven in those salad days for thinking the Internet held the capacity to upend the traditional organization of science.

Most assertions of bald technological determinism turn out to be flawed, and these particular claims were no different. While we shall observe all sorts of phenomena being attributed to the imperatives of Moore's Law, there has been no empirical demonstration of commensurate falling costs of research or learning activities, beginning with the famous dispute over the inability to observe productivity gains attributable to the computerization of all sorts of business activities over the course of the later 1990s (Gordon 2000). All sorts of things seem easier on the Web, but this productivity enhancement did not show up in simple indices of the economy. This foreshadowed the first hint that the commonplace argument for the public good character of "information" or "science" might have been precariously perched on shifting sands. As for the vaunted end-to-end architecture, the actual technical state of affairs in the infrastructure was never as clear-cut as the social analysts had wanted to make out (Gillespie 2006). Lessig was correct, however, to argue that, soon after the decision was taken by the government to commercialize the Internet in the mid-1990s, with the prior military infrastructure split off into a parallel net, there would be pressure to change the Internet architecture in the direction of enhanced control (and censorship) of what gets sent under what circumstances by telecoms, Internet service providers, and manufacturers of routers. Indeed, much of this deformation of the architecture occurred in the interim (Zittrain 2008). While the Internet could potentially have fostered decentralized cooperative research over proprietary research, it hasn't happened—at least not yet. There never was any inherent or intrinsic political or economic order to the architecture of the Internet: What you find there, by and large, is what has been put there by intentional activities of the major players. Be that as it may, it nevertheless became commonplace to suggest that the Internet fostered and *required* certain political or economic reforms, especially when it came to institutions like the university and the corporation. Viridiana needs to appreciate just how much the Internet became invested with a supposedly inexorable mandate for the commercialization of intellectual life.

The assertion of the Internet as external agent of reform has seemed plausible because once communication architectures became congealed into certain network pathways, it becomes almost impossible for individual participants, like, say, individual scientists, to have substantial leverage to counteract the impersonal pressure of the Web. Yet, what at first seems a technological imperative often, with a second look, turns out to have been more directly the product of a hidden corporate imperative. This phenomenon is one aspect of what some have dubbed the rise of "the audit society" (Power 2003). One thing the audit society excels at is keeping track of profitability. One manifestation of those Internet imperatives relevant to science policy has been citation analysis when deployed as performance indicators for individual scientists, individual journal titles, departments, universities, and so on. Just like the Internet itself, citation indexes started out as a neutral bibliographic tool to help researchers, but they have ended up as a bureaucratic means of surveillance, evaluation, and control. Most users don't realize that citation data, for instance, are proprietary and have been completely defined and controlled by a single corporation, Thomson International (later Thomson Reuters), since 1992.[17] If there were a Science Panopticon erected over the last three decades, it was constructed under corporate auspices but justified as a technological imperative.

The Great Transformation in the Legal Regime of Intellectual Property

The next salient socioeconomic development was the corporate push to fortify and extend the purview of intellectual property in the period following 1980. In the legal and business literature, this is now regarded as one of the major watersheds of recent modern economic history.[18] The fact that this "enclosure movement" has not been confined to the United States and Europe, but has been in intent and execution a global initiative, speaks volumes about how the status of knowledge has become a source of naked political controversy in the modern world.[19] Mostly in this volume we must take this phenomenon as exogenous and given, and yet it is difficult to overstate its importance for our present topic. Although the fortification of patent protection has tended to receive the lion's share of attention (Jaffe and Lerner 2004), the conduct of science has been equally impaired by alterations in the terms and conditions of copyright (Lessig 2001) and extensions of tort law into novel areas of "research tools" (Eisenberg 2001), and the imposition of nondisclosure through contracts. It will shortly be helpful to keep in mind that it was not the university sector that sought in any material sense to bring about this Great Transformation: This was someone else's notion of "reform."[20]

The proximate causes of individual components of the intellectual property revolution are fairly well understood: the creation of a special Court of Appeals for the Federal Circuit in 1982 (Jaffe and Lerner 2004, 10); the Bayh-Dole Act, which encouraged universities to commodify their research conducted with public funding; the Sonny Bono Copyright Extension Act of 1998, which extended the horizon of U.S. copyright to the author's lifetime plus seventy years; and TRIPS (trade-related aspects of intellectual property), which was instituted by the Uruguay Round negotiations in the GATT talks, and its attendant imposition of fortified intellectual property (IP) law upon member countries starting in 1995 (Sell 2003; Drahos and Braithwaite 2002). But what tends to be missing from this litany is an appreciation of the fact that each of these forays, across the wide gamut of forms of intellectual property, occurred within a relatively short space of time, and moreover, sometimes involved many of the same corporate protagonists—certain select pharmaceutical, computer, and entertainment companies, in particular, tended to pop up over and over again. Once one realizes the stupendous scale and scope of the interventions that were required to transform highly technical public policy toward numerous forms of IP, and not just in one country but across the world, then the idea that this watershed could be adequately explained by a simple change in the exogenous costs and benefits of pursuit of stronger IP due to technical progress in the computer industry becomes thoroughly implausible.[21]

The enduring presence of legal counsel in scientific research programs is one major defining attribute of the modern regime of science funding and management. Due diligence no longer means attentiveness to experimental controls; rather, it now signifies scrupulous prior allocation of IP between the multiple claimants to the products of research. Faculty and administrators ineffectually spar over whether "conflict of interest" really kicks in at payments of more than $10,000. And while universities at the turn of the millennium were being forced to police their own student bodies to make them stop downloading free music and videos over the college server, now vigilance seems to dictate the dissolution of the campus library as a core university function and the privatization of proprietary academic information in scientific publications conveyed over the Web.

The Restructuring of the Corporate Form and the Outsourcing of Commercial R&D

Economic historians, legal scholars, and management theorists all generally trace the metamorphosis of corporate structures in many developed countries back to roughly 1980.[22] Viridiana may find this utterly boring, but it was far more consequential for her life than the things she normally attends to. The trigger was said to be the widespread conviction that the United States had lost ground to international competitors during the oil crisis and economic slowdown of the later 1970s. Although there was substantial disagreement over the causes of the supposed sclerosis, an imposing array of initiatives were crafted to defeat the diverse culprits thought to be sapping America's economic dominance. One major candidate for economic reform was the organizational structure of the large Chandlerian corporation.[23] Various market participants had become convinced that the huge managerial conglomerate had become too unwieldy to effectively compete in the world market in the 1970s, and thus the 1980s initiated the era of hostile takeovers, leveraged buyouts, and shareholder attacks on the top management of large corporations. In response, there was a significant retreat from diversification within firms, with one calculation suggesting that by 1989 firms had divested themselves of as much as 60 percent of acquisitions made outside of their core business between 1970–1982 (Bhagat et al. 1990). Consequently, there ensued a retreat from previous levels of vertical integration in industries like automobiles, computers, telecommunications, and retail. Corporations began to equate agility with repudiation of hierarchical managerial control of process, and with it relinquishment of the multidivisional paradigm, and thus sought to reengineer the supply chain to depend to a greater extent on market coordination. Networks of subcontracts began to displace ownership ties as modes of organization; venture capital began to channel investment into start-up firms. Labor-intensive heavy manufacturing began to be concertedly outsourced to low-wage countries. Moreover, the roster of America's largest corporations experienced severe shakedown, after having enjoyed relative stability for the previous sixty years. The lumbering giants were prodded into defensive action, which was widely interpreted as a return to market methods of coordination (Langlois 2004).

For the reader to understand everything that follows, it is significant to appreciate that the in-house corporate R&D laboratory had been a major component of the Chandlerian model. Starting in the early twentieth century, a substantial proportion of all PhD scientists were employed in well-funded separate research divisions within many large corporations. The breakdown of the Chandlerian model of the hierarchical integrated firm then prompted this nagging question: Why integrate R&D into the firm when you can buy it externally and reduce costs by doing so? But that question presumes that R&D is a distinct fungible commodity in a well-developed market, one so competitive that it can lower the costs through outsourcing relative to doing it yourself. In one important sense, the reengineering of the corporation implied the commodification of science.

A very common bromide is that science has "always" been market-oriented, in one way or another, so qualms about the modern commercialization of science are misplaced. However, this overlooks the fact that market-oriented science assumes very different formats throughout history. No matter how commercialized science may or may not have been, this state of affairs—the ability to outsource commercial research as an external for-profit proposition—until recently was uniformly absent. The strengthening of intellectual property, the weakening of both domestic antitrust prosecution and the ability of foreign governments to counter corporate policies, the capacity to shift research contracts to lower-wage and easier regulatory environments and therefore engage in regulatory arbitrage, the availability of low-cost real-time communication technologies, and the presence of an academic sector that was willing to surrender control of research to its corporate paymasters: All of these were necessary prerequisites to seriously countenance the corporate outsourcing of research on a mass scale.[24]

The globalization of corporate R&D is one of the primary hallmarks of the modern regime of knowledge production. Of course, multinational companies headquartered in smaller countries like the Netherlands and Switzerland have long internationalized their R&D activities essentially from their inception; but the more striking trend is the sharp rise in international outsourcing of research across the board since the 1980s (Reddy 2000, 52). Global outsourcing to foreign low-cost performers has tended to be concentrated in a few industries, such as pharmaceuticals, electrical machinery, computer software, and telecommunications equipment. Nevertheless, surveys within these industries reveal a sharp increase in research carried out beyond the home country's boundaries from the 1960s to the 1990s (Kuemmerle 1999). A later survey by the Economist Intelligence Unit reveals the globalization of R&D gathering apace over the 1990s, with over half the respondents indicating they would expand their overseas R&D investment in the next three years. An UNCTAD survey of the world's seven hundred largest corporate performers of R&D found an average of 28 percent of total R&D expenditures sited outside the nominal home country in 2003, with 69 percent of firms indicating that their proportion would increase in the future (OECD 2008, 21). When respondents were queried as to the major considerations governing their decisions, their most popular responses were strong protection of intellectual property, lower costs, and the tapping of indigenous research capacities (Lieberman 2004). It is the access to lower-wage labor in the context of an academic infrastructure, ***disengaged from any corporate obligations to provide ongoing structural support for local academic infrastructure,*** which helps explain the shift in research funding to countries like China, India, Brazil, and the Czech Republic (Economist Intelligence Unit 2004, 9). Another way to cut costs is to disengage the firm from nationalist appeals to help support scientific infrastructure, accompanied by improved opportunities to further reduce or avoid corporate taxation.

The range of research involved in offshore outsourcing is daunting; most people, including Viridiana, might be surprised to learn how much work is done in this way:

> Even Wall Street investment banks such as JP Morgan, Lehman Brothers and Bear Stearns [prior to their demise, of course—P. M.] export work in financial analysis . . . Radiologists in India and Australia interpret CT scans for patients in American hospitals. Fluor Corporation employs thousands of engineers and draftsmen who work on architectural designs in the Philippines, Poland and India. Statisticians in Bombay process

> clinical research data for American drug companies . . . Engineers in Russia design parts of Boeing's airplanes . . . Intel's China Software Lab in Shanghai, one of four Intel research groups in China, works on projects to enhance Linux technology for Intel based servers . . . Scientists at Texas Instrument's research center in India design next generation mobile phone chips. (Lieberman 2004, 13)

Politics and economics factor into this trend to outsource R&D, but they also govern our ability to find out just how prevalent the trend might be. After all, we are dealing with multinational corporations, one element of whose *raison d'être* has been to evade the inconvenient politics of individual nation-states. Because nothing would seem to refute the neoliberal raptures over a new knowledge economy more dramatically than a rapid and highly visible loss of science base, the quality of the data on offshore outsourcing of R&D has become very badly degraded over time. The government office charged with measuring such activity was located in the Department of Commerce in the Bureau of Economic Analysis (BEA). In the 1990s, it was becoming apparent from Commerce BEA statistics that employment in R&D and engineering within U.S. parent multinationals was being shifted abroad: for instance, by their figures in the period 1989–1999, while R&D employment grew in the United States a paltry 1.3 percent, R&D employment in foreign affiliates grew 9.8 percent. The comparable figures for engineering and architectural services were −0.6 percent and 7.3 percent (Landefeld and Mataloni 2004). Corporate R&D expenditures abroad quadrupled from $4.6 billion in 1986 to $17.5 billion in 2000. But after 2000, this data became contentious and not a little embarrassing, although for a short time it was possible to use the existing database if you had the access of a political insider like Senator Joseph Lieberman, to isolate just where and when corporate R&D was being shifted abroad.

For instance, it became possible to pinpoint the grand inflection point of the offshore outsourcing of corporate R&D to China in the years 1998–1999 (see Figure 1.3). This ability to focus on specific diversions in the channels of world commerce and investment became a flash point in the 2004 election; consequently, in a dynamic we shall repeatedly encounter in this volume, the facts were changed to protect the not-so-innocent.

The National Science Foundation (NSF) had long signaled that the data on R&D expenditures by private firms have shown signs of corruption from the 1990s onwards. Hence a number of vain attempts to "correct," revise, or otherwise redefine the data ensued over the last decade, which have effectively rendered comparative statistics like those reported in *Science and Engineering Indicators* useless. But the past few years of tinkering have now finally

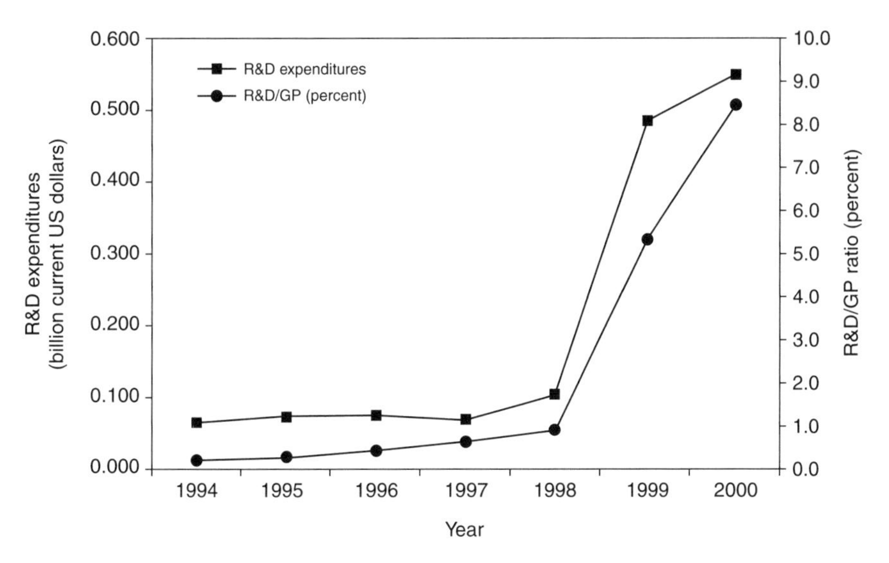

Figure 1.3. U.S. corporate R&D spending in China. *Source:* Lieberman 2004.

been brought to an abrupt halt; the decision was finally made to shoot the messenger:

> Budget pressures will force BEA to stop collecting detailed data on the financial investments being made by multinational companies both in the United States and overseas. And the agency will stop gathering data to measure how spending on research and development is impacting U.S. gross domestic product. The decision to stop collecting the data was made because the agency does not have enough money, say BEA officials . . . By not knowing what is happening with the globalization of the U.S. economy, those concerned about the sell-off of America won't be able to raise a fuss . . . The R&D data that will be dropped was started two years ago as a means of measuring the impact research is having on the economy. (McCormack 2008)[25]

We are living through a profound transformation of the American science base: Offshore outsourcing is the bulk of privatized scientific research, but we are bereft of dependable means to gauge its grossest outlines. Indeed, one consequence of the commercialization of knowledge has been to leave us in the dark concerning such trends that really matter to scientists. This induced ignorance leaves us with little more than neoliberal assurances that the free flow of investment across national borders can never possibly harm a nation.

Approaching the commercialization of science from this angle profoundly revises the usual narrative of the privatization of modern academic science as a relatively straightforward case of cash-strapped universities simply following the money, albeit beset by a few nagging qualms concerning the propriety of telling corporations only what they want to hear.[26] Rather, an alternative account might be proposed where many of the nascent institutions of globalized privatized research were first pioneered *outside* of the academic sector per se, as adjuncts to modification and reengineering of the modern corporation. Once these innovations were well underway, only then were universities lured into incorporating these exemplars of the new globalization regime through their own internal restructuring of scientific research.

The Withdrawal of the State from Its Role as Science Patron and Manager

Of all the trends that impinge on Viridiana's day-to-day activities, this is the one that is treated as blindingly obvious by her peers. The percentage of national R&D expenditure provided by the U.S. federal government has been falling since roughly 1967, while that emanating from private industry has been rising. Federal budgets for R&D have been essentially flat in real terms since the 1980s, as can be observed in Figure 1.4. Science policy

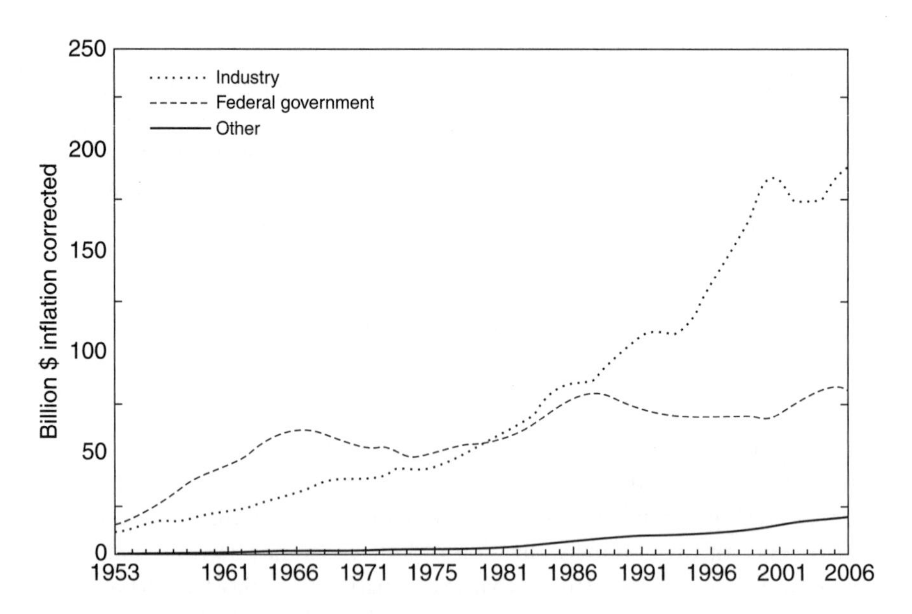

Figure 1.4. Sources of American R&D expenditures.

experts know this wasn't only a matter of straitened federal budgets but was also caused by a dramatic reversal of principles that had become ingrained in the political psyche in the years just after World War II. During the Cold War, it had become a matter of strategic necessity for the United States to maintain an advanced national science base, for both military superiority and ideological reasons. For about three decades, the military had been recruited to serve as science manager and to organize patronage (contrary to some impressions, the civilian NSF never loomed large in this period), but around the time of the Vietnam War many imperatives began to encourage the military to divest this function. One scenario might have been to shift the function wholesale to some other civilian arm of government, but outside of areas of biomedicine covered by the National Institutes of Health (NIH), this by and large didn't happen. By 1980, the proportion of R&D accounted for by industry surpassed that coming from the federal government, and thus the problem of who or what would take over the job of science manager in charge was starting to be settled by default. Various individual states were tempted to mount their own initiatives (Douglass 2006); Geiger and Sá 2008, 85–115) in response to the federal pullback, but these tended to be targeted to very narrow sub-subfields of the sciences and remained small.

Far from being some idiosyncratic aspect of American history, much the same thing happened in other Western nations with large military sectors, although with a variable lag as this was conditional on the previous character of their own indigenous university systems.

The slow unwinding of the military/academic complex wended its way through many constituencies, but the one that has received the most attention of late has been the Bayh-Dole Act of 1980. The first wave of commentary portrayed the act, which initially allowed universities and small businesses to own patents on discoveries made with government R&D funding, as freeing up a tsunami of commercial-grade science, which had been supposedly dammed up in ivory towers by earlier government regulations. This was followed by similar legislation in many other Western countries, chasing the dream of cornucopia locked up in their own universities (Powell, Owen-Smith, and Colyvas 2007). In 2002, the *Economist* notoriously called the Bayh-Dole Act "possibly the most inspired piece of legislation to be enacted in America over the past half century. More than anything, this single policy measure helped reverse America's precipitous slide into industrial irrelevance."[27] A second wave of commentary pointed out the real story was not very much like that at all; nonetheless, the lines of causality became even murkier in the interim, as arguments degenerated into disputes over whether U.S. universities had actually asked for the Bayh-Dole legislation or whether universities had already enjoyed the ability to patent government-funded research before Bayh-Dole, or indeed if the net results were just a plethora of lower-grade patents by universities that had never before sought them out.[28]

The Bayh-Dole Red Herring has diverted attention away from the fact that it was simply one entry in a whole sequence of legislation in the United States, each act devoted to extricating the government from previous Cold War science management, as a prelude to turning responsibility over to the corporate sector.[29] No single piece of legislation could have done (or has done) the trick. Government revisions of policies with regard to intellectual property or educational subsidy may have constituted incentives, but they could not unilaterally impose the fully elaborated structure of the new regime. Governments could give a onetime windfall to research performers in the form of IP they had not paid for, but that did not mean they intended to continue providing such massive implicit subsidies as a matter of course. Hence universities should not have seriously expected that Bayh-Dole was going to spell their ultimate deliverance from penury or even prove to be a dependable source of research subsidy. While legislation such as the Bayh-Dole Act was enabling certain novel behaviors, it should not be confused with the *cause* of the privatization of science, which was instead attributable to the larger societal shift in the nexus of science management and funding.[30]

Retreat from the Premise That the State Should Be the Primary Provider of Education for Its Populace

Viridiana, who teaches at a state university, can't help but shrug and roll her eyes when she hears her state's governor wax lyrical over how her campus is going to miraculously drag her state out of its economic doldrums. The level of misrepresentation only starts with the fact that her state has reduced its subsidy to such an extent that only about a fifth of the university's budget comes from state appropriations.[31] This just skims the surface of what has become a standard yet utterly incoherent justification for the very existence of her university.

There have been some interesting analyses of what has been happening to higher education emanating from the beleaguered departments of education, themselves caught in the crosshairs of the business consultants hired to slim down and streamline the modern university.[32] By pungent contrast with the economists, these analysts approach what has been happening to science and higher education as a subset of larger political movement, one that seeks to reengineer democracy by privatizing one of the largest sectors of state expenditure—namely, the provision of education—and to turn universities into competitive global service industries. In short, universities are being exhorted to become more like corporations—to regard their products as "information" and "human capital," to treat their students more like customers—as a prelude for the state to withdraw from all responsibility for the provision of education. The problem remains that existing universities cannot wholeheartedly embrace complete conversion to the corporate model, for a raft of historical reasons. For instance, much of the accounting structure of existing universities is oriented around the legal presumption of their nonprofit status; over time, subsuming more for-profit activities would undermine their tax-free status, their claim to special exemptions with regard to certain classes of IP (as in the very important case of *Madey v. Duke University*),[33] and much else that renders them distinctive. And it is not just the uneasy relationship to the government that is strained through imitation of the corporation. Many universities depend on the prospect (whence university presidents spend most of their waking hours) that their alumni will donate generously to their alma mater, but who in their right mind leaves a bequest to Coca-Cola, GE, or Microsoft? The more the university opts for corporate behavior, the more it relinquishes one of its most important (though certainly uneven) sources of sustenance.

But far from being something that can easily be redressed, this transformation of education has turned out to be an intentional goal of much federal policy. One can trace it back to the decision in 1972 to shift government aid to individuals, away from the universities involved. Student grants were converted to student loans not simply to save money, but to hasten the redefinition of the student as consumer, so that the university might act more like a retailer. In the 1990s, many tax credits and federal programs for higher education were explicitly extended to for-profit universities. At that point, huge distance-education diploma mills like the University of Phoenix, DeVry University, and Universitas 21 Global sprang up across the landscape.[34] Viewed dispassionately, the ultimate objective of U.S. education policy is two-pronged: to preserve a few private legacy institutions (such as those in the Ivy League) for the affluent seeking that boutique diploma; and, conversely, to convert most of the rest of the educational opportunities to low-cost for-profit options for the great mass of the population. The great public state universities are being slowly divested and phased out: Appropriations are cut except when the sports teams are doing well; permanent faculty are replaced with contract labor; business models are imposed top-down on academic units in the name of "accountability"; and public/private identities are blurred. Indeed, one model for this process could be the establishment of the University of Illinois "Global Campus" in January 2008, a distance-learning operation run entirely by a separate administrative structure, employing part-time and nontenured faculty. Originally intended as a freestanding for-profit subsidiary, a faculty revolt at Illinois led to a more ambiguous status, allowing "market-based" pricing and "full-cost recovery" policies.[35]

It has not gone unnoticed that there is very little room in this model for an elaborate or extensive research capacity (J. Johnson 2009). The cheap for-profit "schools" providing distance education have explicitly renounced any such functions as being inconsistent with the business plan; the state universities lose their ability to maintain a diversified base as tenured faculty are phased out; and the gold-plated private schools pour most of their own resources into areas of the natural sciences that are able to attract private money, while starving out everything else. The net result can only be that, wherever research is to be conducted, it will only be supported under conditions of commercialized joint ventures with external corporations.

And here we observe the individual trends described above begin to converge into a novel system. Everywhere Viridiana turns, things that used to be cheap (if not free) are now occasions for making a profit. Rich is the new smart. In the new knowledge economy, a dollop of high-class human capital is an offer you cannot afford to refuse, so you should be willing to pay dearly for it, including student loans that stretch out well into your working career. If you don't happen to be one of the fortunate few born with a silver spoon or a golden credit rating, then the Internet will supply a lower-quality version of the commodity on the cheap in the form of distance education. Because education is no longer about the formation of citizenship or character, all that really matters is that some bureaucratic entity sanctions that you purchased the stipulated commodity—one reason for the popularity of the MBA and the undergraduate business major. The worldwide strengthening of IP foists a version of this knowledge economy on the entire globe, under the rubric of the free trade of global service providers. It exhorts arenas recently voided of manufacturing employment to seek to live off the tribute of far-off others. Since the whole idea of an academic peer group loses its rationale, while information shades off imperceptibly into infotainment, knowledge becomes defined in a circular manner as whatever the market will pay for. And just when the modern corporation seeks to outsource its R&D functions as part of its restructuring, voilà, universities everywhere just happen to be vying with each other to attract and accept contract research. Because information can be digitally transferred, owned, and controlled far outside the bounds of the nation-state, the university as research provider finds it must compete with both nonacademic and foreign academic units,[36] imitating the prior global reach of the transnational corporation.

Is it just an accident that such far-flung trends have come together into something that looks very much like an integrated political economy?

The Marketplace of Ideas as Core Neoliberal Doctrine

Although the agents responsible for each individual trend identified above—the knowledge economy, the spread of the Internet, the strengthening of IP, corporate outsourcing of R&D, and the withdrawal of state provision of education—may have had little directly to do with one another, they all were inspired by a particular vision of the economy and the polity, one that we will associate with the set of propositions concerning society called neoliberalism. Viridiana should develop a passing acquaintance with neoliberalism so that she can comprehend how all the trends that bear on her science eventually meshed together into a single coherent system.

There is a small literature that now seeks to connect contemporary transformations of the science base with neoliberal politics.[37] The key neoliberal doctrines that abide at the heart of each of the trends identified in the previous section are the reification of the ideal economy as a "marketplace of ideas" and the conviction that the state as an actor can never measure up to the ability of the abstract marketplace in both conveying existing ideas and in summoning forth further innovation in ideas. This doctrine, contrary to untutored impressions, is itself relatively recent, dating at its earliest from the 1930s. It is eminently a *political,* and not simply a cultural phenomenon, because it was developed as part of a concerted effort to counteract the rise of planning and other market-skeptical movements that grew out of the Great Depression and the experience of World War II. A brief history of the transformation will assist our understanding of its contemporary hold on science.

As Dan Schiller once perceptively queried, "Why wasn't the status of information a major topic in economic theory in 1700, 1800, or 1900? Why was it only in the postwar period that the economic role and value of information took on such palpable importance?" (1988, 32). The easy retort that it was inadvertently overlooked until World War II simply will not stand up to scrutiny. Rather, there were very specific conditions that prompted social scientists blithely to situate knowledge on a par with knishes or narcotics. Although a serious survey of the relationship of knowledge to society would require a thick detailed history to do it justice,[38] I shall point toward two major developments salient to the sciences: the rise within the natural sciences of a concern with "information" as an analytic tool (which includes the rise of computer science), and the transformation of theories of the market from a simple allocation device for material goods to the ultimate in information processing. In both, the natural sciences inadvertently provided some inspiration for neoliberalism. Later in this book I will explain how neoliberalism then turned full circle to bite them back.

Our initial point of departure is the phenomenon of the "cognitive turn," or cognitive revolution, taken in most social and natural sciences in the period immediately following World War II in America (Mirowski 2002; Baars 1986; Gigerenzer and Murray 1987). The immediate implication of such a dramatic shift in orientation across the board is the heightened likelihood that economics would be caught up in the enthusiasm, as indeed it was. "Information" became a topic of mathematical models starting in the 1920s, and it began to spread as a theoretical entity throughout the *natural* sciences during World War II. Claude Shannon's "information theory" started off as an attempt to theorize cryptography (Mirowski 2002, 68–76) and rapidly became conflated or confused with a formal theory of "intelligence." A concern over military intelligence led to the founding of a science of military decision theory, carried out under the rubric of "operations research." This development was crucial for all of the social sciences, full of implications for the ontological status of information versus knowledge, but we must merely take it as background here. The other corollary was the construction and stabilization of the first electronic digital computers in World War II (Edwards 1996). The physical instantiation of "machines who think" (whatever may have been intended by such a locution) provided irresistible metaphors for cognitive activities, many of which would be taken up to various degrees in all of the social sciences during the postwar expansion of academic research. These physical traditions concerning the sciences of information set the stage for a later reconstruction of economics and politics, rendering some attempt at accommodation nearly inevitable.

Just as the physical world began taking on many of the characteristics of mind and thought, discourse concerning the social world began to shed much of its fascination with the tangible and the intractable Otherness of the physical world. Indeed, in retrospect, both sallies might be considered two aspects of the same metaphysical quest for the unity of science. Although one might observe that this unity was never actually attained in any substantive sense, a working rapprochement was forged, making use of the novel tool, the computer. Starting with attempts in midcentury to reify the computer as a metaphor for the operation of the market, the notion slowly grew that what markets did was *reconcile participants' mental states* through the computation of prices, rather than simply shift around physical goods between people who desired them with greater or lesser urgency. The Market was therefore the mother of all computers. At first the conceptual innovation seemed minor, especially for neoclassical economists, who had based models on individuals maximizing their psychological utility. But in prewar neoclassical economics, nobody really had to *think*. Hence the notion that markets could coordinate cognition really turned out to be a dramatic departure for a vast array of the postwar social sciences.[39]

Neoliberal intellectuals love to point to early prewar work anticipating the metaphor of the "marketplace of ideas," especially those found in the realm of American jurisprudence.[40] However, they rarely admit that such metaphoric flights were nowhere to be found in formal neoclassical economics, or in social science generally, at least until they began to surface in the writings of the later generations of the Austrian School. The fruits of the intellectual ferment in Vienna in the 1930s for philosophy and economics, from the logical positivism of the Vienna Circle to the mathematical innovations of the *Mengerkreis,* are well known in the historical literature. What has been less surveyed is the role of the Austrian School in elevating discussion of the notion that the fundamental role of the market was not the static allocation of things betwixt traders but the processing and conveyance of knowledge. In other words, the "marketplace of ideas" was not merely a *façon de parler,* but it existed as a real phenomenon.

The pivotal protagonist in this transmutation was Friedrich Hayek. At one time, it was commonplace to portray Hayek as a figure who was defeated midcareer by John Maynard Keynes and that Hayek was consigned to exile outside the legitimate economics discipline. Recent research has reversed this verdict.[41] Not only is Hayek now regarded as the godfather of the rise of the neoliberal movement in social thought, and the entrepreneur behind the establishment of the Chicago School of Economics, but he is also now conceded to be the locus of real intellectual inspiration behind the cognitive revolution in American economics. The mid-twentieth-century intoxication with the marketplace of ideas constitutes the fire within and the fuel beneath the innovations described above, and hence it becomes crucial to understand the extent to which the notion was intimately bound up with the innovation of neoliberalism in the social sciences in America and Europe. The origins of the marketplace of ideas in such postwar Hayekian projects as the Society for Freedom in Science, the Mont Pelerin Society, the University of Chicago economics department, Chicago law and economics, think tanks from the IEA to the Heritage Foundation and beyond (Cockett 1995), and his own connectionist theory of mind and its echoes in artificial intelligence and cognitive science (Dupuy 2000), will provide some indication of the political valence of the catallactic approach to knowledge.

Neoliberalism as it developed after World War II diverged from classical political liberalism in a number of ways; it renounced the passive notion of a *laissez-faire* economy in favor of an activist and constructivist approach to the spread and promotion of free markets.[42] The starting point of neoliberalism is the admission, contrary to classical liberalism, that its political program will only triumph if it becomes reconciled to the fact that the conditions for its success must be *constructed* and will not come about "naturally" in the absence of concerted effort. This had direct implications for the neoliberal attitude toward the state, as well as toward political parties and other corporate entities that were the result of conscious organization, and not simply unexplained "organic" growths. In a phrase, "The Market" could not be depended upon to naturally conjure up the conditions for its own continued flourishing. As Milton Friedman once joked in a letter to Friedrich Hayek, "Our faith requires that we are skeptical of the efficacy, at least in the short run, of organized efforts to promulgate [the creed]" (Hartwell 1995, xiv). The ultimate purpose of institutions such as the Mont Pelerin Society and the Chicago School of Economics was not so much to revive a dormant classical liberalism, as it was to forge a neoliberalism better suited to modern conditions. The conundrum of right-wing intellectuals in the Depression and in World War II was to explain why people in general could not comprehend the political truth of the dangers of a planned economy, and thus reconcile their dismay with their commitments to personal freedom. Hayek's response to the conundrum was to insist that "intellectuals" were at fault: They were "secondhand dealers in ideas" and had promulgated false and misleading images of what the market was and what it did. The man in the street was, of course, free to believe whatever he liked; but the neoliberal conviction grew that something about the way ideas had gotten promulgated was corrupting the very essence of Western knowledge.

It may have seemed incongruous then, but the resolution of Hayek's problem situation was to be found in the promotion of the idiom of the marketplace of ideas to pride of place in neoliberal theory. As he wrote in his much-quoted 1945 address "The Use of Knowledge in Society,"

> The economic problem of society is thus not merely a problem of how to allocate "given" resources . . . It is rather a problem of how to secure the best use of resources known to any of the members of society, for ends whose relative importance only those individuals know . . . it is a problem of the integration of knowledge which is not given to anyone in its totality. (1948, 77–78)

This revision of "the economic problem" was the embryo from which both neoliberalism and the quest for an economics of knowledge would germinate. But why couldn't your average denizen of the Cold War democracies feel the force of this notion? One might have drawn the lesson from such a position that conservative liberalism had already proffered its wares in this agora of the mind, and it had failed to find a buyer; but this was emphatically *not* the lesson extracted by Hayek and his followers. On one level, Hayek believed the correct apprehension of the nature of the market had not been as yet offered up on the auction block; all that had been previously available were debased and tawdry wares, such as "rationalism," Marxism, and (yes) static neoclassical equilibrium analysis. The primary task for his neoliberal cadres was therefore to produce a novel set of doctrines, including (from their perspective) a more valid economic theory. Another would be an attack on the state-dominated structure of education, which would be brought to better approximate the ideal marketplace of ideas if it were privatized.[43]

On a different and more disturbing level, Hayek posited that no one would ever be capable of fully grasping the operation of a marketplace of ideas *by its very definition:* It processed information in ways that any human mind would be stymied in attempts to imitate, such that no central planner could ever mimic its operation. Therefore participants would always languish unsatisfied: Neoliberalism preached "the necessity of the individual submitting to the anonymous and seemingly irrational forces of society . . . the understandable craving for intelligibility produces illusory demands which no system can satisfy" (Hayek 1948, 24).[44] This was just another instance of the neoliberal insistence on how The Market could not be left to its own devices to foster its own successful operation.

In the interests of relating a manageable version of what eventually became neoliberal creed to our immediate concerns—namely, the commercialization of science and the university—I have risked oversimplifying here its tenets into a discrete set of ten grossly telegraphed propositions.

[1] The Market is an artifact, but it is an ideal processor of information. Every successful economy is a knowledge economy. It knows more than any individual, and therefore it cannot be surpassed as a mechanism of coordination.

[2] Neoliberalism starts with a critique of state reason. The limits of government are related to intrinsic limitations on a state's power to know, and hence to supervise. These limits are not fixed for all time, however. Nevertheless, the Market always surpasses the state's ability to process information.

[3] Neoclassical economics is a good first pass at the representation of the capacities of the market as information processor. Here the Chicago Program tended to diverge from the writings of Hayek himself. This is related to a profound change in neoclassical economics located just after World War II (Mirowski 2002).

[4] Politics operates as if it were a market, and thus dictates an economic theory of "democracy." This supports the application of neoclassical models to previously political topics; but it also explains why the neoliberal movement must seek and consolidate political power by operating from within the

state. The "night-watchman" version of the state ends up repudiated. This tenet justifies alliances with the powerful in order to push the neoliberal agenda, and it reinforces right-wing suspicions concerning the virtues of what they consider radical democracy (that is, political action outside a market framework).

[5] Governmental institutions should be predicated on the government of the self. Freedom is not the realization of any *telos,* but rather the positing of autonomous self-governed individuals, all naturally equipped with a neoclassical version of rationality and motives of self-interest. Foucault (2004, 2008) is strongest on the role of these "technologies of the self," which involve an elaborate reassessment in concepts of human freedom and morality.

[6] Corporations can do no wrong. Competition always prevails. This is one of the most pronounced areas of divergence from classical liberalism, with its ingrained suspicion of joint stock companies and monopoly. It underwrites a "degovernmentalization of the state" through privatization of education, health, science, and even portions of the military.

[7] The nation-state should be subject to discipline and limitation through international initiatives. This was initially implemented through neoliberal takeover of the International Monetary Fund (IMF), World Trade Organization (WTO), the World Bank, and other previously classical liberal transnational institutions. This pacification of the state began as advocacy of free trade and floating exchange rates but rapidly became subordinate to the wider agendas of transnational corporations, to whom it became attached. Neoliberal "reforms" can be imposed outside of standard political channels by supranational organizations.

[8] The Market (suitably reengineered and promoted) can always provide solutions to problems seemingly caused by markets in the first place. Monopoly is eventually undone by "competition"; pollution is abated by the trading of emissions permits; McCarthyism is mitigated by competition between employers (Friedman 1962, 20). There is no such thing as a "public good" but only a series of problems handled by different governance structures, themselves determined by relative transactions costs (Coase 1960, 1974a).

[9] Redefinition of property rights is one of the most effective ways the state exerts neoliberal domination. Once such rights become established, they are then treated thenceforth as "sacred." Neoliberal economics often presents property rights as though their specific formats were relatively unimportant for the operation of the Market, but simultaneously they admit that, once created, they are very difficult to reverse. The best way to initiate the privatization program in any area that had previously been subject to communal or other forms of allocation is simply to get the state to institute a new class of property rights. Political self-interest often will take care of the rest.

[10] "Freedom" is recoded to mean only one narrow version of economic freedom within the new reconceptualization of The Market. Neoliberalism purports to value freedom above all else: "Economic freedom is an end in itself" (Friedman 1962, 8). However, it will be indispensable to grasp that neoliberalism did *not* prescribe the abolition of all controls over human action—in this respect, it was the most marked divergence from previous anarchist or libertarian doctrines. In fact, controls were not to be banished so much as recoded and reconfigured into a new version of "competitive order." Chicago neoliberalism transcends the classical liberal tension between the self-interested agent and the patriotic duty of the citizen by reducing both state and market to the identical flat ontology of the neoclassical model of the economy. (The situation at the Mont Pelerin Society, and particularly Hayek's relationship to it, was much more complex over time.)[45] "Freedom" thus becomes recoded to mean the capacity for self-realization attained *solely* through individual striving for a set of necessarily unexplained (and usually interpersonally ineffable) prior wants and desires. Isaiah Berlin ([1958] 1969) captured this innovation with his distinction between "negative" freedom, which was a state of immunity from encroachment by others upon the realization of given desires, and "positive" freedom, which viewed the agent as being encouraged to engage in a process of finding those desires in an environment that actualized that quest. Berlin (and the neoliberals) linked positive freedom with totalitarian movements. Once this new lexicon was firmly set in place, it became virtually impossible within this realm of discourse to regard any economic transaction whatsoever as coercive (V. Smith 1998, 80), which was a massive divergence from classical liberal doctrine.

The Commercialization of Science Is the Apotheosis of the Neoliberal Program

Finally, after this winding detour through the recent history of politics and economics, we can return once more to Viridiana and proceed to a preliminary diagnosis of her malaise.

Diagnosis 1: If only she had been equipped with a sufficient background in the neoliberal project, she would at least come to appreciate that all her beliefs about science being conducted "for the public good"; education as existing to shape moral, civic, and intellectual character; and knowledge as the embodiment of intrinsic virtue as part of its constitution, are all hopelessly *passé.* They have lost all cultural authority at the dawn of the twenty-first century. By this I do not mean they have merely fallen out of fashion—no, they no longer are grounded in the constellation of institutions she holds dear: the state, the university, the scientific journal, the NSF, the high-class media, the hospital . . . The neoliberal marketplace of ideas has come to be inscribed in the laws, the funding agencies, the evaluation structures, the communication vehicles, the computer on her desk, and the very language she uses to teach her students. There are, as always, nodes of resistance: Viridiana has flirted with the idea of having her papers distributed under the "creative commons" copyright license, for instance.[46] But in her darker moments she thinks, isn't it just like a lawyer to believe that one minor change in one corner of IP law can counteract decades of reification of knowledge into a thing that has come to be defined by its relationship to the marketplace? The last time Viridiana checked, she didn't even own the copyright of anything she had written over the last three years, because the journal editors made her sign it away as a precondition of publication.

Diagnosis 2: Viridiana feels disaffected from a few of her colleagues who constantly threaten to organize the faculty into a union, but beyond the stock retort that professionals are not blue-collar workers, she has never been able to articulate why it seems so futile. As we can now see, each of the trends identified above militate against collective action and collective bargaining. The deindustrialization of the West was a direct consequence of the neoliberal push for reducing international trade barriers, particularly when it came to the search for cheap foreign labor that might destroy their political *bête noire,* the trades unions. Once the manufacturing sector had evaporated, what was left over was the overarching culture complex of the "marketplace of ideas": Only the flexible and nimble would survive, with their fickle fealty to stodgy institutional arrangements and fleeting employers. Unions had been broken by the capacity to outsource all forms of production. The rise of adjunct temporary labor in Viridiana's university was a preemptive market that blocked unionization and permitted outsourcing of many professorial functions. Indeed, the knowledge society was merely a bowdlerization of Hayek's reconfiguration of The Market as the ideal information processor, once one takes into account the neoliberal precept that there is "no such thing as society" outside of the market. Neoliberals have built a world where organization is futile.

Diagnosis 3: Viridiana's skepticism toward the blogosphere and uneasiness with techno-utopianism may portend something more than fuddy-duddy Luddism. The computer and the Internet have also become inextricably tangled up with the neoliberal project. For many, the Web exemplifies the neoliberal belief in the "wisdom of crowds" and Google's mantra that "nobody is as smart as everybody," with the market as a stand-in for the hive mind. Reading someone like Benkler (2006), for instance, one encounters a crusader for the freedom of information who has lost all faith in the state as a political actor, and therefore invests what remains of his utopian fervor into the prospect that "the networked environment makes possible a new modality of organizing production: radically decentered, collaborative and non-proprietary . . . [the] commons leaves individuals free to make their own choices with regard to resources managed as a commons" (60–63). One therefore substitutes a thin negative nondevelopmental conception of freedom for something as positive as tenure. The neoliberal Nobel economist Ronald Coase (1974) or the neoliberal legal commentator Richard Posner (2005) could not have put it better. As Robert Bork (1963), a neoliberal theorist and failed Supreme Court nominee, suggested, the First Amendment protection of free speech was only "intended" to apply to political argument, not to literary or scientific discourse. The wisdom of crowds sees very little use for the older hierarchical conservation structures of the organization of science.

Diagnosis 4: Viridiana feels as though her students don't relate to the grand archive of human scientific thought the way she used to. The massive fortification of intellectual property is one of the stunning success stories of the neoliberal project. It has been the primary method deployed to transmute the marketplace of ideas from a dream scenario into quotidian reality. The dogma that no one would think, or at least be bothered to convey their thoughts to others, unless they somehow received market recompense for their labors is a tremendous slander on the history of science and culture, but nevertheless it has carried the day to become folk wisdom in the modern academic order. Students have to be taught not to steal digital music files, even though within their lifetimes making copies of music had been treated as fair use under copyright law. Likewise, students have to be taught not to steal research that they may have carried out themselves, unless they have secured permissions from everyone from their thesis advisor to their university technology transfer office.[47] Because the marketplace is deemed the greatest information processor known to humankind, there is no way that a few extra property rights imposed here and there might actually throttle the further production of knowledge and culture. To suggest otherwise would be to contradict every foundation on which faith in the knowledge economy was built.

Diagnosis 5: Viridiana feels wary around her colleagues who own start-up firms that grew out of their own research. She needs to come to appreciate that this is an artifact of the reengineering of the modern corporation, another triumph of the neoliberal project. Early on, neoliberals had worried that the separation of ownership from control and the bureaucratic character of the large corporation might actually mitigate or otherwise thwart the operation of market forces in the very heartland of capitalism. The neoliberal solution was to make the inside of the corporation work more like a virtual market. The "reforms" that were imposed from the 1970s onward ranged from the award of lavish stock options to upper management to more correctly "align their incentives" with the owners of the firm, to the attack on the M-form structure, which we have cited as the "breakdown of the Chandlerian corporation." One of the most important aspects of this marketization of the corporation was to treat knowledge not as the special possession of the employees of the firm, or as inherent capacity of a social unit borne out of shared experience, but rather to acquiesce in the power of the market as an ideal information processor. Firms should not bother to nurture talent internally over time but instead buy what they need from other firms. The upshot was to render corporate R&D fungible and to outsource it to the lowest-cost producer. Ambitious faculty members are also obliged to turn themselves into little firms, avid entrepreneurs of the academic self. Obviously, the fortification of intellectual property helped render this a viable proposition.

Diagnostic 6: Viridiana harbors a nagging suspicion that pumping more government money into science via the NSF or NIH would not really make things better. She is aware that the NIH budget was doubled from 1998 to 2003, and that in retrospect, this was deemed a failure (Couzin and Miller 2007). Viridiana needs to think harder and with greater subtlety about the promoters of such negative assessments. It should have become apparent by now that it is neoliberal dogma that there is no way that the nation-state should act as any kind of science impresario, since the market will always outperform the state as conveyor and nurturer of knowledge. This was indeed the site of one of the very first battles fought by Hayek and his comrades, even before he became known as the patron saint of neoliberalism.[48] The marketplace of ideas has always been Exhibit A, trotted out whenever the state has endeavored to withdraw from its Cold War functions as patron and organizer of scientific research. Because the marketplace of ideas has progressively become confused with the very definition of freedom itself, it has become almost impossible for the state to resist the implication that the communal promotion and management of science verges on the antidemocratic. This constitutes the half-submerged subtext to modern battles over the politicization of scientific research.

But there is one more shadowy aspect of the neoliberal regime that would itself go a long way to clarify Viridiana's unfocused discombobulation. Neoliberalism preaches that one must *actively construct an ideal market,* not just wait for it to appear on its own; obviously, this applies to the very marketplace of ideas as well. Hayek and the early neoliberals set out to forge the post–World War II institutions they believed were called for by founding a nested set of institutions to propagate their cause. At the center lay the Mont Pelerin Society, the semiprivate debating society, which restricted its membership to a select few true believers vetted by Hayek and other members. This core set was able to discuss the development of neoliberal doctrine with a "blue sky" attitude, absent the scrutiny and critique of those opposed to their politics. Moving outward, the next circle consisted of a limited number of academic departments that were taken over by the like-minded: the University of Chicago economics department and law school, the Freiburg School, the Swiss Institute of International Studies, and others. Their job was to endow the ideas concocted at Mont Pelerin with a veneer of academic legitimacy. The circle after that consisted of a set of purpose-built think tanks, such as the Institute for Economic Analysis, the Hoover Institution, and the American Enterprise Institute, which served to provide homes for neoliberal scholars to produce targeted policy documents in specific national contexts. Another outer circle consisted of foundations like the Heritage Foundation, set up to intervene in political controversies and newspaper op-ed pages and to provide TV talking heads and a presence in other popularized venues in a timely fashion.[49] The point of the neoliberal marketplace of ideas was not (as most still seem to believe) to simply let a thousand flowers bloom on a level playing field, to permit any and all criticism free play, and eventually induce the truth to come out of its own accord. It is instead geared to submit all ideas to the refining fire of dollar votes, within a consciously structured interlocking set of economic markets. This was deemed the only way to counteract those "second-hand dealers in ideas" about whom Hayek was so filled with disdain in the 1950s.

As one can observe, the predominant areas where neoliberals staked out their fledgling institutions were initially the *social sciences*, especially in economics. However, it would be a mistake to think that the model could not or has not been extended in the interim to the *natural sciences*. What Viridiana has observed over the last two decades in climate science, evolutionary biology, pollution ecology, health policy, clinical pharmacology, and any other hot-button area of the natural sciences is the concerted construction of parallel Russian doll structures of the neoliberal blueprint for a vibrant "marketplace of ideas" responsive to corporate concerns. Where does Viridiana think the Discovery Institute, the "Advancement of Sound Science Coalition" (hosted by the Cato Institute), the National Quality Forum, the Lavoisier Institute, the Ewing Marion Kauffman Foundation, the Tobacco Institute, the Heartland Institute, and a veritable brigade of others come from, anyway (Oreskes and Conway 2010)? To blame it all on a few idiosyncratic rich cranks, or better yet, upon the silly season of postmodernism that has supposedly swept like a virus through academe, is to read from the script that the neoliberals themselves have so conveniently supplied.[50]

Americans, with their habitual self-absorption and political myopia, have tried to characterize this as a "Republican war on science" (Mooney 2005) but, as usual, this misses the forest for the trees. Indeed, the Republican

Party is merely a Johnny-come-lately to a phenomenon with deeper roots. What we are living through is a transnational program for the spread of the neoliberal marketplace of ideas to every nook and cranny of human intellectual discourse—or, at least, to every area that holds at least some prospect of making a buck. It was inevitable that neoliberal successes in the social sciences would then be extended to the natural sciences, for as every science studies scholar is aware, the latter fields have at least as many political implications as the former.[51] What this means is the creation of the full panoply of think tanks, activist sites, and echo chambers for the promotion of a specific kind of commercialized natural science, again, conveniently, just as universities render themselves open to the commercialization of university research. The entities funding these novel institutions conveniently remain hidden behind the veil of anonymity, appealing innocently that they are just another democratic participant in the open agora of intellectual discourse. If you try and follow the money, well, then, they are shocked, just shocked, that you would dare to accuse them of such crude interest politics (McNeil 2006). They are just trying to help mankind by getting useful research out the door that much quicker.

Viridiana is not used to getting much for free in the new commercialized regime of science, but she checks her mailbox at school one day and discovers—*mirabile dictu*—a free journal, sent unsolicited to her and other faculty members. It is called *The New Atlantis: A Journal of Technology and Society,* and it is published by something called the Ethics and Public Policy Center out of Washington, D.C.[52] As she leafs through it, there are articles on biotechnology, evolution, cloning, space science, and even the Turing test—and in each and every case, the article hews to the neoliberal line described above. Of course, you might aver, she could just chuck it out, just like she tosses the *Chronicle of Higher Education;* no one is forcing her to read it. She is free to ignore the fact that the cultural agenda for discussion is progressively being set more and more by these broadsides emitted from some indeterminate coordinates vaguely outside/inside of academia—that is, at least until it finally impinges upon her own research funding.

Finally, Viridiana should keep in mind that the extraction of the state from any obligation to provide education to its citizens has constituted a neoliberal tenet since the earliest days of the movement (Friedman 1962, Chap. 6). In the marketplace of ideas, education is the commodity *par excellence,* and the early neoliberals had every reason to think that it was the state stranglehold on education that had created the hostile political atmosphere in which they had found themselves stranded in during the immediate postwar era. Of course, they never advocated full withdrawal cold turkey—instead, they preached choice in education, vouchers, creeping privatization of the school system piecemeal (as in the for-profit classroom TV Channel One, promoted by Edison, an operator of for-profit elementary schools), enforced standardized testing, enabled home schooling, and incessant demonization of the public teaching profession. Twenty years of success in these and other initiatives can only be understood as artifacts of the neoliberal project to render the marketplace of ideas as the true and only legitimate font of education, and to reduce everything else to a crippled surrogate. In this brave new world, it is not anything even remotely approaching a joke that Donald Trump can form what he calls Trump University and promote it with a money-back guarantee![53] If you think that is a cheap shot, then check out some of the course offerings of the fastest growing providers of distance education at the college level, such as the University of Phoenix or Universitas 21 Global, the self-designated "world's premier online Graduate School."

A fundamental premise of my book is that the major objective of the neoliberal program is to totally decouple most functions of scientific research from the educational functions to which they had been wedded during much of the twentieth century. Previously, universities had "bundled" many diverse functions together into one institution, but as they become progressively commercialized, neoliberals insist they will of necessity find they have to become unbundled and sometimes spun off, to better concentrate on their core markets. Science must fragment so as to become more accountable. In this it will have to follow the earlier example of the Chandlerian corporation, which opened its doors to the market gales of creative destruction to blast away the vestiges of nonmarket operations.

I

Why We Should Not Depend Upon the Existing Content of an "Economics of Science"

2

The "Economics of Science" as Repeat Offender

Suppose that Viridiana was coaxed to embark on a program of reading in the literature that one of her colleagues designated as the "economics of science," in, say, 2005. Here are four quotes from well-known economists she might have run across, just as I did:

Fischer Black, inventor of the famous Black-Scholes options formula(in Mehrling 2005, 73): "People love to create ideas, just as actors love to perform . . . We don't have to pay researchers either."

Viridiana muses: Glad we got that settled (though there is the nagging consideration that Black was spectacularly wealthy at Goldman Sachs by the time he said that).

Suzanne Scotchmer, a Berkeley economist and specialist in "innovation" (2004, 2): "Creation and discovery are mysterious processes. But whatever else is required, economists are reasonably certain that incentives matter . . . The only fundamentally new incentive scheme of the past 400 years is intellectual property."

Hmmm . . . these quotes seem to contradict one another at the most basic of premises. But let's move on . . .

Michele Boldrin and David Levine, game theorists at Washington University and crusaders against intellectual property (2005, 1252): "Common legal and economic wisdom argues that competitive markets are not suitable for trading copies of ideas, as ideas are intrinsically different from other commodities. For the most part, these arguments are incorrect . . . ideas are *not* different from other commodities, and those few dimensions along which ideas are different do not generally affect the functioning of competitive markets."

Viridiana squirms: Well, gosh, doesn't that seem kind of extreme? . . . I know Bentham said poetry is as good as pushpin, but wasn't he a fusty Victorian windbag? . . .

Dominique Foray, a prominent French science policy expert (2004, 2–5): "Knowledge is above all a cognitive capacity . . . Information, on the other hand, assumes the shape of structured and formatted data that remains passive and inert until used by those with knowledge . . . The economic problem of information is essentially one of protection and disclosure, that is, a problem of public goods . . . knowledge requires the mobilization of cognitive resources."

OK, that *definitely* contradicts the previous quote. Viridiana would probably be getting a little irritated at this juncture. Maybe she starts to regard her economist colleagues as falling short as bona fide experts on the knowledge economy. (In 2008, she leaves disparagement of their macroeconomic prowess to others.) But suppressing her spleen, she is willing to venture a little further to begin to grasp why economists seem such a letdown when it comes to science.

Beware of Economists Bearing Gifts (of Knowledge)

The root of the problem seems to be that "science" is not just another widget that can readily surrender to the economists' "toolkit"; the nature and significance of science cut to the very quick of the ambitions and competence of economists, to the very self-conception of economics as a valid form of inquiry. Science has resisted and continues to resist being confined to the status of a stable object of economic analysis. Without modern science, the set of contemporary doctrines and practices huddled together under the rubric of "economics" would make no sense whatsoever. This, then, is the first reason why the "economics of science" cannot be treated with the same freakonomic insouciance one might approach the phenomenon of an economics of housing, or agriculture, or monetary policy, or "the family," or sumo wrestling: here the ambitions are a little too debilitating.

The economics of science turns out to be intractably obstreperous precisely because economics and science share so much DNA in their life histories. For instance, most major schools of economic thought from Marx onward owe their very existence to trends in the natural sciences more or less contemporary with their own intellectual trajectories. Early neoclassical economics derived its mathematical model from the prior formalisms developed in energy physics; the American Institutionalist school of economics was predicated upon specific doctrines of biological evolution popular in its heyday; the postwar American orthodoxy in economics owed many of its characteristic themes to the "cyborg sciences" that had grown up around the computer after World War II, as did Carnegie-Mellon–style "behavioral economics."[1] Without access to the authority and integrity of the natural sciences to provide secure templates for intellectual inquiry, modern economics would find itself quite at sea, not knowing what would constitute a "good model," or how to begin to formalize "information," or how to integrate theory and empiricism, or even what sort of activities would qualify for the honorific of "science." For that reason alone, economics continues to be denied a privileged Archimedean point to stand "outside" of science in our culture, banned from a Throne of Judgment from which to pronounce ukases. It is a little like the inability to totally eradicate a childhood accent: it matters not a whit whether the individual economist is aware of it or not.

Some would aver that this caveat smacks of the genetic fallacy: after all, many of the natural sciences themselves grew out of natural theology, but that has not prevented them from nurturing separate and distinct modes of inquiry over time, and even in some instances coming to train their analytical skepticism back upon religious concepts and doctrines. While this analogy displays some intriguing parallels, the critical difference in the case of the natural sciences is that while scientists' research does overlap with the explanatory ambitions of many religions (speaking, for instance, of the origins of the universe, the place of humanity in the natural order, the last days of the Earth, and so forth),scientists do not (beyond a few zealots like Richard Dawkins) generally take it upon themselves to instruct the various religions how better to do their own jobs. Economists, by contrast, have unself-consciously anointed themselves as uniquely competent to pontificate upon the "best" ways to organize and prosecute scientific inquiry. There may be something of a whiff of hubris in this attempt to reprimand the very legators of their own tradition; it would be as if Jimmy Page had endeavored to instruct Mstislav Rostropovich or Jacqueline du Pré on fine points of musical technique.

Perhaps something is lurking here of greater consequence beyond the mere complaint that those ham-fisted gradgrind economists are at it again, engaging in the same old imperialistic slash and burn that they have gotten away with in the past, or as Wade Hands so aptly put it, "Take a game-theoretic model from industrial organization; change firms or players to 'scientists'; add the adjective 'epistemic' in a few places . . . and suddenly you have a philosophical model of scientific knowledge" (2001, 373).The problem, as Hands proceeds to point out, is that proclaiming the existence of a consensus "economics of science" leads one fairly abruptly into the treacherous quicksands of reflexivity. This sometimes pops up in the form of near-Freudian slips, as when the economist Robert Dorfman (1960) once "joked" that he hoped no one would set about calculating the "information content" of his own journal article! However beady-eyed and jaundiced they might be with regard to the sordid motives of others, whenever they start reducing science to the self-interested drives of strategically savvy individuals, they confront the ultimate *tu quoque:* How about you, Mr. Economist? We don't hear you proposing a policy to scale back on the "production" of economics in the interests of efficiency, now do we? Why do economists warrant a special dispensation when it comes to being motivated primarily by "the truth" and the general welfare, when they don't believe it about anyone else?[2] I am still poleaxed that, in the late Science Wars, every sober, lion-hearted "realist" rushed to convict those wicked postmodern sociologists of science of committing the "relativist's fallacy" (*So you think scientific knowledge is a sociological phenomenon? Then how about you, claiming to know the "truth" about science?*); to my knowledge, not a one bothered to turn their disdain upon the economists, who so easily fall prey to exactly the same fallacy.[3]

So maybe the application of economics to the phenomenon of science comes studded with a plethora of pitfalls, waiting to ensnare the confident captain of catallaxy. Indeed, this may help explain the curious fact that a self-aware "economics of science" made a rather delayed appearance on the academic scene. Economists had indulged a predisposition to concoct theories of technological change at least dating back to Charles Babbage and Karl Marx in the nineteenth century, but "science"— that was left as the great unexplained First Mover in economics until late in the twentieth century. It was almost as if economists knew in their bones that an "economics of science" would be a bridge too far, and hence hesitated to cross. Given the reticence, it now becomes of some interest to inquire why what had previously been taboo has currently become not only acceptable but even fashionable in the twenty-first century. Had contemporary economists inadvertently mislaid their sense of decorum? Had they lost all capacity to become embarrassed by their own hubris? Or, more likely, was there some external set of cultural determinants that opened a space for economists to rush in?

The previous chapter proffered glimpses of an explanation; here I treat the matter from the vantage point of intellectual history. The primary factors in the encouragement of the debut of an economics of science have been the across-the-board rise of the neoliberal characterization of the market as an information processor and the subsequent deindustrialization of the advanced Western economies. A resort to such external structural accounts when it comes to the change of heart of economists is nearly unavoidable, because there is no serious evidence that contemporary economists have made any noteworthy theoretical breakthroughs, which would by themselves have summoned forth the newly proclaimed "area" of the economics of science from the 1980s onward. The worst transgressions in this regard have been perpetrated by journalists who have signed on as credulous cheerleaders for the modern profession:

> Thanks in part to our recent experience of the ways new information and communications technologies, biotechnologies and others are affecting our societies—economists have made huge strides recently in analyzing the process of innovation . . . Economists are explaining the process of patenting innovations, the best way to protect intellectual property in order to encourage innovators, . . . the paths through which research in universities is put into commercial practice by entrepreneurs, and many other empirical details of how it is that technical change turns into economic growth. (Coyle 2007, 63)

The idea that technical change somehow induced economists to become experts in how science works and how knowledge should be nurtured is implausible, for reasons I shall shortly enumerate.

Work that appears under the contemporary rubric of the "economics of science" has in fact been distributed over a curious range of classes of models, diverse types of concerns, and levels of analysis. For convenience, let us sort them into three broad categories. First, there is a literature that concerns itself with approaching the actions of the individual scientist as a question of epistemic optimization of something or other; in many instances, these analysts think they are saying something cogent about the brain. The relevant communities here might range from so-called neuroeconomics to philosophers enamored of rational choice theory to decision theorists to garden-variety microeconomists. While I shall briefly touch on these issues below, by and large I bypass consideration of these groups in this volume.[4] Then there is a second cadre, shading over from the first, that tends to be much more concerned with what might be called "social" models of science, remembering that economists treat society in a rather fragmented manner, often as an aggregation of individuals. A major theme of this literature is whether one can or should treat science as a simple marketplace of ideas, and whether there might be cause to imagine alternative social structures relative to those that currently organize science. I shall have plenty to say about this version of economics of science throughout this book. Finally, there is a third group that approaches the issues from a much more concertedly macroeconomic perspective, linking it to concerns about growth, technical change, and welfare economics. This last group receives special attention in the current chapter. Of course, sometimes individual authors range rather intrepidly over all three domains, so it can be at best a rough and ready taxonomy; but no matter who these economists are, they all think they know some deep things about knowledge.

People like Viridiana looking for a primer on the economics of science are therefore put at a great disadvantage: Almost every book or article on the subject tends to retail *one* modern theory from a single category detailed above, and proceed to treat it as the *only* relevant language for discussing science. The same holds for key journals like *Research Policy* or *Journal of Technology Transfer*, and generalist outlets like the *American Economic Review* (*AER*). This has resulted in mass confusion on the part of outsiders, who have little time or interest in keeping up with all the twists and turns of economic doctrine in the modern era. Here is one example: Rebecca Eisenberg is one of the most insightful legal thinkers about the modern problems of science in America, and her work is mentioned many times in the rest of this book. And yet, in a moment of distraction, she wrote the following about the two famous papers by Richard Nelson (1959) and Kenneth Arrow (1962):[5] "These early sketches of the economic problems that lie behind intellectual property have held up remarkably well over the past 45 years" (2006, 1013). Well, not to rain on her parade, but bluntly: no, they haven't. Coming to comprehend how these "market failure" approaches to the economics of science have themselves suffered intellectual decrepitude is another major task of the present volume. If someone as wise and as discerning as Eisenberg can go so perilously off the rails, well, then there isn't much hope for the average scientist like Viridiana. Or, rather, there won't be if we don't clarify the important proposition that *there is no single orthodox approach to the economics of science.*

My immediate task is to inoculate Viridiana (and you, the reader) against acquiescence to commonplace sentences such as "*The* central economic problem of science is *X*," or "The solution to the problem of the efficient production of information is *Y*," or (more concretely) "The logic of the public good rationale is precisely that of the Bayh-Dole Act" (Geiger and Sá 2008, 119). This will help substantially lower the temperature when you are later confronted with sharply contradictory statements by economists, such as those that introduced this chapter. Here I will survey the "linear model" of science, science as a public good, science in growth theory. While this does not comprehensively cover the waterfront, it does hit most of the high points that would be encountered by the interested layperson trying to figure out what economists have said about science in the recent past. Without getting too bogged down in the details, which would only add dyspepsia to disorientation, I will provide a sketch of each position, roughly arranged in the chronological order in which they first appeared.

Things That Economists Say about Science

Neoclassical economists frequently insist they all freely subscribe to the same core set of principles; but the more you learn about them, the more you appreciate it just isn't so (Mirowski and Hands 2006). Perhaps the experience of watching Nobel-winning macroeconomists savage each other over the past few years in the wake of the economic crisis has rendered this point a tad more plausible. As with so many other elaborate doctrinal orthodoxies, it turns out to be much more important for the faithful to discipline the doubters and expunge heretics than it is to get their own stories straight. This is especially the case when it comes to economic accounts of science. While schism and sectarianism potentially can fissure without limit, the last half of the twentieth century saw a limited number of positions regarding knowledge in orthodox economics, which I will survey below.

The Linear Model

It has become a cliché to begin most courses on science policy and the economics of science with something called "the linear model of innovation," only to rapidly disavow and disparage it, almost to the point of anguish. Some rhetorically wonder how it was anyone ever got bamboozled by such an implausible account of the relationship of science to the economy. Luckily, there is now a second generation of literature that asks instead: What did it really signify, and who could have been said to have believed it? Not unexpectedly, the role of economists in this tale looms especially large.

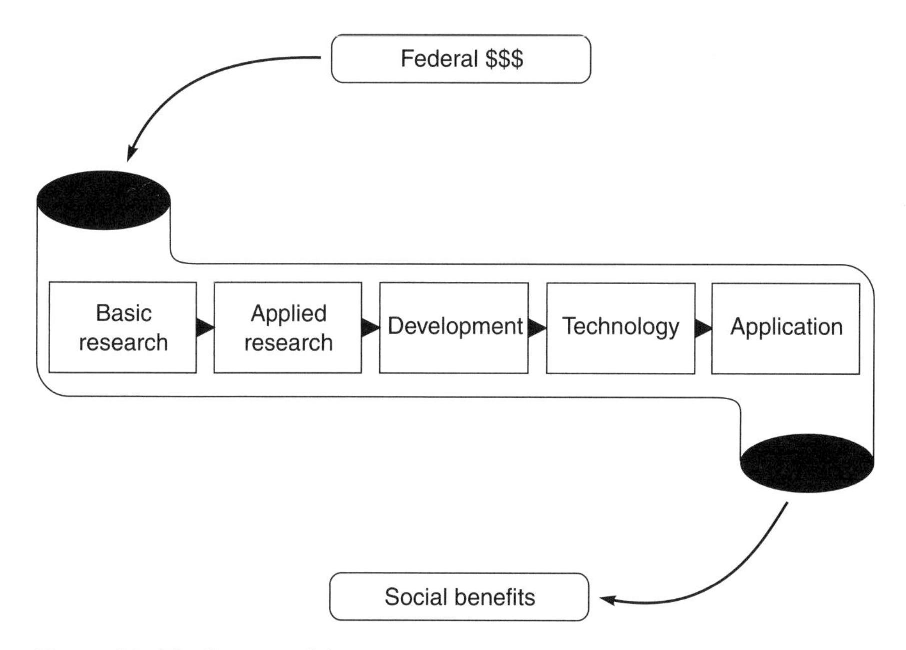

Figure 2.1. The linear model.

Figure 2.1 conveys a number of the salient aspects of the linear model in a pithy fashion: first, it has something to do with nationalist ambitions, given that the $$ at the wellhead are identified as coming from the federal government; there is also an assertion that the world of science can be divided up into "basic," "applied," and technological components; furthermore, that there is an orderly or "linear" progression to these subcomponents, such that each subsequent function in the pipeline requires the preceding activity as a necessary prerequisite; and that the end product of the sequential progress will result in social benefits outside the pipeline of the science pursued, most frequently manifested as general economic growth. Just like water, knowledge naturally finds its level by trickling down from basic research to application. While it has become *de rigueur* to critique most of the intermediate propositions of this linear model, it has perhaps not yet sunk in sufficiently that the real consequence of the recent demise of this model has been to call into question the overarching idea that science directly causes economic growth.[6]

Let us unpack the linear model, starting with the idea of the necessary sequence of basic → applied → technology → commodity. Historians have reminded us that these categories were not timeless but tended to be of more modern provenance. Clearly, distinctions between "pure" versus "applied" contexts were being made well before the twentieth century, but one has to treat the notion of "purity" gingerly.[7] Michael Dennis correctly points out that when later nineteenth-century American figures made their pleas for "pure science," they did not refer to some notion of disembodied science carried on for its own sake nor to an imaginary autarkic scientific community defending its prerogatives, but rather to a pedagogical ideal of a species of hands-on higher education where teaching and research were combined in a setting relatively sheltered from commercial considerations. At that stage there was no necessary implication of a requisite lockstep sequence of innovative activities. "Purity" was inevitably defined relative to the debased, but the activities so disparaged changed dramatically over time, depending on the current institutional structures taken to be the Platonic ideal. Hence in the nineteenth century, "purity" might have been code for freedom from state intervention or else designate something that occurs *ad naturam* free from human determination, or else signify an activity situated outside the orbit of industrial regimentation. At the turn of the twentieth century, "purity" became all bound up with the push to distinguish shop culture from school culture, and the drive to construct a separate curriculum for engineers, as compared to fledgling academics. Yet by the 1930s, the terminology itself underwent further deformation, with the gradual displacement of "pure" by terms like "basic" or "fundamental" when describing the types of science the author in question sought to highlight. Once the terminology shifted, it seems that quite a few writers strove to indicate that there was some species of inner teleology at work, especially relative to that other neologism dating from that era—"technology" (Schatzberg 2006). Only then do we find statements like the following: "No matter how 'pure' the laboratory discovery may be, it issues sooner or later in practical applications, and these applications find commercial sale through their exploitation by entrepreneurs"(Gray 1935, 539). Thus, by the first third of the twentieth century a lockstep primacy of "basic/pure" over "applied" had started to be taken for granted in some circles, and furthermore, it was frequently further hitched to a commodity caboose that was pulled along in train. By some accounts (Forman 2007), the presumption of the logical and temporal primacy of basic over applied science had become nearly ubiquitous by midcentury.

At that juncture, an ideologically loaded dispute elbowed its way onto center stage, one that persists in informing much of the theory of science in economics. Given that there came to be stipulated several distinct classes of "science" (basic/applied/development), which putatively served as inputs to their descendants, it was proposed that concerted organization and funding would be required to keep the pipeline primed and in good order, such that there would be no disruption of the flow of novel goods at the final commodity phase. It was here that economists first became closely involved in the elaboration of the linear model. Many date this development from the Social Relations of Science movement in Britain in the Depression, in conjunction with its nemesis, the Society for Freedom of Science, and the publication of the "Bush Report," *Science—the Endless Frontier* (1945), in the United States. Let us briefly visit each in turn.

If historical justice ever triumphed, then the British crystallographer J. D. Bernal would be credited with the invention of the economics of science, but this will never ever happen, because Bernal was an unreconstructed Marxist.[8] Indeed, his breakthrough book *The Social Function of Science* begins with this statement: "Any discussion of the application of science necessarily involves questions of economics" (1939, xiv). Bernal was involved with the political organization of scientists in the Depression to oppose fascism, to reform the structure of British science, and to promote the government funding and organization of science. He popularized the diagnosis that British industrial failure (relative to Germany) was correlated with the lack of government involvement in the promotion of science. He helped form the Division of Social and International Relations of Science as a subset of the British Association for the Advancement of Science, and he penned *Social Function* as its manifesto. Within it, he claimed that isolated academic researchers were inefficiently pursuing the science needed for the future and that industrial labs had also been ineffective in producing breakthroughs (1939, 107).

He opined that more than half of the existing money spent on science was wasted (120) and that "the vast bulk of scientific publication is . . . of unequal value; a large proportion of it, perhaps three-quarters, does not deserve to be published at all, and is only published for economic considerations" (118). Significantly, he was one of the first to actually try to quantify some of these magnitudes, and, along with his comrades in the movement, he sought to provide estimates of the relative sizes of basic science, applied science, and development (Godin 2006a, 649–650). Bernal thus pioneered the persistent, modern bad habit of equating the sums of money spent in a sector with the amount of research conducted within that sector, and he constituted the inspiration for a slew of subsequent attempts by various governments to quantitatively measure their science base by resort to monetary proxies (Godin 2005). The mere fact of defining and collating such statistics subsequently helped invest the linear model with a modicum of credibility.

Bernal's movement provoked an almost immediate counterattack on his proposals to plan science, not to mention ridicule his fulsome praise for the Soviet Union. In 1940, John R. Baker, Michael Polanyi, and Friedrich Hayek formed the Society for Freedom in Science (SFS) to counter the Bernalists. While it seemed that the SFS did not enjoy anything approaching the immediate popular success of the Social Relations movement (McGucken 1984, Chap. 9), not least because most of its members did not opt to ally themselves with the military during World War II, it did prompt a number of significant longer-term consequences. The most important from our current perspective is that the SFS marks the earliest entry of some of the major protagonists of the nascent neoliberal thought collective into the arena of science organization and funding. It is no accident that both Hayek and Polanyi were soon thereafter present at the creation of the Mont Pelerin Society (MPS), with its mandate of reconsideration of the role of knowledge in society. For the SFS, the state was portrayed as the greatest threat to the ongoing health of science, rather than as its premier promoter, as it had been for Bernal. Another salient consequence is that it later became standard practice to date the inception of the disciplinary specialty of "science policy" to the Bernal-Polanyi dispute, and further, to bracket the duo with the historian Thomas Kuhn (by no stretch of the imagination a love triangle) in telling an origins story of the birth of the social studies of science and technology.[9] If we include the economics of science in this crowded incubator, then Bernal and Polanyi, not to mention the neoliberals, surely have a lot to answer for.

The experience with the SFS led *both* Hayek and Polanyi to compare the practices of science to their respective visions of the operation of the market soon thereafter: Hayek with his marketplace of ideas, and Polanyi with his republic of science. But hewing to our current concerns, I want to stress it also led them both to conceptualize a version of pure science that they could support against the depredations of the Bernalists. As John Baker wrote to Joseph Needham in July 1942,"I think that a man who has a real aptitude for pure science wastes his talent if his social conscience drives him to become a technologist"(quoted in McGucken 1984, 267).Hence it was within the confines of the SFS that a countereconomics of science was incubated, one that initially disdained the linear model in favor of something closer to treating the pure/applied distinction as a transcendental category existing fully independent of social structure, untethered to any considerations of efficiency or economic consequence. In Polanyi's case, it inspired an elaborate philosophical system in which the ineffability of something he called "tacit knowledge" both militated against the rational planning of science and dictated that *only* the abstract natural sciences (and no other sphere of human inquiry) deserved copious state subsidy with no concomitant accounting for performance or economic success.[10]

Most modern commentators on science policy tend to skip over Bernal, Polanyi, and Hayek, and instead they erroneously attribute the origins of the linear model to Vannevar Bush in 1945. His text *Science—the Endless Frontier* has been subject to repeated close scrutiny over the years as the supposed harbinger and blueprint for postwar science policy in the United States, because it is replete with ringing quotes (here is an example):

> Basic research leads to new knowledge. It provides scientific capital. It generates the fund from which the practical applications of knowledge must be drawn . . . A society which depends upon others for its new basic scientific knowledge will be slow in industrial progress and weak in its competitive position in world trade, regardless of its mechanical skill. (1945, 18)

Like any other really seminal document, Bush's report has been variously interpreted as the first influential statement of the model (C. Freeman 1996; Martin 2003), as "not really" containing the linear model (Edgerton 2004; Balconi et al. 2009), as insisting the "one dimensional graphic image is one he himself almost certainly never entertained" (Stokes 1997, 18), and conversely as merely replicating a set of presumptions already "in the air" (Godin 2006a, 645). Another interpretation treats the linear model as a tactical feint, a way for Bush to thread his way gingerly through a political situation he mostly despised (Reingold 1995). Luckily, we needn't adjudicate authorial intentions in this regard or laboriously recapitulate his problem situation at the end of World War II, since it has been recently acknowledged that Bush did not actually compose the passages conventionally identified as containing the linear model. Instead, they were ghostwritten by the economist Paul Samuelson.[11]

This small fact goes some distance in clarifying the precise nature of the linear model and why it held sway. Here, I agree with Godin (2006a, 2008) in that it seems the entire discourse concerning the linear model got hijacked by economists at midcentury, and therefore is better understood as an artifact of internal discussions and controversies within neoclassical economics, than as a model of science actively promulgated by working scientists.[12] This does not deny that many scientists back then did (and perhaps some still do) believe that "science" is a distinct entity with exogenous causal powers, existing autonomously with its own purity of logic in a space removed from social or economic considerations, only acting upon society from without. Conversely, a substantial number of scientists employed in corporate labs remained unpersuaded throughout the period by the simple primacy of science over technology (Shapin 2004, 2008b). It also is worth noting that a majority of historians associated with the Society for the History of Technology were staunchly united in their opposition to the linear model from the 1970s onward.[13] The linear model only thrived in certain specific locales, but that did not render it any less important. For example, it was recruited to play a major political role in explaining why the general populace should be kept out of decisions regarding science planning and funding after World War II (Dennis 1997).

Beyond the landscape of rejection and acceptance, our current concern is to insist that the *specific format* of Figure 2.1 in portraying how science got produced was forged as a necessary accessory to the postwar development of neoclassical economic theory, and it was concocted almost without any empirical inquiry into the relevant histories of science and technology.[14] The Bernal-Polanyi connection guaranteed that the linear model would prove inseparable from the political question of the appropriate role of governmental intervention in a capitalist economy: Advocacy of the linear model from thenceforth tended to be associated with a pro-statist position, especially in postwar American economics. In that guise it spread to business schools, government statistical bureaus, the military, and beyond (Godin 2006a).

One indication of the extent to which the linear model had become a Trojan horse for one particular neoclassical approach to the economics of science could be found in some pre-1980 attempts to resist it. One of the more famous incidents was a landmark in the history of science policy in the United States, Project Hindsight (and its twin, Project Traces).[15] The Department of Defense had been suffering some criticism by the mid-1960s for funding research not directly or obviously related to its military mandate, and therefore it commissioned a lavishly funded study called Hindsight to gauge the extent to which "basic or undirected science" had contributed to twenty high-tech weapons systems, including the Polaris and Minuteman ballistic missiles, the AN/SPS-48 frequency scan radar, the Starlight night vision scope, and the FADAC field digital computer. Merely posing the problem in this manner began to reveal that it might not have any single coherent answer; however, undeterred, the principal investigators pushed ahead, admitting, "Since the challenge was essentially an economic one, the answers would have to be based on economic benefits"(Sherwin and Isenson 1967, 1571). In an early attempt at scientometrics, the team identified 710 "events" that preceded the final technological achievement in a twenty-year retrospective frame, and decided that only 0.3 percent of those events could be classified as basic or undirected science. This was widely perceived as an attack on the linear model, when in fact it might have been better understood as a symptom of the incoherence of existing abstract economic pictures dictating how science and technological innovation meshed. Instead of exploring that possibility, the National Science Foundation (NSF) quickly commissioned its own opposing study, using a different methodology, extending the time frame and choosing different end-products of the pipeline. It was really no big surprise that the NSF-sponsored Project Traces (1968) found that basic (now simply equated with academic) science loomed much larger in thrusting new commodities out the pipeline, pushing the relevant proportion of "non-mission oriented" research estimate up to 70 percent. Everyone in the science policy community then breathed a sigh of relief and unleashed an army of economists who set about calculating astronomical "rates of return" to public R&D, itself blithely equated with state-funded academic science, more "basic" than "applied" and all predicated upon the linear model.[16]

The quest to quantify "rates of return" to nonmarket research bore within itself a fundamental contradiction, one that took decades to finally provoke a rejection mechanism within economics. The linear model held forth a promise that apparently nonmarket, nonaccountable (except possibly in some transcendental search for truth) activities could nonetheless cogently be tamed through a cost/benefit calculation, by "backward imputation" from the empirically observed value of the final goods to the virtual value attributed to noneconomic activities that had initially set them in motion. This notion of temporal imputations had been a calling card of the early twentieth-century Austrian School of economic theory, especially in its treatment of production; but in postwar America, the profession had been rapidly sloughing off any such Austrian doctrines in neoclassical production theory. Hence the clash between these virtual "rates of return to basic science" with contemporary neoclassical production theory, not to mention the awkwardness of the whole notion that one could justify nonmarket activities by boldly pretending they were priced, is one major reason why the profession rapidly shifted its allegiance to the whole theory of "public goods," covered in the next section.

The oddest aspect of the festival of hubris was not so much that economists had wrought a risibly rudimentary model of how science actually worked and then foisted it upon an unwitting public; it was rather the stark rapidity with which the linear model fell into disfavor after 1980. Of course, numerous neoliberals had been unhappy with the portrait all along, dating back to the Depression complaints of Polanyi, but they hadn't managed to make a dent in the overall consensus until roughly the fall of the Berlin Wall. The historian Paul Forman (2007) has rightly posed this about-face as being itself a phenomenon requiring historical explanation. Many, if not most, intellectuals simply took both the priority and primacy of science over technology as gospel before 1980; just as confidently, most disavowed it afterward. It is remarkable the extent to which the linear model abruptly collapsed, and not just in economics, but in such unrelated areas as the philosophy of science, science studies, and the popular culture, all roughly simultaneously. Some, like the science studies icon Bruno Latour, went so far as to claim there had existed only one ontological entity called "technoscience" all along, while all those prior distinctions along the lengths of the pipeline had been pipedreams. Clearly, it was not simply a matter of empirical disconfirmation: stories about the indispensable importance of steam engines in prompting the laws of thermodynamics, or the role of aniline dyes (and not clinical biology) inducing the birth of synthetic pharmaceuticals, or the role of artillery in the prompting of Galilean mechanics had been familiar to historians of technology all along. New stories, like the one about how the technology of railway clocks linking distant locales served as conceptual preconditions for the abstract theory of relativity (Galison 2003), gained enhanced cultural cachet in the new environment.

The sudden loss of faith circa 1980, just like the post–World War II elevation of the linear model, had very little to do with deep changes in our understanding of science per se, and everything to do with the economy and the preoccupations of economists. As Godin documents, "Over the period 1961–2000, the central aim of science policy has been to bend science and technology to economic ends" (2005, 287). Changes in the economy were therefore far more important for the fate of the linear model. The demise of the linear model maps far too neatly into the neoliberal ascendancy in intellectual life and the modern commercialization of science for it to have been merely coincidence. Most of the six structural trends outlined in the previous chapter predated the loss of nerve; and a select number have played a direct role in its demise. The deindustrialization of the West has wreaked havoc with the idea that a strong competence in "pure science" was a prerequisite for a vibrant manufacturing base; the Vannevar Bush quote above has been flatly refuted in China, Japan, Malaysia, Korea, and elsewhere.[17] Even more important was the progressive withdrawal of nation-states from their previous positions as primary patron and science manager in charge. The interventionist politics normally attached to the linear model, once a major source of its credibility, had now become a liability, once the state no longer wanted reasons to directly prime the pump of innovation. And it was no accident that it was precisely at that juncture that certain neoliberals, inheritors of the mantle of Polanyi, launched their most savage attacks on the linear model, both from inside and outside the economics profession.

One of the most prominent neoliberals inside the economics profession, Milton Friedman, was rather quick out of the gate, perhaps heartened by the installation of Ronald Reagan as president. He wrote:

> Basic science in the United States prospered prior to large-scale government research by combining research with teaching in colleges and universities and by soliciting grants from private individuals and foundations . . . Given the undisputed success of private financing in the past, the burden of proof that the benefits of government financing exceed the costs surely rests on those who support such financing . . . I favor major cuts in NSF grants to all disciplines as a step toward the abolition of the NSF. I now brace myself for brickbats from my injured fellow economists, as well as the members of the other disciplines represented in the National Academy of Sciences. (1981)

Amazingly, Friedman did not suffer any of the feared brickbats. Instead, the National Academy, in conjunction with President Bill Clinton, awarded him the National Medal of Science in 1997. Another Chicago economist and MPS member went on record suggesting that the private firm Celera Genomics did a better job mapping the human genome than the government-sponsored Human Genome Project (Becker 2000). A neoliberal biochemist made an even bigger splash, if only by rendering the superfluity of the linear model more manifest:

> Is technology dependent on science as Bernal believed, or is the opposite true as Gillespie and Henderson claim? Is technical innovation largely directed at saving labor or at other costs? How important is the rate of diffusion of new technology, and what determines that? . . . These and other questions are fascinating, but for the science policy-maker or for the politician they are irrelevant. *Laissez faire* meets all needs . . . [the] linear model and [attendant] *dirigisme* have failed . . . The Market Place does not worship false Idols, it makes empirically correct judgments. It is

the government funding of science that is an Idol of the Tribe. (Kealey 1996, 344–345)[18]

Soon, all manner of commentators, even those who would strenuously deny any latent neoliberal tendencies, began to sneer at the linear model. People who would not normally be caught dead as fellow travelers of the Reagan or Thatcher Revolutions nonetheless began to express nagging doubts about whether "basic" research had really ever existed, or in any event, should be considered a necessary prerequisite to keep technological change flowing freely.[19] They might couch their reservations in the fashionable STS language of social construction, or perhaps instead in the rather more prosaic language of "the funeral for the old assembly-line view" of science and technology, but it all pretty much came down to the same thing.[20] Subtle epistemological gradations between science and technology were out; the brave new world of technoscience had arrived.

The final proof that the linear model had become démodé was when some economists who were most closely associated with the invention and promotion of the linear model in the American profession felt impelled to speak ill of the deceased and recant on this topic. "In principle there is a clear divide between science and technology. In practice there isn't" (Nelson 2004, 455). "Industrial innovation emerges from a complicated process, where fundamental research need not play an initiating role, *or at times, any role*" (Cohen, Nelson, and Walsh 2003, 109, emphasis added)."Everyone knows that the linear model of innovation is dead" (Rosenberg1994, 139).

Science as a Public Good

Most outsiders to the economics profession, if they have come across even a sampling of the literature we are in the process of surveying, frequently find themselves attracted to the notion of science as a "public good." Although the idea is more than a half-century old, there is something piquant about the earnestness with which tyros set out time and again to explain the slippery notions of nonexclusivity and nonrivalry as if they constituted some deep insight into the eternal nature of knowledge. The idea that something about science is inherently public, often yoked together with the notion that knowledge subsists as a durable entity that you can use without using it up, seems to strike a chord deep within.[21] The disenchantment comes when they learn that the concept turns out to be much more an artifact of the twists and turns of postwar neoclassical economic theory than it is some philosophical nugget of the Wisdom of the Ages. Furthermore, as the concept has slowly been drained of much of its cogency within the modern economics profession, its utility as a weapon to "defend" science has correspondingly lost its potency in the modern rough-and-tumble of commercialized science. Worst of all, it mostly distracts attention from the way science as a process actually works in the modern regime.

The literature on public expenditures in neoclassical economics dates back almost to the 1880s, but the specific notion of the public good was first formalized by Paul Samuelson in 1954. He conceived it to formally justify the economic legitimacy of government intervention in the economy while still pledging allegiance to the iconic neoclassical model of the free market. Samuelson sought to argue that there was a special class of traded goods that did not exhibit the property commonly assumed in the presumption of "commodity space": the idea that if I use the good, it does not prevent your (simultaneous) use of the good as well—this later was dubbed the property of "nonrivalry." Samuelson (1955) identified national defense, police officers, public safety items like lighthouses and highways, the courts, and public education as public goods. Soon, this definition became confused and conflated with the existence of "externalities," namely, beneficial or detrimental effects of the commodity that have impacts beyond the immediate buyer and seller. In "welfare economics," the generic problem of externalities was redefined as a problem of nonexclusion, such that all the relevant aspects of the commodity were not correctly priced by the market. Kenneth Arrow transported this version of the public good into the arena of general equilibrium theory, insisting that "the problem of externalities is a special case of a more general phenomenon, the failure of markets to exist . . . [There are] two possible causes of market failure: (1) inability to exclude; (2) lack of necessary information to permit market transactions to be concluded" (1985, 513). Information deficiencies were conflated with public goods qualifications from thenceforward.

Over its history, there have been many flaws identified in the neoclassical model of markets, but one distinctive penchant of postwar American economics was repeated recourse to changing the definition of the commodity to make it seem as though one had "fixed" the problem at hand. This was a house specialty of the economists billeted at the Cowles Commission in midcentury. The reason this happened was that the mathematical definition of the commodity space was widely viewed as the analytical wild card in which there was unlimited freedom to let the imagination soar.[22] Whether or not the wild card was actually warranted, it certainly served as an immensely effective defense mechanism, shifting attention from the flaws of the model to interminable "technical" disputes over the true ontological character of the commodity in question and the right way to amend the model to capture it. One observes this *rifacimento* in the standard repertoire, with the blame for "market failure" in Arrow's account being shifted from what would colloquially be considered *failure* of the market process itself to a eschatology over whether some purely virtual markets were "missing" or not: The bald presumption that The Market was always more powerful than any other possible allocation mechanism was barely hidden in such prognostications. Markets worked, except, conveniently, where they didn't.

Thus by the late 1950s the artifice of the public good was one (but not the only) conceptual attempt within the tradition of neoclassical economics that sought to justify intrusion by the government into the marketplace by insisting that there were a few anomalous 'commodities' that did not possess the standard attributes presumed in orthodox economic models. In particular, these goods could be produced at "zero marginal cost," which would suggest that standard equilibrium pricing (where price = marginal cost) would lead to the so-called underprovision of the good, or worse, it not getting produced at all. Consequently, the iconic public good was saddled with further anomalous characteristics identified above—that is, nonrival consumption and nonexcludability (the producer could not prevent you from also using the good through standard property rights). Both characteristics were cited to buttress the stipulation that markets would fail in providing adequate levels of the good relative to public demand. The "solution" to this shortfall in production was to propose government provision of the good, although the means and modalities of provision were left to the mysteries of welfare economics.

This is not the place to pursue the strange piebald career of the public good in economic theory in the latter half of the twentieth century; instead I will narrowly track how it played out in the economics of science. The historian David Hounshell (2000) has shown how all roads led from the pivotal attempts to apply the public good concept to the defense of the postwar subsidy of science in the United States, especially after Sputnik, as manifest in the papers by Nelson (1959) and Arrow (1962). These major proponents of neoclassical market models of science were part of a group of economists primarily associated with RAND, the Air Force think tank during that era.[23] It was they who introduced the now-pervasive habit of treating the genesis of scientific knowledge as if it were production of a "thing," on a par with any other commodity, except for the fact that basic science was said to exhibit the characteristics of a public good. In other words, it was they who reified science as a conflation of object and process as a prelude to encompassing it as a subject of economic analysis. It was also they who dominated the construction of early statistics of national science policy and they who sought to combine the linear model with the theory of public goods to account for science as an input to economic growth (Godin 2005, Chap. 15).

Although these papers are regularly treated as "seminal" in the economics of science, there were warning signs of weakness right from the very start. Glen Asner (2006, 32) and others point out that both papers never presented any actual evidence that the United States was underinvesting in basic research in the 1950s relative to some efficient benchmark—indeed, the papers say very little about corporate R&D whatsoever. Instead, they just relied on the fact of the Sputnik scare to presume the reader was already on their side, and then they came up with versions of the public good story that might explain in a post hoc manner why it seemed that America was falling behind Russia in science. Further, there was no serious consideration of how one might measure the degree to which firms had trouble "appropriating" the benefits of basic research, since neither paid any attention to property rights. The prime objective seemed to be cooking up an excuse for the government (read: the military) to boost its investment in not-for-profit university science, shouldering aside objections of those who worried that lavish government subsidies might actually harm the conduct of science.

In attempting to frame an economics of science as a subset of welfare economics, Arrow and Nelson repeatedly suggested that the greatest flaw in market provision of research would be failure in arriving at Pareto optimality; that was a static (and perhaps totally empirically inaccessible) notion of welfare that had not been a pressing concern of writers on the subject since at least Bernal and Polanyi. When most people worried about the funding and organization of science, they focused on phenomena like Joseph Schumpeter's creative destruction, or the possibility of state or corporate perversion of the types of knowledge produced, or that the advance of science might outstrip the ability of society to use it wisely, or that science was inherently communal and should therefore not be subject to appropriation, or the fact that dynamic cultural and economic progress would be stymied for lack of appreciation for the endemic problem that the quest for knowledge that seemed to have little direct payoff to most participants in the process. Pareto optimality was the furthest thing from their minds and had nothing whatsoever to do with any of their urgent concerns; therefore, again, economists managed to change the subject by tying all discussion down to the Procrustean bed of the Walrasian model. Furthermore, these models could only be said to be concerned with "knowledge" in the most stilted and artificial sense, since they took their rational agent to be logically omniscient and fully cognizant of every possible state of the world, and, incongruously, a passive price-taker in perfect competition. After all, when you "learned" something, couldn't your preferences be said to have been changed? How this would relate to the communal discovery of truly novel facts in the natural world was something none of these economists ever bothered to elucidate. But, beyond all that, the most perverse aspect of these papers was hidden from their readers, since it was omitted from all published versions.

We now know it was no accident that the public good model of science was incubated at RAND in the 1950s. The subtext behind the rise of an economics of technical change at RAND, as explained by Hounshell (2000), was an in-house dispute over whether there really could exist a rational science of war. The systems analysts at RAND believed not only could they plan an efficient war, but they could also create the advanced weapons systems needed to prosecute it. This conviction came to a head in the mid-1950s with the controversy over how to design and build the intercontinental ballistic missile (ICBM). The systems analysts ridiculed the idea that you could wait to let the marketplace innovate the intricate systems comprising the ICBM, and they proposed that the program of research and engineering could be run as a top-down procedure, over which the military could maintain control and secrecy. A cadre of neoliberal economists at RAND, including Armen Alchian, Reuben Kessel, and Jack Hirshleifer, set themselves in opposition to this program, proposing instead that many concurrent R&D programs be contracted out to private corporations, with the military entering in at a relatively late stage to procure the most likely candidate weapons system. The anti-Bernal line that you just cannot plan science was their strident refrain. Arrow took the side of the system analysts at RAND, and in his paper of August 1955, RAND D-3142, explained why you should not leave research into weapons systems to the vicissitudes of the market. This paper, sanitized by the removal of most references to the dispute over weapons systems, became the famous 1962 paper described above. Likewise, Nelson—also at RAND at the time—composed his 1959 paper to explain why the neoliberal proscription wouldn't work. The rivalry with the Soviets lurked below the surface: The attraction of the public good story would be to explain why "we" were committed to markets in a way "they" were not but would nevertheless justify the free hand that had been dealt to the military in orchestrating and subsidizing science in the immediate postwar period. This is a significant reason we will henceforth refer to the public good vision of science as the premier Cold War account of the economics of science. The neoliberals lost this particular round, as the Air Force went ahead with the top-down plan and (apparently successfully—see MacKenzie1990) produced the ICBM, but that didn't mean the neoliberals surrendered in the larger war.

The public good account came to dominate discussions of science policy in the West by the 1960s, and it is informative to observe how the neoliberals struggled with that fact. Especially at the University of Chicago, various representatives grappled mightily with the problem of an economics of science as part of their larger opposition to the ascendancy of welfare economics within American economics. For instance, Harry G. Johnson admitted that for his peers, "Science as a problem for social theory can for simplicity be identified with disembodied knowledge considered as a public good" (1972, 15), but he then spent much of the article muddying up the simplicity with all sorts of considerations mitigating against treating science as a subset of welfare economics.[24] Theodore W. Schultz opened with a salvo in 1979 against the kind of economics he thought was being encouraged by state sponsorship of science and later continued by trying to square the circle of endorsing the public good concept but nevertheless insisting that "the public sector in its financing of scientific research is beset with many distortions and that they have been increasing over time" (1980, 645). Simon Rottenberg (an American Enterprise Institute fellow) expressed the curmudgeonly attitude then current at Chicago, complaining, "Too much should not be made of the defence for an active and public science policy in terms of science as a public good" (1981, 49). Rottenberg knew where he wanted to end up, though he was a little confused how to get there: "If the consequences of research are known and the financial gains from research can be obtained by its discoverer, the knowledge that is discovered is not a public good, and subvention by the state is not clearly called for" (51). Fritz Machlup (another MPS member), after spending a whole chapter on public goods in his three-volume work on knowledge, came to the rather anodyne conclusion that the sphere of application of public good theory was rather paltry and small (1984, 162). It was pretty apparent the neoliberals knew what they didn't like; they were just a bit hesitant in coming to terms on what to do about it.

Ultimately, a different cadre of neoliberal economists cut through the confusion by pledging their troth to two principles:(1)there was no such animal as a public good, once you looked at things properly, and (2) all knowledge was always and everywhere adequately organized and allocated by markets, because the market was really just one super information processor. The former tenet was established in the neoliberal camp by Ronald Coase, George Stigler, and James Buchanan, and the latter found its roots in Hayek but was successfully promoted by Aaron Director (1964) and Richard Posner. Coase had perhaps the most brilliant inspiration: If it had been permissible to tinker with the commodity definition to produce a public good, why not use jiu-jitsu and tinker with it once more to dissolve the legitimacy of the public good? (The cavalier stance of Cowles economists with regard to commodity space thus came back to haunt them.)If the commodity was just reconceived as a bundle of legal rights, said Coase (1960), then potentially there was no problem of externalities, since unbundling the rights would allow the market to correctly allocate each. Of course, in practice, it might be difficult to accomplish this (Coase dubbed such practical complications "transactions costs"), but that was just a secondary consideration. In reality, the message came through loud and clear that "public goodness" did not justify state provision of any such services.[25]

In a later paper (1974a), Coase had a field day ridiculing Samuelson's example of a lighthouse as providing a public good; he claimed if you knew a little English history, it would have become apparent that lighthouse provision was largely privately provided, and efficiently so, thus empirically refuting the very notion of the public good. Of course, one should always be wary when a neoliberal economist begins lecturing us about "real history"; in fact, the success of the British system was neither as uniformly private nor as sterling a success as Coase claimed (Bertrand 2005). But perhaps more germane to our present concerns, Coase did not actually refute the entire public good edifice altogether, since he only attacked the nonexclusivity component of the definition. Nevertheless, he did subscribe to the other main neoliberal tenet: that when it comes to the distinction between the market for goods and the market for ideas, "there is no fundamental difference between these two markets and, in deciding on public policy with regard to them, we need to take into account the same considerations" (1974b, 389). With his usual aplomb, Coase used the occasion of this lecture to accuse his political opponents of living a lie, since they used the First Amendment to protect what he considered the freedom of the marketplace of ideas but that they had no qualms abridging freedoms when it came to the marketplace for goods. Since his opponents did not yet openly subscribe to the marketplace of ideas, this was another instance of Chicago jiu-jitsu. Although he did not mention public policy toward science in particular, Robert Bork, his colleague at the law school, did, and on at least one occasion Bork (1963) opined that First Amendment protections should not extend to scientific research.

Richard Posner (another MPS member) had been willing to take these ideas to their logical conclusions, and even a little beyond. In a popular article called "Bad News," Posner argued that there was no way that news and other journalistic reportage should be considered a public good. He is worth quoting at some length, since it directly relates neoliberal doctrine to the issue of "truth," which presumably would equally apply to science:

> People don't like being in a state of doubt, so they look for information that will support rather than undermine their existing beliefs . . . Being profit-driven, the media respond to the actual demands of their audience rather than to the idealized "thirst for knowledge" demand posited by public intellectuals and deans of journalism schools. They serve up what the consumer wants, and the more intense the competitive pressure, the better they do it . . . Giving a liberal spin to equivocal economic data

> when conservatives are in power is, as the Harvard economist Andrei Shleifer points out, a matter of describing the glass as half empty when conservatives would define it as half full . . . The public's interest in factual accuracy is less an interest in truth than a delight in the unmasking of the opponent's errors . . . The blogosphere has *more* checks and balances than the conventional media; only they are different. The model is Friedrich Hayek's classic analysis of how the economic market pools enormous quantities of information efficiently despite its decentralized character, its lack of a master coordinator or regulator, and the very limited knowledge possessed by each of its participants. (2005, 9, 11)

These notions that most intellectuals are generally nothing more than crass, self-serving spinmeisters, working for themselves and not for the public, and that neoliberals are the only thinkers equipped with sufficient intestinal fortitude to stare that fact in the face, have become the hallmark of the rejection of knowledge as a public good in economic circles. Hayek's disparagement of "second-hand dealers in ideas" has now spawned consequences extending well beyond his original disdain for the socialist tendencies of the intelligentsia. To a new generation, it might have even appeared wickedly *subversive* to suggest that scientists were as debased in their self-interested motives as you and I, but that in the context of sufficient market organization, those tendencies are rendered harmless and even efficient—no need for public subsidy there. In philosophy, arguments of this ilk predicated upon James Buchanan's public choice theory applied to the scientific community began to crop up. In public choice theory, scientists have been portrayed as banding together to influence government so they can bilk the public.[26] Similar arguments signaled a distinct change of attitude in the orthodox economics literature by the 1990s. For instance, Austan Goolsbee argued that government-subsidized R&D mostly just artificially drove up wages of scientists and engineers, and thus "the conventional literature may overstate the effects of government R&D spending by as much as 30–50 percent. In this sense, R&D may be less about increasing innovation and more about rewarding the human capital of scientists" (1998, 298). The "evidence" for this was desultory (after all, why shouldn't corporate science have the same effect?), and the regression analysis justifying the claim was exceptionally sloppy, but in a sense none of that mattered—here stood another fearless stalwart Chicago economist bashing away at yet another Idol of the Tribe. The fact that Goolsbee later became chair of Barach Obama's Council of Economic Advisors might also call into question the widely trumpeted rejection of neoliberal policies toward science that supposedly characterized Obama's "enlightened" Democratic approach to the "knowledge economy."

The repudiation of public goods led in turn to another characteristic neoliberal practice of suggesting that government subsidy of nonprofit R&D would, in an exact reversal of the public goods story, "crowd out" private R&D spending that might otherwise have been provided by the market.[27] Since this was essentially a reprise of the already massive literature in macroeconomics claiming that activist fiscal policy would drive out private investment, it was a foregone conclusion that the arguments would be tiresomely replicated in the economics of science and that every cookie-cutter empirical paper "finding" crowding out would be countered by another equal and opposite one rejecting the hypothesis, with surplus econometric wizardry displayed by all. The whole spectacle became just another tactical exercise to get another article published in an orthodox economics journal. The comprehension of science, on the other hand, just languished.

One of the more cynical reversals of the public good story by economists was the paper by Jensen and Thursby (2001), which became a *de rigueur* justification of the ways that technology transfer offices (TTOs) were transforming the university landscape in the 1990s. The paper lifts a standard model from labor economics and changes the names of a few variables to argue that "the vast majority of inventions licensed are so embryonic that technology managers consider inventor cooperation in further development crucial for commercial success . . . Our theoretical analysis shows that development would not occur unless the inventor's return is tied to the licensee's output when the invention is successful" (255). From this perspective, university scientists don't produce generic public goods but instead targeted boutique goods that must be tailored to the individual entrepreneur. In other words, it was OK for the university TTO to try and soak the external commercial firm for all they were worth, since the latter would just naturally want to have the original academic inventor on board as an indispensable part of the deal and would be willing to pay a pro-rated proportion of the profits for the privilege. Those who bothered to familiarize themselves with the technology transfer process would immediately recognize that (outside of the wooing of venture capital firms, and temporary considerations of window dressing for IPOs) this is precisely the last thing the external firm wants. As David Mowery (2005) noted, "Academic research rarely produces 'prototypes' of invention for development and commercialization by industry."[28] Academics more often just get in the way during that treacherous stage between nascent concept and effective development of a marketable project; like children at a garden party, they should be seen but not heard. Maybe it is not "linear," but in participant testimony commercial development really does look different from academic research. But the historical record meant little or nothing at the *AER*, which was more interested in promoting a neoliberal economics of science.

Nevertheless, the neoliberal economists had succeeded in discrediting the public good account of the economics of science within the larger economics profession at the turn of the millennium. This is not to say that some of their criticisms had been altogether unjustified. "Although underinvestment phenomena are the rationale for government subsidization of R&D, the concept is poorly defined and its impact is seldom quantified . . . R&D policy has not adequately modeled the relevant economic phenomena and thus is unable to characterize, explain, and measure the underinvestment" (Tassey 2005, 89).During the Cold War, it had merely been sufficient to wave in the direction of "basic science," suggest that it naturally qualified as a public good, and then proceed to plead for government subsidy: The quality of argumentation had always left something to be desired. But around 1980, the neoliberals had chipped away at this notion to such an extent that they got both the courts and the Congress to extend property rights to all that novel knowledge that was previously supposedly so hard to appropriate, as even Richard Nelson (2004) was forced to concede. The next generation of economists, even those who would be offended if you called them neoliberal, in effect simply absolved themselves of dependence on the public good concept: "Appropriability conditions most often have only a limited effect on the pattern of innovations" (Dosi et al. 2006, 896); "the conventional market failure justification for the public subsidy of science is weak"(Martin and Nightingale 2000, xxiii). This was the death rattle of the notion of science as a public good.

A curious sideshow to the rise and fall of the public good occurred in the field of "science studies." As early as 1969, George Stigler had called for an economics of science to displace the baleful influence of the sociology of science (of a Kuhnian vintage) (1982, 112), and as the Science Wars hotted up, the calls became more plaintive and insistent (Wible 1998, 17). For all the reasons so admirably summarized by Hands (2001, Chap. 8), many were inspired to hope that the relativist implications of science studies would be countered and banished by the development of a neoliberal economics of science; however, the joke was on them, since one major branch of the field known as Social Studies of Knowledge was already busy at work reconciling their relativism with neoliberal trends. The so-called Paris School of Actor-Network Theory (or ANT for short) had flirted with quasi-economic terminology and ideas since the late 1970s;[29] but by the 1990s found itself in the awkward position of competing for research funds within European governmental circles with many of the protagonists of the economics of science. At that juncture, one of ANT's major representatives, Michel Callon, sought to differentiate his product by—*mirabile dictu*—attacking the concept of the public good! In his 1994 article, he does a credible job of summarizing the literature surveyed here, including pointing out that economists have a bad habit of treating knowledge as a "thing"; that they tend to collapse knowledge to information; and that the proposition that "the market" was uninterested in funding science was belied by "news from the front streaming in every day cast[ing] doubt on this seductive argument" (401). But then his article took a vertiginous turn, arguing that the ANT brand of sociology of science "shows that there is nothing in science that prevents it from being transformed into merchandise" (402). Because nonexclusion was reportedly not proving to be such a problem, and nonrivalry did not really exist in science, "scientific knowledge does not constitute a public good as defined in economic theory" (407). Far from economists politically displacing sociologists, it seems neoliberalism had taken root in the most avant-garde precincts of science studies, gussied up with the seemingly noneconomic terminology of actants, rhizomes, networks, and parliaments of things (Latour 2005). In case the message had gotten lost in his luxuriant verbiage, Callon was induced to clarify the consequences a decade later:

> There are two positions we have to abandon. The first is the idea of critique of the hard economists, which is intended to show them they are wrong. And the second position is to describe markets just to say they are more complicated than economists or political decision makers believe . . . Let us stop criticizing the economists. We recognize the right of economists to contribute to performing markets, but at the same time we claim our own right to do the same but from a different perspective. (in Barry and Slater 2003, 301)

Indeed, the forces bearing down upon the economists, the philosophers, and the science studies scholars must have been considerable, given that they all ended up saying pretty much the same things about science by the turn of the millennium.[30] The neoliberals had won: Science no longer qualified as a public good.

Science as the Ultimate Cause of Economic Growth

This was the locus where the neoclassical economists' fascination with the linear model and the postwar quest to provide an account of knowledge as a public good have converged, only to collide in a train wreck. If one could inspect the tablets etched with the Knowledge Economy Creed, undoubtedly near the top would be inscribed the conviction that more money spent on science invariably gets you more technological knowledge, and that increased applied knowledge eventually buys you more economic growth. Furthermore, should you ask an economist what it was that constituted the ultimate cause of economic growth, in the sense that one might abstract away intermediate-term macroeconomic processes of investment, production, and consumption, more often than not they would reply with some synonym for "knowledge."[31] This message—that economic growth owes its genesis to the consequences of science—is a watchword on a daily basis in popular magazines, in newspapers, on TV, and (especially) on the Internet. Journalists wax lyrical over the near-Hegelian promise of the triumph of the spirit over mere base matter:

> So what is economics all about? Land, labor and capital, with technology considered as a force apart? Or people, ideas and things, with the production and distribution of knowledge as a matter of central concern? Scarcity? Or the countervailing forces of scarcity and abundance? . . . It is the growth of knowledge that is the engine of economic growth. (Warsh 2006, 342)

In the midst of this profuse enthusiasm, it may come as a bit of a shock to realize that the actual evidence behind the hype is, well, a little thin. Let us sample some quotes in chronological order:

> How far the organized pursuit of knowledge in the form of science as it exists today can be credited with responsibility for the economic basis of contemporary civilization, and how far society gains from public support of science, remain open questions. (H. Johnson 1972, 18)
>
> The link between patents and productivity or other economic growth measures has not really been worked out. (Griliches 1978, 172)
>
> One of the things that all knowledgeable people supposedly "know" is that technological change has been the critical variable in accounting for the spectacular long-term growth of the American economy . . . And yet, when scholars of a quantitative turn of mind have attempted to link the story of growing productivity of the American economy to some of the better-known facts and benchmarks of our technological history, that story has turned out to be a remarkably difficult one to tell. (Rosenberg 1982, 55)
>
> As yet, no empirical study proves that technology has been the engine of modern day growth. (Grossman and Helpman 1994, 32)
>
> We have shown that a nation's scientific infrastructure can be characterized by two important dimensions: scientific research and scientific labor force. In our models, scientific research has a substantial negative effect on economic growth over the period 1970 to 1990 . . . On the other hand, a nation's skilled scientific labor force has a substantial positive and significant effect on economic growth. (Schofer, Ramirez, and Meyer 2000, 882)

> After nearly fifty years of studies, one still looks in vain for hard data on the links between science, technology, and [economic] productivity. (Godin 2004, 687)
>
> The bottom line is that resources devoted to research have exhibited a tremendous amount of growth in the post-war period, while growth rates in the United States have been relatively stable. The implication is that models that exhibit strong scale effects are inconsistent with basic trends in the aggregate data. (Jones 2005)
>
> Historians and others have assumed that Germany and America grew fast in the early years of the twentieth century because of rapid national innovation . . . So powerful has this innovation-centric view been, especially in its nationalistic versions, that all evidence to the contrary has been studiously ignored. It was known in the 1960s that national rates of economic growth did not correlate positively with *national* investments in innovation, research and development, or innovation. It has *not* been the case that countries that innovate a lot, grow a lot. (Edgerton 2007, 108–109)

I anticipate the reader might just now be experiencing a minor bout of vertigo. Isn't it just *obvious* that science begat electricity, which begat the dynamo, which begat the transistor, which begat the computer, which begat Microsoft, which begat Google, which begat the American boom of the later 1990s? Isn't that the alpha and omega of any economics of science? I have arrayed these quotes chronologically to show that doubts concerning the supposed tight link between knowledge and economic growth have actually been around for decades and cannot simply be dismissed as some artifact of one or two intemperate naysayers. Indeed, reservations have always been manifest at many frequencies and amplitudes but have nevertheless been drowned out by those whose faith has proven stronger than their auditory acuity. Edgerton is correct to point out that many nationalistic historians have built entire epics upon vague characterizations of the native ingenuity of the peoples they describe, but the real culprits in this grand dissimulation have been the science policy experts and the neoclassical economists.

The field of science policy has simply taken it as axiomatic that should there exist no close dependence of the economy on scientific research: Then they would be compelled to pack up their tents, abandon science to the kindness of strangers, and find another occupation. "The conventional pairing of policies to promote science and technology as 'science and technology policy' formalizes a fiction, namely, that technological and economic progress *necessarily* depend on scientific progress"(Kraemer 2006, 49). There, the links are built in to the very definitions deployed. The first release of *Science and Engi-*

neering Indicators in 1972 was transparently "an attempt to dramatize politically the anticipated costs should declining trends in public support for science continue" (Ezrahi 1978, 310). Even though it has since grown into a multivolume monster chock-full of numbers and tables, the actual fact of the matter is still that "the indicators adapted in *SI-72* are neither directly constituted from scientific knowledge generated in [science studies], nor designated as aids to scholarly research" (289) Aggregate science statistics are not produced in a format to pose the question we seek here, much less to answer it.[32] But for now, it should be sufficient to point out that these statistics on national R&D are constructed upon the unexamined presumption that science delivers the goods. Even given this bias, it still turns out to be a stretch to make use of them to demonstrate that scientific knowledge actually causes economic growth. Mostly, the numbers have been constructed by promoters and politicians for cruder purposes: the Technocratic Creed cannot be impugned.

And this brings us to the other important constituency beset with a certain deafness: the neoclassical economists. It so happens that three-quarters of the quotes above were selected from well-respected orthodox economists; yet, I venture to guess that none of those economists actually believe that science is not in fact the true first cause of economic growth. The most common ploy is to baldly assert all the individual caveats don't matter: "Regardless of whether one adheres to the more narrow old theories or the broader new theories, the evidence is overwhelming that technology drives economic growth" (Feldman, Link, and Siegel 2002, 24). Unfortunately, neither the Solow-style nor self-designated new growth theory demonstrates any such thing. The persistent gap between neoclassical economists' empirical concessions and their deeper beliefs has quite a bit to do with the history of neoclassical growth theory, and the ways in which it has permitted the neoclassical economics of science to flourish under adverse circumstances.

The history of growth theory in economics has been recounted numerous times,[33] so we can make do here with a relatively streamlined account. Although neoclassical growth theory can be traced back to the implications of Keynesian economics drawn by Roy Harrod and Evsey Domar, almost all modern writers trace the theory's lineage back to the work of Robert Solow (1956, 1957). Solow was concerned to reconcile certain Keynesian macro relations, like the consumption function, with his belief in the long-run stability of the capitalist system. He also displayed a particularly physicalist notion of economic growth, portraying it as the technical relationship between a capital substance, labor, and a generic output measure. It is important to stress the latter point, because so much of what immediately follows will concern the neoclassical "production function," purportedly a *technological* relationship, usually written abstractly as:

$$Y = F(K, L).$$

In the standard notation, Y stands for total aggregate output; L stands for labor measured in some temporal or corporeal unit; and K stands for aggregate capital, measured in ways that no economist wants to explore in detail. I don't mean to be flippant about this last point, because it is significant. Right when Solow was engaged in inventing neoclassical growth theory, he was simultaneously engaged in a knock-down brawl with a cadre of economists located primarily at Cambridge in the United Kingdom; this dispute was precisely concerning the logical consistency of neoclassical capital theory and the dependence upon the production function (Harcourt 1972). I may personally think Cambridge, Massachusetts lost this particular battle on points, but that doesn't matter because eventually they won the war. The stakes for victory or defeat were made clear even before Solow published his model: "The dominance in neoclassical economic teaching of the concept of a production function . . . has had an enervating effect upon the development of the subject, for by concentrating upon the question of the proportions of factors it has distracted attention from the more difficult but more rewarding questions of the influences governing the supplies of the factors and the causes and consequences of technical change" (Robinson 1953/1954, 81). This diagnosis was prescient, because it was exactly what ensued.

Solow invented the fundamental neoclassical model that sought to identify the conditions under which steady state growth could occur, given a fixed saving rate s, a fixed rate of population growth n, and imposition of the macro condition that savings = investment. The "steady state" was an attempt to identify what equilibrium might look like in a growing economy, settling upon the condition where all relevant variables grow at a constant proportional rate, such that nothing of consequence changes. The steady state notion suggested to Solow that the production function should exhibit "constant returns to scale," that is, proportionate increases in all factors lead to an equiproportionate rise in output. Solow presumed for his modeling purposes that the savings:income ratio was identical to the investment:output ratio,[34] and he then demonstrated that all these conditions imply (where $k = K/L$, and $f[k]$ is the production function rewritten in per capita terms):

$$\Delta k = s\, f(k) - nk.$$

Because the change in the capital:labor ratio Δk this "period" will feed back into the equation changing k next "period," then the steady state is defined where $\Delta k = 0$, and thus steady state growth happens at $sf(k) = nk$. The substitution between factors in the production function brings about the steady state growth rate, which is set by the savings rate and the population growth rate. Of course, this was the most rudimentary version of the model working in terms of one lone "output" (one of its attractions for economists), and they proceeded to complicate it without end by incorporating bells and whistles like depreciation on capital, multiple "sectors," turnpikes, optimal consumption paths, and a whole raft of other wrinkles that have now been mercifully forgotten.

No one should take this as a serious description of what economic growth ever was like, and indeed, most of Solow's generation spoke in terms of "parables" and other semimystical oracular entities. What should be noted, however, is the extent to which the model was supposedly cast in "materialist" terms, dictated by technology, demographics, depreciation, and the like. In a sense, it was a quest to take the phenomenon of economic growth into the Realm of the Timeless, with timeless physical relations dictating timeless natural growth rates with smooth adjustments to timeless steady state ratios between key variables. No wonder so many theorists in Britain and elsewhere were suspicious of this brand of social science. But the quest was flawed from the very start, since permitting "technical change" back into the model was the Revenge of the Repressed, the intrinsic reinsertion of the temporal and the social into the explanation of growth.

The story of how Solow himself brought this about is well known. In seeking to "verify" the model, he sought to plug in some actual growth rates of empirical variables for the United States in 1909–1949, which roughly approximated the theoretical variables of the model. It was all a little fast and loose, or, as he admitted in the paper, "either this kind of aggregative economics appeals or it doesn't" (1957, 312). The problem was that not even Solow really thought the production function had stayed constant over the relevant period, so he threw in the possibility of a shift of the whole function over time t, in the form of an additional variable A(t):

$$Y = A(t)\, F(K, L).$$

While one might be tempted to call the shift term a "fudge factor," Solow instead opted to call it "technological change," even though he admitted it was "a shorthand expression for *any kind of a shift* in the production function" (1957, 312). This might not have been so consequential if he had not found that 90 percent of the growth of U.S. output per worker over the period was *not* accounted for by growth in factors of production, but rather by shifts in the production function captured by the A factor. Now, there would be two ways to react to this finding. One is to admit that the whole motivation behind the Solow model, namely that growth is governed by timeless physical relationships, is so far removed from anything we actually observe in the world that it would be best to jettison the model altogether. The other is to proclaim a new "discovery" that "technological change" is the primary cause of economic growth. Phrased that way, it was a foregone conclusion which way Solow was inclined but it was hard to ignore the possibility that this construction of "technology" was only a measure of our ignorance about the gross macroeconomic determinants of growth.

Because Solow owes his Bank of Sweden Prize to this brace of papers, he has been rather cagey about the interpretation of the origins of neoclassical growth theory. In a 1998 interview, he protested that he "expected that the main source of growth would be capital accumulation" and was surprised concerning the 90 percent solution. He came close to acknowledging the rabbit thrust into the hat when he further said, "What I meant by saying something was exogenous was that I do not pretend to understand this; I have nothing worthwhile to say on this so I might as well take technical change as given . . . I do not know what the determinants of technical change are in any useful detail" (in Snowdon and Vane 2005, 665, 668). This admission would appear rather startling, coming from one of the main advisors to the National Science Board, the entity that collates the data found in *Science and Engineering Indicators*. Another striking admission is that the whole project of "fixing" the model or "correcting" the data with ever more sophisticated mathematical and econometric techniques more or less hit a brick wall after a decade: "Around 1970 or so we simply ran out of new ideas" (ibid., 666). That is not exactly the way I recall it. Other Nobel Prize winners instead thought the whole episode had been a disaster: "Elaborate aggregative growth models can contribute very little to the understanding of processes of economic growth, and they cannot provide a useful theoretical basis for systematic empirical analysis" (Leontief 1970, 132).

The confluence of events was very strange to live through. The Capital Controversies, now mostly forgotten, were widely seen as having embarrassed the very logic of aggregate production functions in the early 1970s. All sorts of tortured attempts to make "technical change" something more than an empty statistical tautology rapidly grew so baroque that no one had a clue how to judge their relative legitimacy: factor-specific technical change, alternative functional forms for production functions, distinct "vintages" of capital, putty-clay models, different species of "neutral" versus biased technical change, and so forth (Harcourt 1972, 69 et seq). Further, crucial colloquial meanings of technical change had become ruled out by the artifact of reducing everything to "shifts" of the production function (H. Jones 1976, 178–179). Thus, it was not that the economists had run out of ideas; rather, there were too many ideas floating around to pretend growth theory could be adequately organized around the Solow model. The empirical fits were poor, and everyone had his own favorite idiosyncratic remedy. As so often happens in economics, the main protagonists just declared victory and gave up the ghost: Growth theory courses evaporated in the major graduate programs and growth theory chapters migrated to the end of macroeconomics textbooks, where they never quite made it into the curriculum. Growth theory shriveled back to a dead branch of economics from roughly 1970 to 1985 (Snowdon and Vane 2005, 586). Not to worry: Research programs in neoclassical economics don't suffer defeat; they just evacuate Saigon. Yet, strangely enough, this was not the fate of Solow models in science policy: They were just gearing up for a long and radiant future.

By the 1960s, the combination of the linear model, the public good, and Solow-defined technical change became cemented together in the science policy community as the Cold War explanation of choice. Science policy experts did not want to waste their time in extensive historical or microlevel accounts of how technological progress did or did not come about; they just needed a broad-brush aggregate story, one relatively light on detail, that they could use to justify the continuation of the largesse of government subsidy of science in that era. Because the Solow results could be read as suggesting the predominance of technological change, this became the 400-pound gorilla of economic growth, crushing all other factors, and the linear model could be accessed to argue that a prerequisite of technological change was basic science—all that was needed was the insertion of the public good parable to nail why it should be the government's preordained role to guarantee the starter fuel of economic growth. The "Solow Residual" could be shamelessly commandeered as a proxy for an actual measure of the rate of aggregate technological progress or productivity growth at the level of the nation-state (and *not* a disconcerting index of the failure of growth theory); science policy mavens would then point to their other favorite statistic: the ratio of total national R&D expenditure to national income, as the relevant target variable for government manipulation. Solow definitely played along with the notion that the true taproots of economic growth were elusive, external, and noncapitalist, rather than a component of some sort of larger economic structure: "There is an internal logic—or sometimes non-logic—to the advance of knowledge that may be orthogonal to economic logic . . . the 'production' of new technology may not be a simple matter of inputs and outputs"(Solow 1994, 52). In his books, there was no need for a dedicated economic theory of science. That suited the Cold War science policy wonks just fine. Public money should be poured into science with little accountability, they said, just as the military had been doing for two decades.

But then the environment changed. Starting in the 1970s, the Solow Residual (now dubbed "Total Factor Productivity") measure of technical progress began declining in the United States. Fears blossomed in the 1980s that the United States was falling behind its major trading partners in competitiveness. Numerous empirical exercises, like the ones quoted at the beginning of this section, failed to show any simple or close correlation between aggregate government R&D expenditures and the Solow measure of technical change. Even more chilling, the military began to withdraw from its postwar role as premier science manager and patron. And then the neoliberals began hammering away at the linear model and public good theory, as documented above. Growth theory had been reduced to a withered appendage in the interim, such that there were very few intellectual resources available to diagnose the ailments that appeared to beset the American economy.

You might think that so many adverse circumstances would portend the demise of neoclassical growth theory, but against all odds, the 1990s marked its revival in the guise of the "new" or "endogenous" growth theory. The main protagonist in this story was Paul M. Romer, and it is no accident he became the new poster-boy for the New Information Economy and science-driven economic growth (Warsh 2006). Romer's (1990) paper accomplished this by reconciling the Solow growth model with the neoliberal ascendancy of the marketplace of ideas: Technological change was reputedly rendered "endogenous" by modeling it as something that could readily be purchased by and produced in the private sector. As Romer admitted in a 1998 interview, "What endogenous growth theory is all about is that it took technology and reclassified it, not as a public good, but as a good which is subject to private control" (in Snowdon and Vane 2005, 681). He accomplished this by breaking down the public good concept in ways already pioneered by neoliberals like Coase: "Rivalry is a purely technological attribute . . . Excludability is a function of both the technology and the legal system . . . What matters for the results is that knowledge is a normal good that is partially excludable and privately provided" (Romer 1990, S73, S85). By more directly addressing the public good notion, and yet subordinating it to the privatization of knowledge through strengthened intellectual property rights, Romer took what had previously been the canonical justification for state subsidy of science and inverted it into a brief for the privatization of science as a solution to problems of flagging productivity and growth. It proved just what the neoliberal doctor had ordered.

Popularizations of this revival (Warsh 2006) treat the cause as a new infusion of "rocket science" or new mathematical tools (Snowdon and Vane 2005, 679), but a quick glance at the model shows it was nothing of the sort.

Instead, it was a return to the most rudimentary style of production function, only now augmented by an index of Gary Becker style "human capital" as an extra argument in the function, over and above labor input. Spending on human capital devoted to research is posited to directly increase the rate of technological change. Technological change increases output but also increases the productivity of human capital in the research sector, leading to "increasing returns" to scale. Larger rates of growth induce more research, and accelerating rates of growth, restrained only by standard diminishing returns on the conventional capital and labor inputs.

A number of economists subsequently observed that there was very little new in this purported new growth theory.[35] Romer himself was relatively unapologetic about this: "When I started working on growth I had read almost none of the previous literature" (in Snowdon and Vane 2005, 679). It is certainly true that in the Romer model there was no steady state determined by timeless conditions like the population growth rate, but neither was there any substantial advance over the Solow nondefinition of technological change. Indeed, the real perplexity comes in seeking to understand its extraordinary popularity in restoring neoclassical growth theory to pride of place in the hearts and minds of orthodox economists. Romer had not truly "endogenized" technical change, since he still made use of the Solow trick of equating it with shifting the production function, only now linking that shift to yet another unobservable variable dubbed "human capital devoted to research" (1990,S83). Splitting one portmanteau variable into two equally inaccessible variables does not make for rigorous clarification. Solow himself has noted this curiosity: when asked what we have learned from Romer-style growth theory, he responded, "Less than I had hoped."[36]

Perhaps more to the point, the Romer-style growth models have faltered in empirical tests just as dramatically as did the original Solow model. The key prediction of these models is the acceleration of growth (or "scale effects") in regimes where investment is plowed back into research and human capital: In most situations, that has apparently not happened. Charles Jones (1995b, 2005) has merely been the most vocal in pointing this out, but there have been numerous other empirical disconfirmations.[37] Postwar spending on R&D investment in the United States had expanded tremendously (at least up until the current economic crisis), but the trend growth rate of the U.S. economy had not accelerated. This just reprises the point made at the beginning of this section: Researchers have repeatedly searched in vain for correlations between R&D spending and conventional national income measures, even absent the spur of endogenous growth theory; equally they repeatedly have failed. If the claim to fame of endogenous growth theory was to finally demonstrate how scientific knowledge serves as the ultimate motor of economic growth, then it failed miserably.

It would be complicated (and frankly a distraction) to diagnose all the tangled conceptual roots of the recurrent failure of neoclassical growth theory, but Joan Robinson certainly put her finger on the most likely culprit: the stubborn dependence upon the analytical construct of the production function. The problem is not simply a secondary matter of "aggregation." The entire motivation behind early production functions was to posit a sharp distinction between factor substitution and something more dynamic and pervasive, which, for lack of a better term, is still commonly called "technological change." The production function, itself modeled upon the original utility function (Mirowski 1989, Chap. 7), was to permit an equal freedom of choice between supposedly coexistent yet distinct means of producing the good in question, the same way the consumer could choose between different baskets of commodities. The comparison was and still is a fallacy, as has been pointed out by numerous critics of the neoclassical approach.[38] No one can effortlessly change production processes as if it were merely leafing through to a new page in a "book of blueprints"; it takes time, new knowledge, and a period of breaking in the new process before one could ever dream of finding that "production possibilities envelope" once again. But a deeper issue lurks here. The production function is supposed to represent the physical/technological boundaries of what can be accomplished in the production process, but examination of the history of the production function in economics demonstrates that the mathematical form of the production function (whether it is Cobb-Doubles, CES, trans-log, and so on) has *never* been dictated by any physical or engineering laws. Instead, it was chosen for other reasons, mostly so that it would exhibit symmetric properties to the "well-behaved" preference function. The reason production functions have never been derived from technological schemas is that *they violate one of the most important laws of physics, namely, the second law of thermodynamics* (Georgescu-Roegen 1976). Movement in input space—changes in technique—is neither reversible nor path-independent in practice. Hence the isoquants do not remain fixed as relative prices change, because every change in technique leads to irrevocable physical losses, undermining the very notion of free constrained maximization. No wonder the use of "well-behaved" production functions in growth theory leads to utter nonsense when it comes to discussions of technological change. Production functions should therefore be banished from a rigorous economics of science primarily because they deny one of the most important principles of science: the entropy principle. You cannot discuss technical change with a tool that violates the laws of physics and pretends "pure" innovation can be segregated from the learning process attendant to any change of the production process. This again reveals the folly of attempting to reduce all economic change to the stasis of neoclassical equilibrium theory.

Of course, almost no one in contemporary mainstream economics sees things that way. Just as in the first phase of growth theory, most economists have not opted to reject the Romer model but rather to amend or augment it in various ways. Saigon beckons. This feels to some like *déjà vu* all over again, because mostly this involves re-imposing diminishing returns to whatever the analyst in question thinks is the true source of technological change, bringing the entire mathematical artifact full circle to Solow's original presumption of constant returns to scale, governed by constrained optimization. From our current perspective, it is not necessary to track down all the mutant strains of this modern literature, because the pursuit of the primary objective, namely, to actually *demonstrate* that science causes economic growth, has not gotten one iota closer to its goal.

Yet economists and journalists alike extol the new growth theory, and there is no denying its popularity in certain circles. What the reader should take from this episode is that, just as Solow owed much of the popularity of his model to the way it was used to buttress and justify the Cold War regime of science organization, the Romer model owes its cachet to the way it seems to validate the post-1980 commercialization of science and the neoliberal romance of the marketplace of ideas. Rigorous mathematics and assiduous empiricism had very little to do with it. However, I do not want the reader to leave this chapter with the impression that, in the real world, scientific knowledge has nothing whatsoever to do with technological change or that science has no causal relationship to economic growth. I would rather hope that the reader would realize that the triad of linear model/public good/growth theory is a badly flawed and expendable set of concepts to structure inquiry into the economics of science. No such simplistic macroeconomic statements concerning science and economic growth have been found to hold up very long.

Notes on the Contemporary Economics Profession

I would not want the reader to get the impression that I had come up with the critique of economics found in this chapter all by myself. As anyone who had spent time around intellectual history knows, ideas always grow out of earlier ideas. But in this particular case, problems with the earlier neoclassical economics of science had grown so glaring and insistent that they practically conjured a sort of counter-orthodoxy from the 1970s onward. While many of the disgruntled stewed in isolation, over the decades a subset of the disaffected successfully gathered together a transnational epistemic community, through international workshops, shared participation in (European) science policy positions, and purpose-built journals. Recently, a few of the main protagonists have taken to explicitly proclaiming the existence of a "Stanford/Yale/Sussex School," thereby striving to claim the mantle of the "new orthodoxy" in the economics of science and technical innovation.[39] I want to start off by acknowledging the extent to which their writings have inspired much that happens in the rest of this book. Whatever the weaknesses of their work, these people have devoted their lives to transcending the superficial approaches to science that have pervaded postwar neoclassical economics. Some of them have been bold thinkers who have sought to integrate the history of technology (and sometimes science), and philosophical approaches to knowledge, with economic models.

I initially thought I would present a survey of contemporary economists who were proponents of what they themselves called a "new economics of science," but the further I got, the more frustrated I became. Although these protagonists have been proud to sport their credentials as evolutionary and historical economists, I eventually came round to the opinion that their work was still too concerned to "save" the holy trinity of the linear model, the public good, and the new growth theory—or to not put too fine a point on it, to be accepted by their orthodox colleagues.[40] But the decision to ignore them here was primarily dictated by the realization that they contributed relatively little to the type of empirical understanding of the modern commercial regime of science prosecuted in the remainder of this book. Economists may dispute whether the current level of intellectual property protection is optimal or they may suggest that the commercialization of the modern university has gone too far, or they may plead that technological change is more complicated than a simple shift of the production function, but concrete propositions on the relationship of science to the economy (and vice versa) seem as remote and sketchy as they did in the days of Bernal. Because many of these economists are located in schools situated at the forefront of the drive to privatize research, and consult for organizations like the National Academy of Sciences and the Ewing Marion Kauffman Foundation (who ardently support such initiatives), their work inevitably avoids confronting the really profound structural problems in modern science. So, regrettably, I must pass by this rather large literature (nonetheless included in the bibliography for the curious reader).

I would not want to leave this chapter on the bleak note that nearly everything economists have said about science has been utterly bankrupt. If there is a bright patch on the horizon, it appears within the "left" wing of Europeans within evolutionary economics, if only because they have had the least invested in protecting neoclassical economics. A number of figures, who either trained in the department of Science and Technology Policy Research (SPRU) at the University of Sussex or have been affiliated with it, have produced a raft of insightful studies on the economics of science (referenced in later chapters), and in particular with regard to the current regime of globalized commercialized research, although it would be an exaggeration to claim that they have as yet produced a working theoretical synthesis. It is noteworthy that they tended to be affiliated with the new model "business schools," which have sprung up across the European terrain in the last few decades, or else find themselves housed in public policy centers. Their remit is the explicit understanding of science and technology in the contemporary political economy, in close consultation with science policy directorates, and not in keeping the economics discipline happy. Perhaps because of the relative novelty of their disciplinary identities, they are more inclined to produce fine-grained empirical studies and small-t theories, and therefore they do not tend to garner the attention they may deserve on this side of the Atlantic. I shall repeatedly refer to their work in the remainder of this volume. Here I shall just briefly touch on the work of three representatives—Giovanni Dosi, Benjamin Coriat, and Paul Nightingale.

Giovanni Dosi is currently Professor of Economics at the Sant'Anna School of Advanced Studies in Pisa, where he also leads the Laboratory of Economics and Management. He is also the impresario behind the attempt at the synthetic packaging of the "Stanford-Yale-Sussex" school, and so manifestly he is a proponent of a Big Tent approach to building epistemic communities, in contrast to the categories and distinctions I have sought to delineate in this chapter. In that same spirit, Dosi has been relatively promiscuous in the past in his appeals to "evolutionary economics" as something that all manner of disaffected scholars could endorse (Dosi and Nelson 1994). He started out back in the 1980s seeking to document the existence of "technological trajectories" possessing a quantum of inertia, motivated by comparison to Thomas Kuhn's "paradigms" (1982). This may sound a lot like Paul David's "path dependence" of technologies, and, indeed, Dosi also pioneered the application of Polya urns to models of technology before David as well (Dosi and Kaniovski1994). However, of late, he has grown much more skeptical of David's amalgam of neoclassical economics and history: "One must sadly admit that evolutionary arguments have been too often used as *ex post* rationalizations of whatever the observed phenomena . . . the relevance of path dependent selection of relatively 'bad' institutional setups and technologies remains a highly controversial question" (Dosi and Castaldi 2004, 21, 23). Increasingly, he has also taken aim at production functions as flawed tools for analyzing technological change and the role of knowledge. As he craftily suggests, the correct answer to the question "How do I bake a cake?" is most emphatically not to respond "Maximize price times output minus price times inputs" (Dosi and Grazzi 2006, 179). He has argued that dependence upon "routines" and tacit knowledge as fundamental theoretical entities in Nelson and Winter had the unfortunate effect of rendering working rules in exaggerated individualist and cognitive formats, and that the role of power relations in knowledge had been neglected (Coriat and Dosi 1998a).

More recently, Dosi has been moving much closer to something resembling the science studies position, although he has not been very explicit about this. For instance, while many economists treat institutions as parameterizing state variables, derived from the rational choice of profit-seeking agents, formed by explicit "constitutional" conventions (Vanberg 2008), he argues instead that institutions shape the cognitive and behavioral identities of agents, with institutional structures existing prior to the definition of agent self-interest. Also he insists that much institutional structure derives from unintended consequences of processes that resemble notions of "self-organization" and emergence found, for instance, at the Santa Fe Institute. Rather than treat firms as a nexus of rational contracts, subject to the frictions of "transactions costs" as has become commonplace in the neoclassical tradition, he stresses firms as exhibiting a structural persistence beyond rational organization, often constituting what sorts of economic exchange are possible or feasible. Most important for this volume, he has taken to insisting that power relations cannot be simply reduced to asymmetric information or other adventitious asymmetries but that they constitute an essential feature of organizations, and as such cannot be portrayed merely as artifacts of economic exchange.

The direct relevance of Dosi for my version of the economics of science consists of two recent projects:(1) the quest to argue that recent initiatives by the European Union to imitate the neoliberal reorganization of research characteristic of the United States will bear deleterious consequences for the future of research in Europe (Dosi, Marengo, and Pasquali 2006; Dosi, Llerena, and Sylos Labini 2006); and (2), a trademark project of the SPRU–Sant'Anna axis to rethink the track record of one of the key sectors of the modern regime of science organization: the pharmaceutical/biotechnology sector (Mazzucato and Dosi 2006). Both are utilized in subsequent chapters.

Another important figure in the European evolutionary economics community is Benjamin Coriat, a professor at University of Paris-XIII. Coriat's connections to the French "regulation" school have predisposed him to be much more willing to think in terms of "regimes" of science organization and governance than many of the other economists, breaking out of the endemic blinkered perspective of so many analysts, and fostering new ways to apply this organizing principle to the modern regime of commercialized science.

His work, along with the work of the Parisian historian of science Dominique Pestre, has been inspirational for this volume. Coriat combines close attention to specifics of the changes in intellectual property rights (Coriat and Orsi 2002; Coriat 2002) with a comparative perspective on the ways in which the biotechnology sector resembles and differs from other nominally "high-tech" sectors (Orsi and Coriat 2005; Coriat, Orsi, and Weinstein 2003). He also was the first to point out that there were severe problems showing up in the biotechnology sector, counteracting the reflex tendency to treat it as a wonderful cornucopia of technology transfer and the wave of the future of scientific research, a stance that has now spread to establish a beachhead within the Harvard Business School (Pisano 2006b). This will be a key theme of Chapter 5.

The third and perhaps most important figure of the younger evolutionary economists is Paul Nightingale, a Senior Lecturer at SPRU. Nightingale was trained as a chemist, but he later became interested in the political economy of science (Martin and Nightingale 2000). Nightingale has revealed both theoretical inclinations, which involve integration of real cognitive science and the social studies of science literature into theories of the organization of scientific research (1998, 2003, 2004), and deft empirical capacity, which has been trained with great effect upon the pharmaceutical industry (2000; Nightingale and Martin 2004; Hopkins et al. 2007). His role in sounding the tocsin over unfounded enthusiasm regarding the "biotechnology revolution" will provide a salutary prophylactic for universities and governments in their unseemly haste to jump on a bandwagon hobbled by a broken axle for at least a decade now.

At a deeper level, Nightingale (2004) revisits the Cold War argument that the inherent unpredictability of science must necessarily lead to a version of the "public goods" story in the economics of science. Instead, he argues for the "moderate-unpredictability thesis—that when dealing with economically important technologies, predictability is typically, but not always, constructed" (2004, 1264). Here the knowledge attained does not exist independent of the social means used to arrive at it. Hence technology can exist as a relatively autonomous body of knowledge distinct from science, "because it is possible to know how to produce effects without knowing how those effects are produced" (1271–72). However, to deploy a technological capacity to produce and maintain a constructive predictability, distinctively scientific capacities and abilities are required. This can be used to explain why firms cannot simply apply science "off the shelf" to their problems and why internal R&D capacity is a complement, but not an alternative, to external market sources. Technological capabilities are costly to set up, have long gestation periods, are generally localized, and are mostly nontraded. They look a lot like what I call in this volume "the science base": an interlocking set of institutions that meld research, education, development, politics, publication, and recruitment. Nothing could be further from the neoliberal notion of a marketplace of ideas in framing and execution. As Nightingale puts it, "The justification for the public funding of science is not based on unquantifiable, abstract theory or market failure arguments about the provision of public goods. It instead revolves around the empirical requirement for the infrastructure needed to produce technology and allow markets to work" (2004, 1278). The rest of this volume constitutes a sequence of empirical illustrations of precisely that point.

One reason Nightingale is so important for this volume is that his insights can equally be applied to the economics profession and its relationship to science, a matter he has recently considered (2008). One lesson of this chapter might be that economists have approached their contemplation of science more as a technological phenomenon than in a scientific spirit. As Nightingale has said, they have been more concerned with producing a preconceived effect (say, a certain policy) than coming to understand how a theory of science, one which would venture into all manner of unforeseen and ill-defined areas, might be built up from multiple acts of empirical scrutiny. For example, Nelson (1959) and Arrow (1962) were more concerned with justifying a big increase in the federal (military) subsidy of what they called "basic" university science in the Cold War than they were with understanding how corporations, governments, and universities might fit together in constituting a flourishing and stable American science base. It was easier for them to simply *presume* science was just another instance of a market phenomenon, full stop, to satisfy their patrons at RAND back then. Later on, David and Dasgupta (1994) produced a model to assuage our fears concerning the ongoing privatization of science; they speculated that we could simply choose how much privatized science from column A and public science from column B that we wanted in any given university, in something like a metamarket for the production of ideas. Likewise, there was no serious attempt to explore how technology (e.g., the Internet) and the economy (e.g., deindustrialization) were feeding back on the very structure and content of globalized, commercial scientific research.

This raises the possibility that there is an unhealthy codependency between the science policy community and the economics of science that needs to be broken before we can really come to understand our current predicaments. Langdon Winner has raised this possibility in congressional testimony:

> Indeed, there is a tendency for career-conscious social scientists and humanists to become a little too cozy with researchers in science and

> engineering, telling them exactly what they want to hear (or what scholars think the scientists want to hear). Evidence of this trait appears in what are often trivial exercises in which potentially momentous social upheavals are greeted with arcane, highly scholastic rationalizations. How many theorists of intellectual property can dance on the head of a pin? (Winner 2003)

Nightingale's message is that a better understanding of the relationship of science to technology today would need to acknowledge that some technological innovation can certainly happen in the absence of a developed science base and that science, technology, and the economy do not interact in any single fixed, lockstep fashion. However, the prerequisite of certain kinds of scientific infrastructure and institutional ties to other power centers (corporations, governments, universities, NGOs) is required to render the world sufficiently predictable so that innovation can itself proceed in a somewhat predictable manner. Without an ongoing, vibrant science base, no one would recognize those random acts of genius for what they really were. The purpose of the remainder of this book is to take that lesson to heart and to describe in an empirical fashion the ways in which the science base and the economy have mutually shaped one another in the twentieth century. In the next chapter I begin by identifying three different historical regimes of organization; then in the following chapters I deal with the period since 1980 in much greater detail.

Of course, there has been a cornucopia of work in the social sciences that will aid in our understanding of the modern predicaments of the economics of science. However, to appreciate it, it will be necessary to clear away the deadwood of "linear models," "public goods," and the "new growth theory." It may even be wise to stop genuflecting to Romer or Arrow or Schumpeter. Only then, in Chapter 7, can we return to this question: What has been the effective role of economists in the modern commercialization of science?

II

A Modern Economic History of Science Organization

3

Regimes of American Science Organization

Claims about the proper method for writing the history of science are simultaneously claims about the relations between the producers and consumers of scientific knowledge.[1]

Money Can't Buy Me Truth?

Viridiana is not especially distressed to learn that the economists are a few cadenzas short of a concerto when it comes to science, but the people who really get her goat are the ones who insinuate that she is pining for a lost Shangri-La that never really existed. They say things like, "The enlistment of science in the cause of commerce and production goes back to Antiquity" (Shapin 2008b, 95). And that from a historian! What really irks her is the suggestion that science has always been commercialized and that therefore in her wistful nostalgia she must be suffering from premature onset of Alzheimer's. Surely there has always been some sort of interdependence between the economy and science, but she doubts it has been as monolithically invariant as the cheerleaders of checkbook science intimate. But Viridiana lacks the time or inclination to search out an alternative take on the economic history of science.

It's not easy to get past the noise, because the vast majority of writers on this topic tend to plump for linear narratives, be they Annals of Decline, or else Ripping Tales of Progress. In the one corner, there mill the motley ranks of Cassandras, who nurture a soft spot for the Good Old-Time Virtues of the Mertonian norms and bewail expulsion from the prelapsarian Garden.[2] Back in the golden day there may have been an invisible college, chorused sweetly in concert in the quest for truth, they lament, but now there are only feckless individual entrepreneurs scrabbling for the next short-term contract: "Who will now defend the virtue and purity of science?" they wail. By contrast, there also parade the massed phalanx of economists, science policy specialists, and historians like Shapin and their bureaucratic allies, who by and large tend to reverse the valences but nevertheless engage in much the same forms of discourse. For them, most scientists in the "bad old days" had been operating without sufficient guidance from their ultimate patrons, the corporate pillars of the economy; but luckily, with a bit of prodding from the government, a friendly nudge from their university's intellectual property officer, plus a few dollars more waved in their directions, scientists have been ushered into an era that genuflects to the compelling logic of "technology transfer." At the risk of caricature, one might summarize their *raison d'être* as the collation of empirical data in order to argue that the expanding modern commercialization of scientific research turned out to be "inevitable," with the corollary that little evidence exists that it has "significantly changed the allocation of university research efforts."[3] Admittedly, many of these purveyors of glad tidings would still regard themselves as defending the preservation of an "optimal" sphere of research reserved for open public science and pure unfocused curiosity: a "nature reserve" where researchers could freely gambol; a "separate but equal" doctrine applied to unspecified portions of the university. The history of science for them is simply divided into an Age of Confusion when "open science" had unaccountably been mistakenly conflated with the whole of science, fostering a lack of understanding of the efficient organization of systems of innovation, and our own current Age of Free Enterprise, when we see the true situation of pervasive ownership with clarity. This kind of crude "before and after" discourse has regrettably come to dominate much of the contemporary science policy literature, redolent with euphemisms like "enhanced technology transfer" and "democratically responsive science," which seek to reconcile the harsh authority of the almighty dollar with the delicate sensibilities of those otherwise inclined to resist the onset of the End of History.

This rather superficial stage 1/stage 2 narrative, be it upbeat or downbeat, has little to do with the actual histories of the sciences (plural). In this chapter we shall observe that the "commercialization of science" turns out to be a stubbornly heterogeneous phenomenon, resisting simple definition, and not much illuminated by the economic theories described in Chapter 2. Consequently, many contemporary discussions of the commercialization of science have proved deeply unsatisfying, tethered as they are to totemic monolithic abstractions of Science and the Market pushing each other around in timeless Platonic hyperspace. Indeed, some historians have long sought to remind their readers of what one anthology (Gaudillière and Löwy 1998) calls the "Invisible Industrialist" who has occupied the interstices of numerous laboratories and frequented the hallways of universities since the middle of the nineteenth century. Yet, in rejecting the false polarities of the neo-Mertonians on the one hand, and the economic apologists for the era of commercial science on the other, it would appear that the denizens of science studies and some historians of science have of late run a very different risk of denying that there has been any significant change whatsoever in scientific protocols; hence important structural differences are overlooked that might be traced to alterations in the ways in which science has been paid for and accommodated within economic systems over long stretches of time. One recent instance of this sort of attitude has been expressed by Steven Shapin:

> Throughout history, all sorts of universities have "served society" in all sorts of ways, and, while market opportunities are relatively novel, they do not compromise academic freedom in a way that is qualitatively distinct from the religious and political obligations that the ivory tower universities of the past owed to the powers in their societies. (2003, 19)

A cruder version of this orientation was captured in interview transcripts with the chair of an electrical engineering department (in Slaughter et al. 2004, 135):

> You have to accept the fact that it [research] is going to be driven by the people who give you the money. [If] the state gives us money, they tell us what to do. [If] NSF gives us the money, they tell us what research they want done. [If] DoD gives us the money, [it's] the government . . . Why is it any different with industry? I see no difference whatsoever.

Strangely, this widespread antihistorical insistence on "the way things have always been" in science in its intercourse with the economy dates back to the supposed godfather of social studies of science, Thomas Kuhn. In a little-read set of comments on a pivotal conference on the relationship of industrial R&D to science held at Minnesota in 1960, he insisted that "the two activities, science and technology, have very often been almost entirely distinct" and, indeed, that "historically, science and technology have been relatively independent enterprises," going back as far as classical Greece and Imperial Rome! As a historian, Kuhn felt impelled to admit that,

> since 1860 . . . one finds that characteristic twentieth-century institution, the industrial research laboratory . . . Nevertheless, I see no reason to suppose that the entanglements, which have evolved over the last hundred years, have at all done away with the differences between the scientific and technological enterprises or with their potential conflicts.[4]

The indisputable fact that scientists and their institutions have always and everywhere been compelled to "sing the prince's tune when taking the prince's coin" in one form or another does nowhere imply that the modern trend toward the escalated and enhanced commercialization of science has not altered the makeup of the supposedly invariant "scientific community," not to mention the nature of the "outputs" of the research process. Furthermore, the growing suspicion that the political economy of the sciences in America has been transformed from top to bottom at least twice over the past century has yet to be correlated with the types of science that have been performed, or, indeed, the ways we tend to think about the successful operation (or conversely, the pathologies) of the "scientific community." This sort of agenda was called for in a perceptive paper by Michael Aaron Dennis back in 1987, but his entreaty has yet to be heeded within the STS community, much less appear on the radar screen of the economics of science.

Close on the heels of the political movement surrounding J. D. Bernal and the Hessen thesis in the 1930s,[5] and the subsequent Cold War anti-Marxian backlash against it, most appeals to economic structures as significant conditioning factors in the production of science simply dropped out of postwar theoretical discourse within science studies and most of the history of science. As Dennis has written about American historians, the manner of "solving the problem of providing for the support of the material foundations of science—salaries, labs, instruments—effectively eviscerated the possibility of anything even remotely resembling the materialist historiographies of science that had developed between the wars" (1997, 16). Something similar seems to have happened in Europe as well. Elsewhere I have speculated that the postwar political shift in the philosophy of science played its part in repressing such questions (Mirowski 2004b). Consequently, as the next great transformation of the organization of research was taking place in the 1980s, science studies was instead turning its attention to microscale studies of laboratory life, ignoring how the laboratory's macroscale relationship to society was being reengineered all around, not to mention the changing identities of the paymasters for all those DNA sequencers and inscription devices.[6] At this late date, the qualitative and quantitative effects of the panoply of market activities on scientific research remain an open issue. In this book I seek to revive that discussion and provide some initial answers.

The alternative approach to the economics of science advocated here explores the possibility that the diverse forms of the commercialization of science actually have indelibly shaped both the practice of research and the contours of whatever it may be that we encounter at the end of the process. A key entry point for this discussion turns out to be an enumeration of the ways in which that protean entity "the laboratory" was appropriated and reconstructed by higher education, corporations, and the government over the twentieth century, a point first made with great brio by Dennis (1987) and further propounded by Pickering (2005). In addition, the inescapable fact of globalization tends to undermine earlier nationalist and parochial approaches to the problem of the economics of science, including the once popular notion that there might persist coherent "national systems of innovation" (Sharif 2006; Hart 2009). Another crucial variable is the way in which the divide between nominally "public" and "private" conceptions of knowledge has shifted in the recent past and how that has fed back on the rationales for various actors in their exercise of the governance of science (Slaughter and Rhoades 2004).

If serious analysis is going to get beyond endless disheartening evocations of Merton's norms and Arrow's "public goods," then we need to make a fateful choice between casting the "constructivist" stance as one treating the entirety of science as everywhere and always just another form of marketing (Woolgar 2004), and the alternative stressing the essential historical instability of the commercial/communal binary as instantiated in actual concrete practice. Following the lead of historians like Dominique Pestre (2003b, 2004), Benjamin Coriat (Coriat et al. 2003), Paul Forman (2007), and Paul Nightingale (Hopkins et al. 2007), I organize an account of the last hundred years of the history of science as a sequence of temporally specific "regimes" of economic and social organization, intertwined with changes in the ecology of the sciences themselves. This is not intended as some Grand Unified Theory but rather as an alternative idiom to those found in the previous chapter, by means of which one can better discuss the political economy of science. Viridiana can relax—there are no overweening ambitions here for some grand Science of Science, just a better grounding in the archival record.

Once the ground has been prepared in the next section by the presentation of an analytical scheme of temporal periodization of the twentieth century (albeit one rooted primarily in the American context), I will explore the differing meanings of the commercialization of science under each individual regime. Although market considerations were never absent from the laboratory or the classroom, the modern commercialization movement can in no way be considered a "return" to anything like the interwar science promoted by Jazz Age captains of industry.[7] In other words, Viridiana is correct in thinking she has lived through an unprecedented major watershed in the organization of science. This latest transformation has had more than a little to do with the rise of neoliberal ideas, as suggested in Chapter 1. Modern science has turned out to be a qualitatively different phenomenon from its predecessors because it has been grounded in equally profound historical transformations in the corporation, the university, and the government, with consequences for their respective initiatives to exercise control in the organization and funding of science.

Three Regimes of Twentieth-Century American Science Organization

Historians of science and STS scholars have been wary of reifying the concept of "science" as a transcultural, transhistorical category, and for good reason. The more observers learn about scientists and their livelihoods, the more we come to appreciate the sheer diversity of their activities, the vast compass of their societal locations, and the multitude of ways their findings have become stabilized and accredited as knowledge. What keeps this daunting multiplicity from defeating analysis is the dominance of certain identifiable institutional structures involved in organizing scientific inquiry in the modern period. Scientists have never subsisted as a purely autarkic self-organized discourse community, contrary to the rhetoric dominant during the Cold War era. Rather, they have always been enmeshed in complicated alliances with and exclusions from some of the dominant institutions of our era: primarily, the commercial corporation, the state, and the university.

The approach I adopt in this chapter has been heavily influenced by a literature in the history of science that has begun to frame changes in twentieth-century science in terms of "regimes" of science organization, funding, and thought styles.[8] Rather than reduce issues in science policy to the cognitive peccadilloes of individuals on the one hand and macro levels of R&D expenditure compared to GDP on the other, as has been the tendency in the older economics of science, these scholars have sought to identify the intermediate level structures that have proven crucial for the operation of science within specific disciplines and specific national contexts. As one scholar has insisted, "The fact that Galileo successively worked in a university, then for the Republic of Venice, and finally at the court of the Grand Duke of Tuscany is of direct relevance to the kind of knowledge he produced" (Pestre 2005, 30; also Biagioli 2006a). Something similar will be argued here for science in the twentieth century. Of course, research can assume a dizzying multiplicity of forms, so a certain prudence is called for to guard against the temptation to descend into an endless morass of picayune cultural, legal, and social particularities, especially when it comes to discussing science.

It is a byword of the economics of science that the quotidian activities of the scientist always presume some social scaffolding of material support, which in the modern epoch has been most frequently built up from corporate, governmental, and educational (CGE) elements. Furthermore, various individual scientific fields will be experiencing their own rhythms of relative growth or stagnation, depending on the particular historical configurations of their own intellectual trajectories, in combination with the levels of encouragement provided by the CGE sectors. To render this set of propositions more concrete, I provide in Table 3.1 a schematic outline of the three regimes of science funding and organization in the United States in the twentieth century, based on numerous contributions of historians of science. To prevent this historical sketch from becoming unwieldy, in this chapter I restrict the table to indications of CGE developments that have had direct bearing on the constitution of the "laboratory" in scientific research; considerations of length preclude simultaneous extension of the CGE analysis to, say, clinical medicine, field sciences, or purely abstract mathematical endeavors.[9] My current purpose is to insist that the corporation, the legal framework, the political status of the scientist, and the structure of the university have not been static over time, and their alterations can be directly related to the ways that scientists have made their livelihoods and pursued research agendas, promoted by their immediate patrons. Thus, contrary to prognostications of contemporary economists, no single "market" governed the evolution of science in America over the last century; rather, there have been multiple formats of provisioning, often embedded within larger social structures.

The designations provided in Table 3.1 for the various regimes are predicated on vernacular characterizations found in the existing historical literature. They are intended to evoke something about what contemporaries thought to be the social phenomena most relevant to their science. The "captains of erudition" regime is so designated in honor of Thorstein Veblen (1918), who wrote one of the earliest descriptions of the American research university as becoming subject to specific corporate organizational principles; it also bows in the direction of the dominant American school of business history based on the work of Alfred Chandler.[10] The subsequent Cold War regime is a label regularly used to designate what many now portray as a fleeting interlude of military dominance over science management in the period beginning in World War II.[11] The terminology of "globalization" for the current regime is not so much an appeal to a fashionable concept in contemporary social theory as it is an insistence on a set of factors indispensable for an understanding of the forces that drive the current wave of commercialization of science.

The Genealogy of the American Laboratory

Laboratories were not something that just naturally appeared in the American landscape: They had to be built and, to be able to subsist as more than ephemeral entities, they had to be integrated into some sector of the

Table 3.1. American regimes of science organization in the twentieth century

Period, regime	Corporation evolving	Government corp. policy	Government science policy	Science managers	Higher education	Pivotal disciplinary science
1890–World War II Captains of erudition regime	1895–1904 great merger movement: Chandlerian firm of "Visible Hand." Innovation of in-house R&D labs to control competition.	Massive expansion of corp. preroga-tives. Corps. become legal agents; patents a major strategic tool. Beginning of antitrust. Employers own research of employees.	Almost nonexistent. NRC formed as trade assoc. lobby for natural sciences. General suspicion of gov't involve-ment. NRE fails. Wartime patent bounty.	Charismatic PhD directs corp. labs. Foundation officers run few elite univ. grant programs (on corporate principles).	Elite liberal arts. Research subordinate to pedagogy. Science not a major priority. Foundations attempt reform. Corporate-style Labs founded.	Chemistry, electrical engineering.

World War II–1980 Cold War regime	M-form, conglomerate diversification. R&D units as semiautonomous revenue earners (due to military contracts). Regulatory capture.	Corporate powers augmented; antitrust strengthened. IP weakened. Military contracts as industrial policy.	Huge expansion. Federal military funding and control. Military promotes basic science to defeat enemies. Nat'l labs. NSF as nonmilitary face of "pure" science.	Military primary science managers for research univs., think tanks, nat'l labs, corporate contract research. "Peer review" a secondary means to assert quality control.	Mass education at expanded research univs. Integrated teaching/research. Mandate: Turn out democ. citizens: academic freedom ideology.	Physics, operations research, formal logic.
1980–? Globalized privatization regime	Breakdown of Chandlerian model. Retreat from vertical integration, diversification. Corps outsource R&D, spin off in-house labs.	Transnational trade agreements expand corp. powers to circumvent national control. Antitrust weakened; IP vastly expanded.	Privatize publicly funded research: Bayh-Dole, etc. Kill OTA. Science just one political resource among many.	Globalized corp. officers control. univs., hybrids, CROs, startups.	Stock up human capital for those who can pay. Only entrepreneurs are free. Sever the teaching/research connection. Distance ed for masses.	Biomedicine, genetics, computer science, economics.

economic infrastructure. American chauvinists like to imagine a prelapsarian state of the "democratization of invention" in the nineteenth century, but in truth, "Very little work of scientists in industry before World War I was of interest to anyone concerned exclusively with science and its progress" (Mowery and Rosenberg 1989, 37). Unlike the situation in Europe, large-scale laboratory science did not originate in the university sector in America. Rather, from the outset, it was very much a commercial initiative.

Some economists seeking to read Cold War structures back into the nineteenth century have cited the founding of the U.S. Geological Survey (USGS) in 1879 as an indication of a favorable inclination of the U.S. government toward subsidy of science (David and Wright 1997), but this profoundly misconstrues both the setting and the politics of the situation. Almost everyone of consequence in postbellum America simply accepted the premise that the government had no place in the organization or promotion of science. The Allison Commission, convened in 1884, was among other things an early attack upon the USGS and its use of public funds to underwrite science, with both John Wesley Powell and Louis Agassiz agreeing that "the government should not undertake to promote research in those fields where private enterprise may be relied upon" (Allison Commission 1884, 1079). They were more concerned to limit military control over the coastal survey, because of previous bureaucratic tussles between the Navy Hydrographic Office and the Army (VandeWall 2007) than they were to promote the advancement of science in America. The lineage of the birth of the laboratory in America was to be found rather in the industrial sector.

The broad outlines of the rise of the industrial research laboratory are now quite familiar.[12] Everyone concedes that its origins are to be found in continental Europe, primarily but not exclusively in Germany, and that it was initially located in large firms engaged predominantly in what has become known as the "second industrial revolution": chemicals, electrical machinery, railways, and pharmaceuticals. In France and Britain, the formal science base (such as it was) had very weak ties to industry and therefore did not partake of this innovation. An earlier vintage of historiography tended to assert that the rise of "science-based industries" simply summoned an implicit exigency to incorporate research activities within their ambit, in both Germany and the United States, but modern historians have long since grown more cautious, realizing that the ingredients to explain the appropriation of what had previously been a specialized pedagogical device for industrial purposes would be found in a strange brew of state policy toward advanced education; ideologies of state-building and political rectitude; the rise of various notions of intellectual property (hereafter abbreviated as IP); the conditions that gave rise to large and powerful corporations in particular national settings; and the ambition to exert control over burgeoning transnational mass markets in clothing, transport, and communications, electrical equipment, and patent medicines. Whereas most manufacturing firms had long made provisions for internal quality control, routine testing, and incremental process improvement, an innovation arose around the 1870s to expand the purview of these specialized corporate arms into patent protection, the bureaucratization of trade secrets, and the generation of novel processes and products. It resembled a phase transition between the periodic use of existing sciences for corporate purposes to something approaching the institution of bureaus dedicated to *doing* science for corporate purposes. The distinction was not always sharp, the results were not often that immediately striking, and the transition was not always conscious.

The rise of the industrial laboratory was the consequence of an American pincers movement: on the one hand, a push to bureaucratize and industrialize (or vertically integrate backward, as economists might say) something that heretofore had been conceived as the ineffable capacity of the individual genius, and on the other, a pull to adapt a purpose-built academic social formation to corporate imperatives that itself had only recently been stabilized in specialized educational settings for pedagogical purposes. Michael Dennis correctly points out that when later nineteenth-century American figures made their pleas for "pure science," they did not refer to some notion of disembodied science carried out for its own sake or to an imaginary autarkic scientific community defending its prerogatives, but rather to a pedagogical ideal for a hands-on higher education where teaching and research were combined in a setting relatively sheltered from commercial considerations. *Pace* Bruno Latour, the issue was not whether the denizens of laboratories or their proxies "circulated" with impunity in the wider world, but rather whether laboratories themselves were a robust phenomenon that could be severed from the nascent research university and successfully grafted onto the multidivisional corporation. The wrenching estrangement of the laboratory from its teaching functions constituted so dramatic a departure from its conceptual origins in the later nineteenth century, that it was not hard to find any number of academics expressing scorn for the newfangled industrial laboratories and their spiritually debased inhabitants, disparaging public confusion of untutored tinkerer-inventors with real "scientists." Yet it would be an anachronism to read these as indicative of some transcendental incompatibility of science and commerce, as Kuhn himself later did. Rather, it makes more sense to approach them as symptoms of conflicts attendant on institutional innovations in the construction of both the public and private spheres, artifacts still in their early stages.

The Captains of the Erudition Regime

One of the most salient differences between the German situation and its American counterpart circa 1900 was that, by and large, the academic research laboratory did not substantially predate the rise of the industrial laboratory in the United States.[13] Higher education in the natural sciences and the social sciences was acknowledged to have been superior in the German setting at the beginning of the twentieth century; it was also recognized as having attained an unprecedented level of state-sponsored centralization. The German university had pioneered both the research seminar and the research laboratory; by contrast, the pedagogical research laboratory had not yet become solidly established in American universities, which were predominantly devoted to moral uplift and liberal arts education for a narrow strata of the elite, although the forms this assumed were widely decentralized and diverse.[14] As David Noble put it, in the nineteenth century "shop culture" was deemed to be opposed to "school culture" (1979, 27); if anything, the universities lagged behind large firms when it came to building and staffing labs. In fact, units that might have been called "labs" in the later nineteenth-century American context were more often than not glorified machine tool shops, which existed in a parallel universe decoupled from formal learning and tenuously attached to the firms that claimed them (Israel 1992).

Far from being transplanted bodily from an academic to a corporate context in the United States as it had been in Germany, the American scientific laboratory was built up almost from scratch, modulo some Germanic inspiration, more or less simultaneously at both sites. For instance, as early as 1881, American Bell Telephone experimented with the location of a new physics laboratory, offering Harvard University the money to build it, as long as "professors could use university laboratories in work for private companies" (Guralnick, in Reingold 1979, 133). MIT's fabled Research Lab for Applied Chemistry, originally intended to carry out industrial research, dated from 1908. Since dedicated university laboratories were rare, the academic/commercial distinction was less than distinct. Yet the siting of industrial research on college campuses often proved less than satisfactory for its patrons, mostly owing to perceived insufficiency of corporate control (Lécuyer 1995, 64), redoubling the formation of in-house laboratories. This made for an unusual political economy of science in early twentieth-century America, going some distance toward explaining a certain impression of "exceptionalism" in the culture of science that one encounters among many commentators (Wright 1999) and one that contributed to the fact that American scientific research achieved an advanced level of one kind of commercialization far more quickly than did any other country by the 1930s. It also coincided with the successful elevation of a subset of the natural sciences to world-class status for the first time in the United States, thereby raising the intriguing prospect of the existence of multiple institutional paths to the fortification of a research base in the course of economic development of national systems of research.

Science in the American university system had gained a stable foothold comparatively late, around the beginning of the erudition regime.[15] The highly decentralized character of the American higher education sector at first posed an obstacle to the development of a scientific curriculum, although it would later prove a boon. While later historians might point with pride to the earlier founding of Harvard's Lawrence School, the Yale Sheffield School, Throop Polytechnic (the predecessor of CalTech), or the Massachusetts Institute of Technology, the impact of these and other educational institutions on actual practices of research and the shape of American science was slim to negligible prior to the 1890s. The impetus for the change in regimes originated instead mostly from within the corporate sector, initially in the creation of a new kind of in-house laboratory for commercialized science but later in the export of corporate protocols and funding structures to some handpicked research universities, by way of the instrumentality of a few activist foundations. Hence our brief overview necessarily begins with a flyover of the relevant background history of the corporation.

American historians of technology have tended to lean on the work of Alfred Chandler, and in particular his book *The Visible Hand* (1977), to provide the framework within which they situate their understandings of the rise of commercialized science.[16] This turn of events has been mildly incongruous, partly because Chandler devoted very little explicit discussion to the role of industrial laboratories in his history, but also because it is sometimes predicated on a fairly outmoded technological determinism (Chandler 2005a). Set against an earlier historical literature that considered the corporation to be a nexus of power growing dangerously out of control, Chandler portrayed the rise of the large American corporation around 1900 as a rational organizational response to technological imperatives of high-throughput capital-intensive patterns of production, found primarily in the newer science-based industries, which could only be made viable through the parallel construction and organization of mass markets on an unprecedented scale. Chandler praised the Jazz Age mega-corporation for adopting centralized bureaucratic managerial structures and vertically integrating backward into inputs and forward into sales, advertising, and market research. Although he did lightly touch on the rise of the industrial laboratory (e.g., 1977, 425–433), it is treated as just another exemplar of the line-and-division managerial structure to which Chandler sought to attribute the success of firms such as Standard Oil, General Electric, and DuPont. Hence Chandler did not so much proffer an explanation of the rise of the industrial research laboratory as mutely point to one necessary bureaucratic prerequisite for its American incarnation coming into existence. Some industries could have sought to "integrate backward" into research, except for the inconvenient fact that in most cases there were no preexistent stable structures in which they could integrate themselves backward.

The Chandlerian narrative as it became manifest in science studies and economic history should therefore be supplemented by legal and political considerations, which Chandler largely shunned. The limited liability corporation, far from being an established fixture on the American scene, had just undergone a period of substantial judicial fortification at the end of the nineteenth century, owing to the infamous Santa Clara nondecision extending Fourteenth Amendment rights to corporations (Nace 2003), the race to the bottom of states to liberalize corporate charters, and the unprecedented merger wave of 1895–1904. This sudden arrogation and consolidation of power had not gone unnoticed and had begun to provoke a countermovement beginning with the Sherman Antitrust Act of 1890 and continuing with the Clayton Act of 1914, which in turn provoked political movements hostile to corporate dominance of the economy in the Progressive Era. The rise of the American industrial laboratory should be situated in this context to appreciate some of its more distinctive characteristics as well as its impact on academic science.

The standard popular account portrays the *fin de siècle* industrial lab as a sort of factory of innovation, churning out gadgets that became new products or improved production processes on demand for the corporate hierarchy. This was the image promoted by the Scripps Science News Service, the very first corporate-backed "public relations of science" initiative, which began in 1921 (Tobey 1971, Chap. 3). But more recent historical literature resists this tendency to frame the lab either as a straightforward invention factory or as some university-science-department-in-exile, and for good reasons.[17] The prime directives behind many of the innovations growing out of the large corporation were the drive to control markets, render unforeseen events manageable, and stifle external competition. As the government began to block direct attempts at market control such as explicit cartels, pools, and other tied arrangements through its antitrust prosecutions, the locus of corporate control began to shift to indirect arenas such as IP, the imposition of technical standards, and the like. One primary reason that large corporations turned their attention to bringing scientific research within their walls in this period is that "invention and innovation were effective defenses against antitrust suits" (Hart 2001, 926) and that patents in particular but IP in general were conceived as the best and most effective means of controlling competition in the early twentieth century (Noble 1979, 89). This trend was actively promoted by certain related U.S. government policy moves, such as the seizure by the Alien Property Administration of German patents in 1919 and their licensure through the Chemical Foundation to American firms under highly favorable terms.[18] As both case law and legislation were slanted in the direction of integrated discrete corporate organizations instead of interfirm cartels (or other features of the German model),[19]

> legal doctrine inadvertently spurred corporate consolidation, and the consolidated corporations in turn, enhanced their investments in R&D. "The birth of the central corporate laboratories in this period . . . [is] therefore in part the product of antitrust law." (Hart 2001, 927)

Legal redefinitions of IP and clearer stipulations as to who might assert claims over the fruits of scientific research were heavily conditioned by the shifting needs of the fortified corporation. In a move with untold consequences for the future organization of science, corporations managed to have the case law with respect to employee inventions shifted away from older labor-theoretic notions of the fruits of individual genius and toward a presumption of employers' ownership of *anything* an employee might do or invent. Prior to the 1880s, the standard default rule was that rights to inventions were vested in employees; first, through the creation of the doctrine of "shop right" in the 1880s to 1910s, and afterward, through a series of judicial decisions that made direct reference to corporate research laboratories, the presumption of ownership was shifted decisively to the firm itself (Fisk 1998, 2009). Corporate initiatives then fed back on general cultural images: By the early 1920s, American court decisions began appealing to the apparently commonly accepted notion that invention and science was a "collective" and not an individual phenomenon.[20] As a sign of the times, Nobelist Robert Millikan began to complain in the 1920s that the German research university did not sufficiently respect the collective character of scientific research (Tobey 1971, 219). These profound alterations in corporate power made it possible for Joseph Schumpeter to assert in the 1940s that "innovation is being reduced to routine" (1976, 132).

However, the convenient notion of the "collectivity" was not to be allowed to exude too far outside the firm's boundaries (as in the writings of Thorstein Veblen) for that might bring back the dreaded world of cartels, patent pools, plunderbunds, and trusts. The communal character of science was never given a broad interpretation in America. The legal bias against cross-firm combinations and joint ventures bore direct consequences for the existence and viability of corporate labs that might try to escape from the tentacles of corporate bureaucracy. While freestanding, independent industrial labs were also founded in this period, they never caught on or expanded to the extent that in-house industrial research did; unlike some of the largest in-house labs, they never conducted any world-class science. Moreover, they undertook contract work that did not mimic that of the big corporate labs but was most often subordinate and supplementary to them.[21] It is difficult to overstate just how important this fact is in distinguishing the broad structure of the erudition regime from the current privatization regime. Thus, before World War II, even though the research process was clearly becoming commercialized in some connotations, it was not rendered so thoroughly fungible to the extent of being freely outsourced by its corporate sponsors. The modular "marketplace of ideas" turns out to be a much more recent phenomenon. Hence, the particular format assumed by contract research in America was (and continued to be) heavily conditioned by industrial policy and IP conventions.

After the first generation of the captains of industry had built or consolidated their massive industrial corporations and retired, or otherwise cashed out some of their gains, they or their descendants decided to devote some funds to philanthropy (or perhaps merely engage in tax avoidance) through the creation of various foundations: The Russell Sage Foundation (1907), the Carnegie Corporation (1911), and the Rockefeller Foundation (1913) are some of the better known. Assistance to higher education had become part of their agenda, but serious questions arose as to the most appropriate way to pursue this goal. At first, grants were patterned on other philanthropic practices, and when it came to academic recipients, they were pitched to essentially provide temporary individual outdoor relief to indigent or otherwise needy scholars. However, just as in the case of IP, by the 1920s the focus on the isolated individual as the monad of science funding had gone out of fashion, and attention turned to the targeted application of funds to provide research endowments for continuing programs, reorient whole disciplines, and build new institutions. It was consistent with this vision that the grants were overwhelmingly channeled to private universities and structured to concentrate "excellence" in a few powerful institutions. As Robert Kohler put it most succinctly, "The large foundations were . . . carrying business methods and managerial values from the world of large corporations into academic science" (1991, 396). In everything from recasting the research grant as a contract that imposed certain standards of bureaucratic accountability, to imposing the line-and-division managerial structure on university administrations and departments, to encouraging the creation of hierarchical teams of researchers, the corporate officers who staffed the large foundations tended to foster the standards and practices of the large American corporation within their targeted flagship research universities. According to E. B. Croft of Bell Labs,

> It might appear that it would tend to destroy the initiative of the individual; that it would make it difficult to properly assign the credit and give the reward to the individual worker. These are all problems of administration that have had to be worked out. First of all we must establish in the individuals a state of mind, which leads them to really believe that their best results are attained through cooperation with others. (in Noble 1979, 119)

Harvard and Chicago would be coaxed and inspired to become the AT&T and Standard Oil of American higher education, surrounded by smaller and relatively insignificant rivals who had not learned the lessons of building a permanent and successful managerial hierarchy (the Chandlerian lesson) and, not inconsequentially, a strong research capacity. Colleges would face the choice of emphasizing liberal-arts pedagogy or else aspiring to technical expertise in research. Consequently, the scientific research laboratory was propagated throughout the academic landscape as the necessary accessory to the mature corporate business plan. "Foundation managers allied themselves with the small but growing numbers of academics . . . who realized that [corporate] organization and management were good ways to keep ahead of the pack in the increasingly crowded and competitive world of basic research" (Kohler 1991, 400).

The fact that so much of the structure of the American academic science laboratory was directly inspired by that of the industrial research lab did not imply that academic scientists uniformly sought to mimic the behavior of their industrial brethren, however. Even as the social structure of laboratories was becoming patterned upon corporate social structures, the academic scientists still lauded the university laboratory as a pedagogical ideal, existing separate and apart from commercial pressures and also free from government subsidy. Yet this quest for "purity" only exacerbated the problem of who precisely would fund and manage the research carried on under that banner. The nagging tension between science beholden to special interests versus science in pursuit of the public interest proved a challenge to those who apprehended the "erudition" dynamic as a danger to democracy, such as Walter Lippmann, Thorstein Veblen, and John Dewey (Mirowski 2004b). The foundations were increasingly targeting their funds to support specific research projects in a consciously limited portfolio, or else professionalized arenas of higher education such as medical schools, and could not be expected to bear the burden of the health of the whole gamut of sciences, much less the careers of the next generation of scientists. The National Research

Council (NRC), established in 1916 as a sort of trade association to lobby for the support of the natural sciences, actually opposed direct government subvention of researchers (Noble 1979, 155). The NRC-backed drive to institute a National Research Fund, a private nonprofit that would derive its endowment from corporate subscriptions, failed miserably in the period 1926–1932 (Tobey 1971, Chap. 7). In 1933–1935, an abortive Science Advisory Board was formed within the federal government to distribute science grants but it was smothered in its crib by the National Academy of Sciences and some economists, like Wesley Clair Mitchell (Auerbach 1965). Robert Millikan was still denouncing federal support for the sciences at private universities as late as 1937 (Lowen 1997, 33); it remained minuscule. Outside of a few private universities favored by the foundations, the problem of sustained privatized care and maintenance of a diversified academic research base was not solved by the supposedly collectivized community of researchers or by its corporate patrons. It would not be solved until World War II.

Nevertheless, American laboratories for the first time in their history were able to produce some world-class science under the erudition regime. Whether the Nobel Prizes were for work originated in the academic sector, as with Theodore Richards's chemistry prize in 1914 or Robert Millikan's physics prize of 1923, or from within the burgeoning industrial sector, as with Irwin Langmuir of GE in 1932 or C. J. Davisson of Bell Labs in 1937, there was a certain American style of research that traced a part of its lineage to the corporate inspiration of the laboratories. European commentators noted a certain empiricist temper regnant, a kind of phenomenological exploration well suited to teams of researchers, infused with an experimental and accounting mentality as contrasted with a rationalist orientation. German world dominance in both physics and chemistry were still widely acknowledged in this period. Electrical engineering, however, found its center of gravity shifting westward by the 1930s. Chemical engineering and mining engineering had already become American specialties.[22] Nevertheless, America's deficiencies with regard to theoretical imagination were a common theme of opprobrium emanating from the older and cultured precincts of Continental Europe. Chemistry, probably the most lavishly supported of the natural sciences in America in this era, itself produced no radical changes in fundamental doctrines (Mowery 1981, 104). One might therefore conclude that the corporate orientation of American science did indeed influence the types of research performed in this era as well as some of the results produced. More to the point, when larger cultural movements felt impelled to come to terms with the world-historical significance of the advancement of science, most frequently it was European science that served as their reference point.[23]

The Cold War Regime

The fact that the American science base was utterly transformed in World War II, and then persisted in that novel configuration throughout the Cold War, is a historical generalization hardly requiring defense at this late date.[24] Nevertheless, it does tend to get confused with another notion: that mostly this was due to the rise of "big science"—the idea that postwar science organization was driven by pure scale effects, in much the same way that Chandler asserted that the structure of the modern corporation has also been driven by scale effects.[25] But concentration on abstract size and its quantification, a tendency often associated with Derek de Solla Price and the scientometric movement, tends itself to succumb to technological determinism and therefore requires an inoculation of economic history. There is no doubt that the constitution of huge teams devoted to the production of a particular weapon or device, such as the MIT Radiation Lab, the Manhattan Project, or Lawrence's cyclotron, could not help but provoke revisions in the way Americans would apprehend the nature of the "laboratory" in the postwar period. Science seemed increasingly to be organized around "gadgets," as the denizens of Los Alamos called the Bomb, and such devices were big along almost any dimension one would care to assess: reactors, accelerators, space vehicles, von Neumann's room-sized computers, and so forth.

Yet, before we become blinded by the shiny surfaces, blinking lights, and long phalanxes of bench scientists, it becomes necessary to direct our attention to some rather more pedestrian aspects of the quotidian prosecution of postwar science, namely, the myriad of ways in which the government, primarily but not exclusively in the guise of the military, transposed and inverted the previous understanding of the relationship between science and industry characteristic of the interwar period. The military, responding to a relative vacuum in science policy in the immediate wake of World War II, moved to retain access to the scientists who had done so much in helping them win the last conflict; then when other governmental agencies were eventually brought into play, the political situation dictated that military innovations and military funding would remain the dominant consideration in science organization.[26] The American government had destabilized the political presumptions that had ruled prior to 1940, and in altering its stance toward both industrial and science policy, it compelled both the corporation and the university to revise the ways in which science would be carried on in their precincts. This became the era of the now derided "linear model"—the linchpin of the older neoclassical economics of science discussed in the previous chapter. Indeed, one historian has claimed, "The linear model served as the dominant paradigm for organizing research and development during the early decades of the Cold War" (Asner 2006, 580). Under the triple imperatives of classification, rationalization, and projection of ideological superiority, the military refined the "purity" of the laboratory in a different crucible. One consequence was the strengthening of the institutional sway of the disciplines within the university: "After 1945, the increase in federal funding and the increasing impact of federally supported research centers threatened the hegemony of the universities and their departments. By the 1950s there was talk of a researcher's allegiance to their discipline, rather than their university" (Reingold 1995, 313). As an even greater unintended consequence, the change in regime underwrote a conviction, almost a dogma, that science and commerce should never mix, even though this flew in the face of a previous generation's experience.

The wartime experience of the Office of Scientific Research and Development (OSRD)/National Defense Research Committee (NDRC) and the immediate postwar debates over civilian versus military control of science have been superbly covered by the present generation of historians, so it need not be recapitulated here. What has been missing from these accounts is consideration of the ways in which the militarization of science had an impact on the previous erudition regime of corporate science, as well as the ways in which the American university was forced to reorient itself to occupy the space cleared for it within the postwar settlement. The most obvious alteration was the intrusion of the government as the third (and now largest) player in the funding and management of science, but this implied something more than slinging largesse at a few favored natural sciences. It involved subscribing to a tenet that the federal government was in the business of picking winners and losers in the realm of technological development by running a sub rosa industrial policy under the auspices of the military, which included promotion of a very different set of practices than had held sway before the war regarding IP and antitrust. Meanwhile, concurrently the corporation was growing in power and reach, given that many of its European competitors had been hobbled by the war. Both the government and the corporations were impressed by the efficacy of science in winning the late war; it was taken as given that it would also play a pivotal role in winning the Cold War.

Some people still find it hard to comprehend just how dominant the military was across the gamut of the natural and social sciences during much of the Cold War. For instance, in 1962, just one program within the Department of Defense (DoD) dedicated to fostering "fundamental research" (dubbed IR&D) disbursed $480 million *just to corporations,* when the entire National Science Foundation (NSF) budget for all sciences was only $84 million (Asner 2006, 296). This neglects the fact that each of the individual services had their own R&D divisions, as did "black" budget agencies like the NSA and the CIA, not to mention the "separate" agencies like the Atomic Energy Commission, NASA, and the Department of Energy, which conducted much military-guided research under a nominal civilian designation. Simple budget figures for federal R&D have therefore always been less than solid. In 1958 the DoD plus the Atomic Energy Commission (AEC) accounted for roughly 91 percent of reported federal R&D, and physics was receiving ten dollars for every dollar spent on all other scientific fields combined (K. Moore 2008, 26). The direct military dominance peaked just after the Sputnik scare around 1960, and there was a notable retreat in the military exercising its prerogatives as science manager-in-chief after the Mansfield-Fulbright Amendment of 1970 and the Military Procurement Act of 1971 (Asner 2006, 343–348). Nevertheless, funding and management did not devolve to a different agent for roughly another decade thereafter, mainly because it involved a wholesale reorganization of science once the military-induced structures were perceived as outmoded and no longer lavishly supported by steady subsidy. It is true that the National Institutes of Health (NIH) subsequently grew dramatically as science manager and funder throughout this period, eventually becoming the 800-pound gorilla of federal science policy, but even this ignores the extent to which the military informed the early practices of its research management as well as its influences on postwar biology (Appel 2000).

Historians of military science policy often point to the military's decisive effect on the universities, but it had even more impact on corporations. The Cold War is now regarded as the golden age of the Chandlerian firm. The line-and-division mode of management had proven its mettle during the war; throughout the 1970s the roster of the hundred largest American corporations displayed amazing stability; since a certain equilibrium had been reached in the control of their core markets, the new watchword became "diversification." Dominant firms in mature industries sought to grow by buying up new product lines and moving into newer industries, and the M-form, or multidivisional bureaucratic managerial structure, spread throughout the corporate sector (Lamoreaux et al. 2003). As corporations became less tied to single product lines or nominally related competencies, the role of the corporate laboratory began to shift. Industrial science still assumed many of the functions it had done prior to World War II, such as routine testing and product improvement. Yet the increasingly multidivisional or conglomerate nature of the firm dictated that each division should become its own profit center and that funds would be allocated within the firm according to criteria applicable across all divisions. Here is where the military takeover of science policy came into play. Not only did military funding come to dominate academic science, but it also reconfigured a major portion of industrial or commercial science (Graham 1985; Choi 2007).

Because the American military did not set out with deliberate forethought and intention to become commander-in-chief of science policy in America, but rather found itself backing into the commitment fitfully and by degrees, it had to be flexible about experimenting with various methods to fund and manage the scientists whom it wished to keep on retainer; in the process, it invented many new configurations of laboratories. Many point to the Manhattan Project as the first decisive American military experiment with science organization. Although the original OSRD contracts were run through universities as the research entities, soon it was decided that the industrial-scale centrifuges and uranium enrichment research at Clinton, Tennessee, and the Hanford site would be contracted out to private firms—in that case, DuPont. Even here, the public/private divide was never as sharp as it seemed in theory. The postwar legacy institutions at Oak Ridge, Los Alamos, Argonne, and Brookhaven were set up as something else that had been resisted throughout the previous regime: government-run "national labs" funded directly by the AEC (Westwick 2003). Other sorts of research were deemed to require something other than a university or corporate setting, and so the Air Force and the Ford Foundation concocted a university campus without students or faculty combined with a nonprofit Santa Monica beachfront resort at RAND in 1948 and thereby conjured the think tank. Finally, in the critical areas of aerospace, electronics, and missile development, it was decided that R&D would best be done on a commercial basis, and there the military took the fateful step down the road of subsidizing corporate R&D in areas where it believed there was a compelling national interest in maintaining supremacy at the forefront of research.[27]

The dramatic reorientation of the in-house corporate lab from an internally oriented product development agency to an externally oriented research contractor had profound implications. First and most significantly, the ability to attract military funds reconciled the corporate lab with the M-form corporation, in that the lab could (and often did) justify its divisional status by capturing its own streams of external revenue. However, for this to happen, the corporate science lab had to be brought into line with the rather different protocols of accounting, control, and IP propounded by their military patrons. Recently, Glen Asner made the fascinating argument that a series of accounting, tax code, and procurement regulations imposed by the military and/or the Congress over the 1950s "provided incentives for the corporations to restructure their research programs on the basis of the linear model" (2004, 2006). For example, the Procurement Act of 1947 effectively perpetuated the wartime innovation of the cost-plus contract in the realm of military R&D. The DoD did not mind funding what would be dubbed "basic research" in the aftermath of World War II, because its regulations concerning overhead would putatively allow it to control the mix of basic and applied as needed. For the first time, the DoD instituted a program (called IR&D), which actively encouraged corporations to pursue federally funded research without a stated practical goal as rationale. Further, the 1954 tax code revisions allowed accelerated write-offs of new investments in research infrastructure, which the DoD sought to encourage. Written into the rules were incentives for corporations to construct new purpose-built facilities, which were consciously sited away from production plants, under the rubric of geographical diversification in the case of nuclear war. Here we observe that the basic/applied distinction, far from mapping preset divisions between universities and industry, was inscribed in the very contracts that propagated it, largely through a myriad of nearly invisible stipulations concerning the economic provisioning of research. Far from being mere boondoggles, these practices had the dual effects of allowing a greater degree of disconnect of corporate lab research from the activities of other divisions of the same corporation, while at the same time allowing the corporate lab to be structured more along the lines of the university. (The fact that the model had historically come full circle undoubtedly rendered the transition easier.) Corporate labs were thus consolidated at locations remote from production facilities on campus-style settings, often justified by levels of secrecy and classification demanded by the military, when not explicitly on grounds of dispersal of capacity in defense against nuclear attack (Asner 2004). Scarce postwar research personnel were often courted with promises of university lifestyles and a fair amount of autonomy with regard to research agendas. Bell Labs, Xerox Parc, IBM Yorktown Heights,[28] RCA Princeton, Westinghouse Churchill, Merck Rahway, DuPont Central Research, and other labs became powerhouses of basic research, often enjoying substantial autonomy in setting their own research agendas. Firms that had long possessed profit-making development labs opened separate dedicated units to "fundamental" or "basic" research (Hounshell and Smith 1988). "A two-class system (military and nonmilitary) developed, with the best and brightest concentrated in the military class" (Hounshell 1996, 49). And the investment began to pay off in indirect ways: between 1956 and 1987 twelve corporate scientists won Nobel Prizes (Buderi 2000, 110). Many of them, like Charles Pedersen at DuPont, received the prize for work that had no tangible benefits to their employer (Hounshell and Smith 1988, 373). Was it therefore so very odd that even the community of corporate scientists came to subscribe to the linear model, since everything seemed inclined to ratify its existence?

Although it had not been the intention of the American military to transform the industrial research lab so that it would more closely resemble the university science facility, it was the military's intent to channel research in such a manner as to conduct what has been sometimes called a "stealth industrial policy" (Hart 1998a, 227–229; Teske and Johnson 1994). Specialists in funding agencies like the Office of Naval Research, the AEC, and the Defense Advanced Research Projects Agency (DARPA) thought they could predict which industries were making use of cutting-edge science to produce the technologies of the future; given the imperative of national security, they could justify their interventions to make their own predictions come true. Their successes in the areas of quantum electronics, solid-state physics, and computers are well known, but there were also significant initiatives in pharmaceuticals, radiobiology, meteorology, and catalysis. Not only did the government back select horses in the departmental derby but they ventured to dabble in equine husbandry as well. Through a combination of intentionally weakened IP rules and fortified antitrust practices, the government sought to breed a fortified corporation better suited to withstand the chill winds of the Cold War.

The American military had publicly pledged its troth to the magic of the market but generally was not willing to entrust mission-critical aspects of weapons development or considerations of national security to the vagaries of the free market. The postwar innovation of systems management was constructed to *plan* invention (G. Johnson 2002). In particular, the Cold War regime witnessed a policy of mitigation of IP rights in areas where the military was directly involved in science management. Starting with the Atomic Energy Act of 1946, the government asserted a policy to retain patent rights deriving from military-funded research, but only to make any such inventions that arose available to American firms on a nonexclusive royalty-free basis.[29] The policy was both chauvinistic, in the sense that national security dictated the subsidy of American firms, and also antimonopolistic, in the sense that national security would be compromised if the military were to become inordinately dependent on any single firm. Such considerations also governed the "second source rule" promulgated by the DoD, which conveyed the IP surrounding critical weapons systems or military technologies to a second competitor firm, so that the fortunes of no single producer would constitute a bottleneck.

A fascinating episode that reveals how IP first came to be mitigated by the military in World War II is the saga of patents on the atomic bomb (Wellerstein 2008). IP is one way to keep something secret; military classification is a different method. With the bomb, the clash between the two systems became apparent. In 1940 the NDRC attempted to impose a policy that all patents resulting from their research funding would revert to the government. War or no war, the corporations resisted this unto their last; the outcome was a jury-rigged system where much ownership devolved to the corporations, except under certain circumstances, designated the "short form." Vannevar Bush wrangled a presidential memo soon thereafter to impose the short form on everything having to do with nuclear weapons (contrary to his own public postwar position that it was acceptable for firms to claim patents on government-funded research). In effect, the military could designate whole classes of patent applications secret and thus prevent the U.S. Patent and Trademark Office (or anyone else) from acting upon them. Amazingly, this system was kept in place throughout most of the Cold War, with a few odd patents filed in the 1940s only granted by the patent office in their own sweet time in the 1970s. Nothing shows better how the military was vested with the ability to exercise a veto over any aspect of IP it deemed mission critical during the Cold War.

One of the main hindrances to understanding the extent of the military's role as science manager in the Cold War is a dearth of evidence as to just how much science was classified during the regime, and thus how extensive was "gray" and "black" science, and how much remains classified long after the regime has passed. Peter Galison (2008), claims that secret science was a substantial proportion of the total, supported by rough estimates of the sheer volume of pages that have been embargoed. In 1978 one estimate of the volume of classified documents during the previous twenty-five years had been on the order of 1.6 billion pages, which should be compared to another estimate that the Library of Congress holds at roughly 7 billion pages. Far from tapering off, since then "about five times as many pages are being added to the classified universe than are being brought to the storehouses of human learning, including all the books and journals on any subject in any language" (2008, 38). Even if much of this is irrelevant to science, it cannot be denied that the natural sciences were the major site of military classification and control and that much of this was intimately related to military subsidy. Hence, from the viewpoint of subject matter, no one has a good estimate on the amount of science produced under military auspices during the Cold War.

Not only was the military skeptical concerning the virtues of strong protection of IP in frontier science, but so too were the economic experts that (for a time) dominated antitrust policy in the United States. In the 1940s the Department of Justice adopted the position that one of the more deleterious effects of monopoly was the suppression of technological innovation, and it filed suits against some of the nation's most high-technology companies of the time, such as DuPont, Alcoa, IBM, and General Electric. Compulsory licensing of patents became for the first time a common element in antitrust settlements (Hart 2001, 928). One lab that was turned in a "fundamental" direction by the pincers action of military inducements and antitrust policy was RCA Princeton. "In 1957, RCA finally agreed to a consent decree that ordered it to grant royalty-free licenses . . . This came as a hard blow to the financial well-being of its labs, which relied almost exclusively on royalty income from its patents . . . One way the labs sought to make up this loss of income was to actively pursue military research contracts. The amount of research contracts supported by the military rose to reach, by one estimate, more than 75% of research done at RCA labs by the late 1950s" (Choi 2007, 776–777). The effect of these policies, in consort with military regulations, was to induce firms to pull back to some extent from acquiring the promising technologies of would-be competitors, or to play down the aggressive pursuit of patent infringement cases against major rivals, and to pour more of their resources into their own in-house labs. The result, under the banner of national security, was an oxymoronic regime of relatively open science hedged round by classification and secrecy.

It is through this Cold War lens that we can better understand the ways in which academic scientists could come to believe in the independence and isolation of the ivory tower. The military was convinced that encouragement of a certain format of higher education was an indispensable complement to the protection of national security. In stark contrast with the prior erudition regime, postwar public policy was aimed at sustained subsidy of academic science beyond the narrow scope of few private universities, although those fortunate few also benefited immeasurably under the new regime. Indeed, it might be suggested that only during the Cold War did all the sectors embrace higher education as an exercise in American nation building, with all that might imply: mass education, a diversified research base, a democratic ideology, open science, and the open propagation of research results. The military played a major role in fostering this system, primarily through the innovation of overhead payments on research grants, but also through more fleeting initiatives such as the GI Bill and the generous fellowships integrated in its grant structures. The objective was to fuse teaching and research together into a single symbiotic system, held together by the glue of generous funding.

It was a fateful decision at the OSRD to keep a high proportion of contract research tied to university settings and to reconcile university administrators to that fact with lavish subsidy. Vannevar Bush arbitrarily proposed overhead payments of 50 percent of labor costs for university research grants (although his real allegiance was demonstrated by the 100 percent rate proposed for corporations). Although the magnitudes of the subsidy were the subject of some controversy during the war, universities learned to deal with the inconveniences of having to subject these payments to bureaucratic accountability and oversight (C. Gruber 1995). Some university administrators were convinced that the postwar period would return rather quickly to the erudition regime's dependence on industrial contract research, but other more visionary captains were impressed by the sheer magnitude of military largesse. As Robert Hutchins of the University of Chicago admitted in a memo in June 1946, "It seems likely that within the next five years the Government will become, directly and indirectly, the principal donor of the University" (in C. Gruber 1995, 265). Those who were willing to go along with the drastic shift in patronage thereby stood a chance of stealing a leg up on their rather more hallowed and prestige-laden competitors. MIT notoriously took advantage of the opportunity to climb the league tables (Leslie 1993). In 1946, Stanford managed to accumulate military contracts that were twice the value of its contract research during the entirety of World War II (Lowen 1997, 99).

As we have insisted, some fields of science flourished and others languished under the new regime. It has been estimated that in 1953, 64 percent of federal funds for unclassified research in university physics departments was allocated to nuclear and allied topics, 10 percent was granted to solid-state physics, and a paltry 1 percent was given to low-temperature physics (Kragh 1999,300). But nonetheless, to be a physicist was to be in heaven:

> Physicists were economically pampered indeed. In 1958, the population of US physicists was 12,702, with the two largest fields being nuclear (2,622) and solid-state physics (1926). In terms of mere numbers the physics community was not all that impressive. In the same year, the US counted 35,805 chemists and 18,015 biologists; even the earth sciences, with 13,701 geologists, counted more than physics . . . Each physicist received an average [federal support] of $11,000, while the corresponding figure for the chemist was $1,900; the average biologist received $4,900, and in geology and mathematics the amounts were $1,800 and $1,700 respectively. (Kragh 1999, 298)

It may seem from this summary that the saga of the Cold War regime could be sketched entirely while bypassing consideration of the role of the private foundations, but this would not be altogether valid. Older foundations continued various programs of academic subsidy, and a few new players, like the gargantuan Ford Foundation, lumbered onto the scene (Raynor

2000). However, a government crackdown on the use of foundations as tax shelters as early as 1950, and further politicized by the Cox Committee thereafter, combined with the fact that even the largest foundations could not begin to match the magnitude of the federal government's impact on higher education and science, meant that most foundations scaled back their ambitions concerning the management of science in this era. For instance, in 1960 the Ford Foundation was channeling more support to American universities than the NSF was, but by 1970 it had all but withdrawn from the support of academic science (Geiger 1997, 171). American foundations became notorious for their fickle initiatives, which could disappear with each executive change or board turnover; they were no longer participants in science management for the long haul.[30]

Hence the American Cold War regime was largely structured as a concertedly nationalized system of science and as one whose ideological significance was so highly charged that it had to be presented as though it were the spontaneous eruption of an autonomous invisible college of stalwart, stateless individuals, disembodied intellects who need pay no heed to where the funding and institutional support for all their pure research was coming from. "Purity" had become conflated with "freedom" and "democracy"; "science" stood as the embodiment of all three states of virtue; and American science organization was promoted as a rebuke to the Soviet machine, but equally it was thought to stand as reproof to anyone who sought to make science submit to an imperious political master.[31] It was only within the Cold War regime that "academic freedom" became invested with sufficient gravitas to actually be deployed in an effective defense of the system of academic tenure—something we can now appreciate in the era of its disappearance. The researcher was said to have only to answer to his disciplinary peers, or in the last instance, to his individual conscience, and to feel an enlightened disdain for the hurly-burly of the marketplace—at least until the DARPA grants officer came to call.

The Globalized Privatization Regime

It may not be quite so necessary to expend surplus effort to describe the modern regime of globalized privatized science, if only because we have already proffered a brief flyover of the high points to Viridiana Jones in Chapter 1. The six fundamental trends—deindustrialization, the structural effects of the rise of the Internet, the egregious fortification of IP, restructuring and outsourcing of corporate R&D, the withdrawal of the state as science manager and patron, and the opting out of the state as the primary provider of advanced education—should all be seared into the minds of any reader who has lived long enough to experience their advent. Nevertheless, we do need to reconnoiter the outlines of the history of the current regime once more, if only to highlight the specific ways that it has altered laboratory science, and to prepare us for subsequent chapters where individual aspects of the modern regime are scrutinized in even greater detail. But even more important, the history in this chapter will directly refute the absurd neoliberal proposition that "there's nothing new about professors selling their expertise off-campus or performing industry-funded research in university laboratories" (Greenberg 2007, 83).

The advent of the globalized regime of privatized science was not heralded in an unmistakable way by war or depression, by contrast with the previous regimes. It might seem that one could seek the watershed at the fall of the Berlin Wall, since, after all, that was the dramatic event that signaled the cessation of the Cold War. However, if we triangulate between corporate evolution, educational transformation, and government policy, the inauguration of the privatization regime in America would have to be located a decade or so earlier.

Economic historians, legal scholars, historians of education and science studies researchers all tell the story in somewhat different ways, but it is significant that each group traces the metamorphosis back to roughly 1980.[32] The trigger seemed to be the widespread conviction that the United States had lost ground to international competitors during the oil crisis and economic slowdown of the late 1970s. Although there was substantial disagreement over the causes of the supposed sclerosis (Asner 2006, Chap. 1), with even specific economists unable to make up their minds,[33] an imposing array of initiatives were crafted to defeat the diverse culprits sapping America's economic dominance. One major landmark slated for economic reform was the organizational structure of the Chandlerian corporation (Lamoreaux et al. 2003, 2004; Langlois 2004). Various participants had become convinced that the huge managerial conglomerate had become too unwieldy to effectively compete in the world market in the 1970s, and the 1980s was an era of hostile takeovers, leveraged buyouts, and shareholder attacks on the top management of large corporations. In response, there was a significant retreat from diversification within firms, with one calculation suggesting that by 1989 firms had divested themselves of as much as 60 percent of acquisitions made outside their core business between 1970 and 1982 (Bhagat et al. 1990). There was also a retreat from previous levels of vertical integration in industries like automobiles, computers, telecommunications, and retail. Consequently, corporations began to equate agility and nimbleness with repudiation of hierarchical managerial control of process, and with it the M-form paradigm; thus they sought to reengineer the supply chain to depend to a greater extent on market coordination.[34] Networks of subcontracts began to displace ownership ties as modes of organization; venture capital began to channel investment into start-up firms. Labor-intensive heavy manufacturing was outsourced to low-wage countries. Moreover, the roster of America's largest corporations underwent severe shakedown, after having enjoyed relative stability for the previous sixty years. The lumbering giants were prodded into defensive action, which was interpreted in some quarters as a "return" to market methods of coordination (Langlois 2004).

Another important initiative that marked the shift in regimes was in the arena of organization and control of international trade. In a far-sighted mobilization, a handful of representatives of corporations located in high-tech industries such as pharmaceuticals, semiconductors, computers, and entertainment, formed the International Intellectual Property Alliance in 1984 for the purpose of linking issues of IP to larger trade negotiations.[35] They succeeded beyond their wildest ambitions, using the Uruguay Round of negotiations over the General Agreement on Tariffs and Trade to impose U.S. standards and levels of IP protection on developed and developing countries alike and to enforce them with trade sanctions through the World Trade Organization. TRIPs (trade-related aspects of intellectual property) came into force on January 1, 1995, and has implanted the basic legal premises of the globalization regime in all corners of the world, refashioning academic and corporate activity in the interim. TRIPs might be regarded as one facet of an even larger concerted political movement to weaken the prerogative of national governments to exert regulatory control over their own corporate entities, all carried out in the name of trade liberalization and the protection of foreign investment. In any event, U.S. manufacturing capacity was shifted to lower-wage countries in search of a quick productivity boost, and manufacturing job losses accelerated beginning in the late 1980s (Burke et al. 2004).

These major restructurings of the corporate sector coincided with a crisis in the sphere of higher education. After 1975, enrollments in U.S. higher education ceased to grow for the first time in U.S. history, while cash-strapped states began to contract their funding (Geiger 2004, 22ff.). The military, under pressure to reduce funding of projects not immediately relevant to its mission, had been attempting to withdraw from many of its commitments to the financial support of academic science in the 1970s, so universities suffered a double deficit, with no end in sight. To maintain graduate enrollments, many departments in the sciences began to admit rising proportions of foreign students (National Science Board 2004, 5–25). While this had a salutary effect on the rather parochial atmospheres of many American university towns, it also had the deleterious effect of revealing the essential bankruptcy of the

Cold War justification of education as serving the objectives of state building. Many of the students in scientific/technical areas were not U.S. citizens, and periodically some politician would demand to know what the universities were doing by training the workforce of potential competitors at American expense.

But more to the point, the whole ideal of an informed citizenry and skilled workforce began to lose salience as more and more production activity was shifted overseas and corporate managerial cadres became ever more multinational. The university was losing its grip on its previous social *raison d'être*, even as it remained the preferred path for individual economic advancement.[36] It also, in an ironic twist, was being revamped in a Chandlerian direction, even as many corporations were fleeing that organizational model in droves. Significant aspects of faculty governance were diminished or dismantled altogether (Geiger 2004, 25) and were replaced with top-heavy managerial hierarchies that multiplied divisions, institutes, and other offices, often in the name of rationalization and cost saving. University finances were more directly addressed by replacing tenured faculty with temporary labor and part-time teachers, reversing the Cold War tendency to unite teaching and research as mutually reinforcing activities. By 2005 part-time and nontenured faculty positions had topped 48 percent of all faculty at colleges that award federal financial aid, according to the National Center for Education Statistics (Lederman 2007). For the first time in 2006, the majority of professional full-time employees in U.S. higher education occupied administrative rather than teaching positions (Jaschik 2008). Administrative hiring grew unabated during good economic times and bad (Newfield 2008).

A hallmark of the privatization regime was the sequence of overt political attempts to bring the hobbled universities more into line with the leaner and meaner corporation. It has become *de rigueur* for commentators on the commercialization of science to identify the Bayh-Dole Act of 1980 as a major turning point in the treatment of IP in the United States, because the act allowed universities and small businesses to retain the title to inventions made with federal R&D funding and to negotiate exclusive licenses.[37] Actually, the historical situation with regard to IP was much more complex, yet the end result was almost a complete reversal of practices under the Cold War regime. First, universities had been permitted on a piecemeal basis to patent federally funded research via individual institutional patent agreements since 1968 (Mowery et al. 2004, 88; Sampat 2006, 778). Nevertheless, some designated areas of research were treated as off-limits to patents: for instance, Harvard prohibited patents by faculty on medical inventions until 1975 (Matkin 1990, 62). Only in 1983 was Bayh-Dole style permission extended to large corporations (their real intended beneficiaries) by an executive memo from Ronald Reagan—the better to fly under journalistic radar (Washburn 2005, 69). Second, Bayh-Dole was only one bill in a sequence of legislation throughout the 1980s that expanded the capacity of corporations to engage in novel forms of collaborative research while capturing and controlling their products (Slaughter and Rhoades 2002, 86). For instance, the Stevenson-Wydler Act of 1980 opened the door to commercialization of research performed at the national laboratories. The National Cooperative Research Act (NCRA) of 1984 shielded corporations from antitrust prosecution when engaged in joint research projects. The Federal Technology Transfer Act of 1986 allowed federally sponsored research facilities to spin off previously classified research to private firms. Over the same period, corporations sought and won numerous amendments to strengthen both patent and copyright, and in 1982 they managed to have a special Court of Appeals in the Federal Circuit dedicated to patent cases. The scope of what had been deemed susceptible to patent in America has been progressively broadened, and challenges to the legitimacy of patents have become less successful. (This is analyzed in detail in subsequent chapters.) The very notion of a public sphere of codified knowledge has been rolled back at every point along its perimeter, often by blurring the lines between public and private property. No self-respecting research university, whatever its denomination, felt it could persist without the full panoply of technology transfer officers after 1980 (Sampat 2006, 781). This hyperrestrictive system of IP was then exported to the rest of the world under the aegis of the World Trade Organization and the World Intellectual Property Organization.

The year 1980 also was a watershed for multimillion-dollar deals between firms and individual universities, which burst the previous firewall between faculty governance and external research support in the name of control of IP. In one famous incident in 1981, the pharmaceutical firm Hoechst signed a ten-year agreement with Mass General Hospital to establish a new department of molecular biology, a development then so unprecedented it provoked a congressional investigation. In another deal, the endowment for the Whitehead Institute at MIT in 1981 allowed the outside board, and not the existing faculty, to appoint twenty new members to the MIT Biology Department (Matkin 1990, 2,43). Now such intrusive arrangements, like the Novartis and BP deals with Berkeley, have become much more commonplace and the concerns of those worried about the integrity of the university are generally ignored (Busch et al. 2004; Washburn 2010).

The concerted fortification of intellectual property was accompanied by the weakening of antitrust policy—another exact reversal of the Cold War regime. Absolution was not just granted in the specific case of the NCRA but, more generally under the influence of the Chicago School of law and economics, monopoly was increasingly downgraded as a source of inefficiency or political danger in the viewpoint of the Justice Department (Hart 2001; Hemphill 2003; van Horn 2007). The doctrine was propounded that monopoly was not necessarily harmful to innovation (even in the case of *United States v. Microsoft*), that size of the R&D budget was not correlated with a demonstrated ability to innovate, and that good products win out in the end, no matter what the industry structure. In any event, defenders could point to the increasing resort to cross-licensing and joint ventures to suggest that there was no return to the bad old days of trusts and patent pools (Caloghirou, Ionnides, and Vontoras 2003). Rather, a fortified and unfettered corporate sector free to contract for research whenever and wherever it saw fit was thought to be one of the best prophylactics against upstart foreign producers and looming national economic decline.

The cumulative consequence of all these convergent vectors was a fateful restructuring of the American corporation and a consequent revision in the organization of science within the regime of globalized privatization with the relative demise of the in-house corporate research labs and the spreading practice of the outsourcing of corporate research.[38] The significance of this transformation cannot be overstated. It is here, and not in any vague shift in the Zeitgeist or narrative of the rationalization of technology transfer, that we find the full expression of the new model of commercialization of science in the twenty-first century. Although the trends identified above were not deliberately attuned in isolation to bring about the destruction of the in-house corporate lab, each nonetheless contributed to its demise. It is important to understand the ways in which the withdrawal of the military from science management, the perceived failure of the Chandlerian firm, the push to globalize the neoliberal Washington consensus, and the crisis of higher education all converged on the corporate lab.

Pundits in business schools often attribute the passing of the large corporate lab to the supposed empirical observation that big in-house research labs don't deliver the goods (Anderson 2004), usually accompanied by reference to some neoliberal doctrine stating that in the long run healthy science resists being planned, but this superficial analysis ignores the fact that the labs had been weaned from their internal parochial commercial orientations by military contracts during the Cold War (Graham 1985). The corporate labs had been permitted to maintain their external orientation and unfettered curiosity in a campus ambiance as long as they were revenue centers for the firm, but when the military withdrew from the organization and funding of basic research, the semiautonomous corporate lab became a liability. In a more forgiving environment, perhaps labs might have been reoriented more concertedly toward the development side of R&D and persuaded to renounce the linear model of technological change, but by the 1990s they ran up against the anti-Chandlerian movement to divest the firm of its extraneous product lines and scale back on vertical integration. In many corporations, the research division was a prime candidate for downsizing or spin-off, and that is precisely what happened throughout the 1990s. RCA Sarnoff was first sold off to SRI International, and soon thereafter it was spun off as Sarnoff Corporation in 1987. AT&T slashed research at Bell Labs starting in 1989, only to spin off the remnant as Lucent in 1996 (Endlich 2004). Raytheon sold off its central lab in Lexington in the early 1990s and divested its Hughes research unit in 2006. Texas Instruments began cutting its internal R&D capacity as early as 1982, only to divest it altogether in 1997 (Anderson et al. 2008). Westinghouse Churchill was first decimated and then sold off to Siemens (Asner 2006). Research divisions disappeared altogether at firms such as U.S. Steel and Gulf Chevron. By 1995 IBM had eliminated a third of its research budget, essentially gutting its flagship Yorktown Heights facility; other units, such as its Zurich laser group, were spun off as separate firms. After the merger of Hewlett-Packard and Compaq and the spin-off of Agilent, the renowned HP Labs were slated for reorganization and downsizing. Pfizer shut down its pharmaceutical labs in Ann Arbor after a decade of job cutbacks, just as it announced major expansions in China.[39] It had gotten to be so ubiquitous a phenomenon that R&D directors would openly admit that Wall Street no longer expected that a successful corporation would have a central research lab (Anderson et al. 2008, 16); rather, having an R&D lab was evidence of managerial negligence.

The historian Robert Buderi, who has been most concerned to document this phenomenon, admits that research directors regarded the reorganization as a "research bloodbath" in the late 1980s and 1990s (2000, 22), but he has sought to paint the bloodletting as a prescription for both corporate and scientific health. The problem with this diagnosis is that it is too narrowly focused on the individual firm in splendid isolation, and it ignores the larger system of the funding and organization of science. Buderi writes, "We now see less basic research going on. IBM does not chase magnetic monopoles anymore, but should it have done so in the first place?" (2002, 249). This presumes someone somewhere else will take up the chase for magnetic monopoles, and someone else will worry about where and how that will happen. But this question of who organizes which science to what ends is precisely the debate that is glaring in its absence in the globalized privatization regime. Some enthusiasts for the new knowledge economy have asserted that "Bell Labs is gone, academia steps in" (Zachary 2007). But that assessment was both too hasty and too superficial.

The downsizing and expulsion of in-house corporate labs has not implied a corresponding contraction of private funding of research and development in America—quite the contrary. In a pattern that has been mimicked with a lag in other countries, in the United States federal R&D expenditures as a proportion of the total R&D has declined continuously since the late 1960s, while the proportion of R&D expenditure originating in the industrial sector has increased from the same period, surpassing the federal proportion around 1980 (National Science Board 2008). This is starkly illustrated in Figure 3.1.

If corporate labs were being slashed and divested everywhere, how could this be? The resolution of these seemingly contradictory trends was that an increased volume of research is being performed outside the boundaries of the corporations funding it. Some research is being performed in other corporations purpose-built for research under the new regime, while the rest is increasingly performed in academic and hybrid settings, like research parks and quasi-academic start-ups. It is precisely at this juncture that the other historical trends described above—the globalization of corporate trade and investment, and the crisis of the research universities—come into their own.

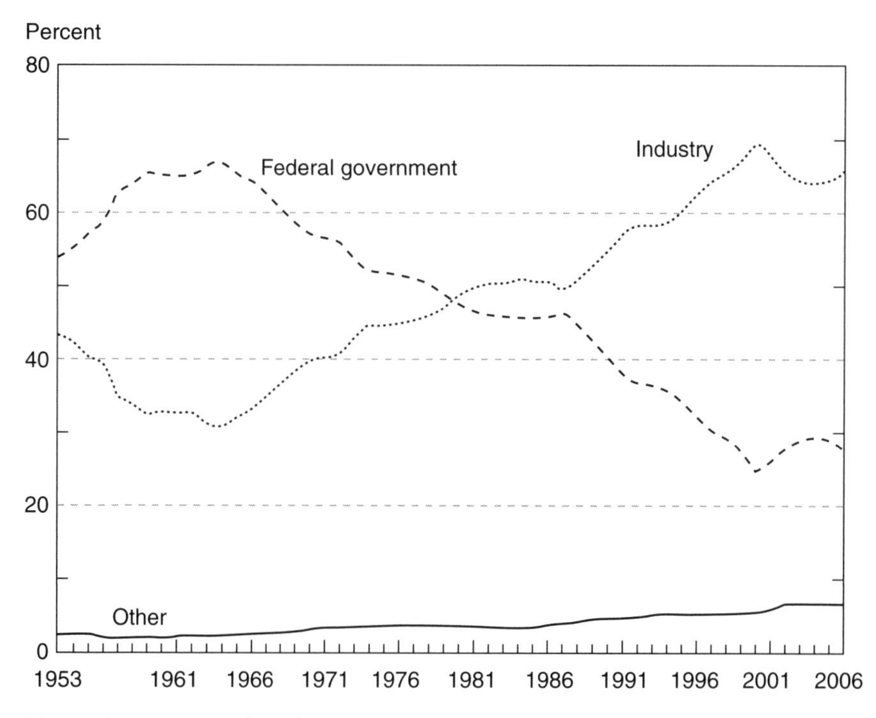

Figure 3.1. Percent of total national R&D by Funding Sector from 1953 to 2006. *Source:* National Science Board, 2008a.

The breakdown of the Chandlerian model of the hierarchical integrated firm has prompted the nagging question of why integrate R&D into the firm when you can buy it externally and reduce costs by doing so? But that question presumes that dependable R&D is a distinct fungible commodity in a well-developed market, one so competitive that it can lower the costs relative to doing it internal to each firm. A major thesis of this chapter is that, no matter how "commercialized" science may or may not have been in the previous American science regimes, until recently it was this state of affairs that was uniformly absent. The strengthening of IP, the weakening of both domestic antitrust and the ability of foreign governments to counter corporate policies, the capacity to shift research contracts to lower-wage countries and easier regulatory environments and therefore engage in regulatory arbitrage, the availability of low-cost, real-time communication technologies, and the presence of an academic sector that was willing to be restructured to surrender control of research to its corporate paymasters—all these were necessary prerequisites to seriously countenance the corporate outsourcing of research on a mass scale.

The globalization of corporate R&D is a characteristic hallmark of the new regime. Of course, multinational companies headquartered in smaller countries like the Netherlands and Switzerland have internationalized their R&D activities essentially from their inception, but the more striking trend is the international outsourcing of research across the board since the 1980s (Reddy 2000, 52; OECD 2008). It became necessary since 2000 to try and define the phenomenon more precisely (United States Government Accountability Office 2004; Huws, Dahlmann, and Flecker 2004), to try and comprehend a dynamic that threatened to outstrip the ability of outsiders to keep track of it. It became the custom to distinguish between "outsourcing," which was the purchase of services outside the boundaries of the firm, and "offshoring" as the purchase of services outside the boundaries of the nation-state in which the headquarters of the multinational corporation was domiciled. A firm might "offshore" R&D by opening an affiliate in Shanghai without technically "outsourcing," if the lab was a wholly owned subsidiary (Buderi and Huang 2006). Conversely, it might "outsource" R&D to a university spin-off without technically "offshoring," if the university start-up was also domiciled in the home country. But here is where the grand reengineering of the Chandlerian corporation came into play to bedevil most attempts to gauge the extent of outsourcing and offshoring.

> In its legal sense "outsourcing" refers to a business activity, involving the production of either goods or services, purchased by an organization from an external supplier rather than internally. It is, in other words, "subcontracting." However, in the current context of rapid organiza-

> tional change, determining what is "internal" and "external" is increasingly difficult. Mergers, demergers, strategic alliances, public-private partnerships . . . are increasingly common. If a company is restructured on the basis of separate cost or profit centers, for instance, should transactions between them be regarded as "outsourcing" or merely as internal accounting flows? (Huws et al. 2004, 3)

In short, by operating across national borders, multinational corporations could evade or ignore individual national attempts to define and measure cross-border movements of funds and employment (through the phenomenon of "transfer pricing") and thus manipulate measures of "offshoring" with relative impunity; by increasingly blurring the boundaries of the firm by reducing it to a "nexus of contracts," a corporation could also alter what would count as "outsourcing." Hence, in the globalized privatization regime, there is no longer any fixed phenomenon called "the offshore outsourcing of R&D"—there is only whatever the corporations involved choose to make of it. And realizing that the topic is a political hot potato, they have taken the neoliberal route of promoting confusion and ignorance around the issue so as to avoid any serious examination of modern trends on the ground.[40] In effect, the quality of most of the data that the National Science Board (NSB) issues concerning American R&D has been utterly compromised, insofar as the great bulk of expenditures are carried out by the corporate sector.

One has to recognize that the NSB follows the procedures set up by adherence to the generally accepted accounting principles enshrined in the national income accounts. Because there is no absolute method to get at the extent of "error" in the national income accounts, statisticians mostly follow double-entry procedures that seek to measure most aggregates in two independent separate ways. For GNP, this means collecting data on incomes and on expenditures, and seeing how closely they match up for individual sectors, and in total. For measures of R&D, a similar procedure is followed where data are collected from two perspectives: from that of the source/contractor and that of the performer of the research. In some instances the two might be the same (as when universities internally fund their own faculty research out of endowment funds), but in many others, such as outsourcing, they are not. You might think that the difference between the two measures might be substantial, due to all kinds of statistical considerations, but as long as it was random and uncorrelated over time, then the net result of aggregate measured R&D would stand as a reasonable indicator of trends and levels of activity. However, as the NSF has admitted for years now, that has not been the case: the "discrepancy error" has been growing prodigiously over time, as revealed in Figure 3.2.

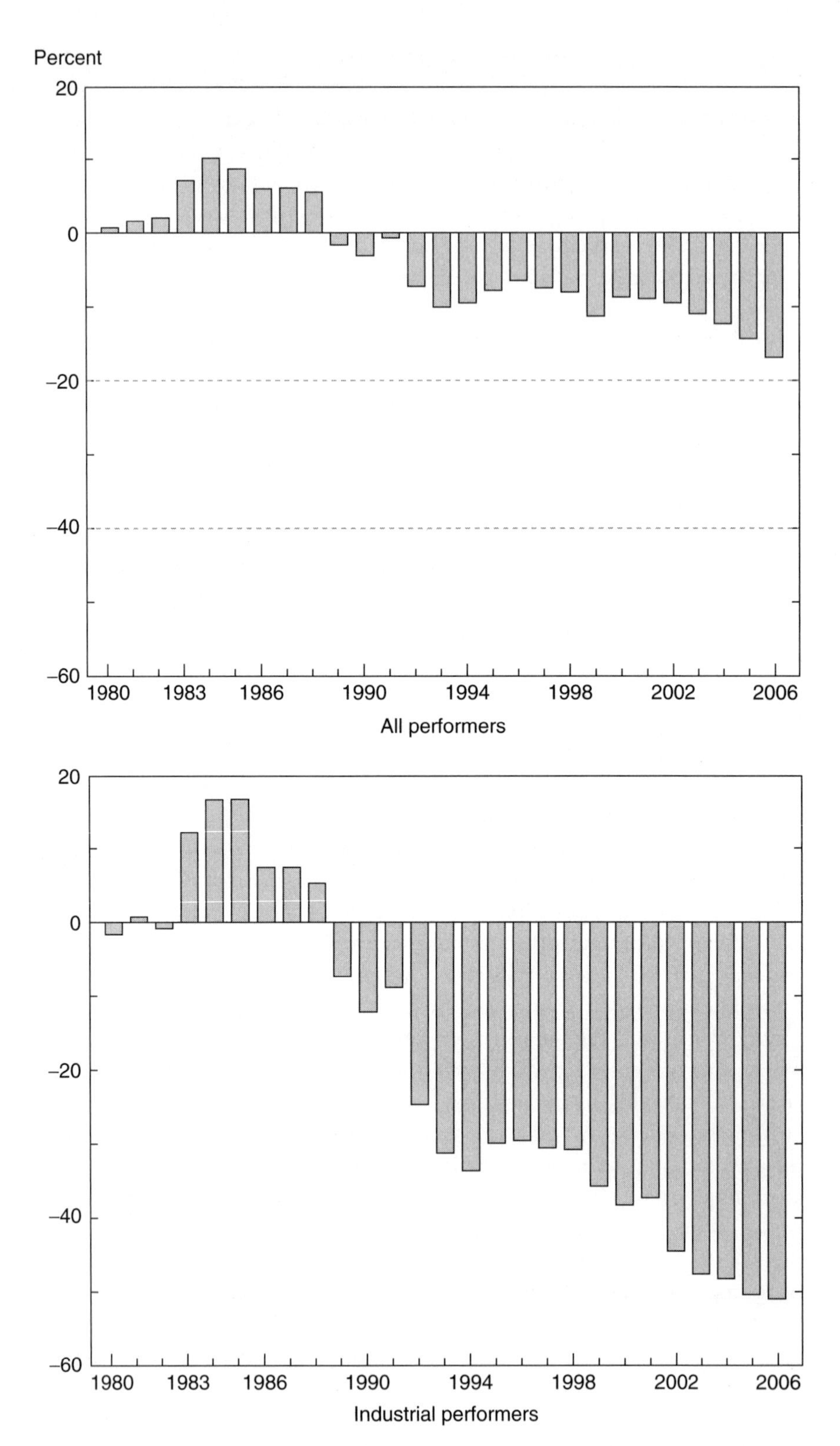

Note: Difference defined as percentage of federally reported R&D, with positive difference indicating that performer-reported R&D exceeds agency-reported R&D.

Justified or not, government statisticians feel more confident about the quality of government statistics on government R&D expenditure than they do on statistics supplied by the corporations (for reasons explored below), so this particular exercise in Figure 3.2 takes as its target *only* federally funded research. It shows a persistent and worsening gap between what the government thinks it is buying and what the recipients report they are doing with the money since 1990. In national income jargon, the "error" is systematically in the red, with performers reporting progressively *less* research expenditures than the government has allotted to them; this should be interpreted as a warning light that the reliability of the accounts is becoming degraded over time. The statisticians have attempted to pinpoint where the discrepancy is concentrated and suspect with some reason (as shown in the lower half of Figure 3.2) that it is primarily emanating from within the private corporate sector. In other words, *the amounts of R&D expenditure reported by the U.S. corporate sector have been increasingly unreliable over the past fifteen years.* The magnitudes being cited here are not insignificant: the size of the discrepancy is *over half the total reported expenditure on R&D* in the corporate sector. While the NSF has been loath to speculate on causes, much less admit to the loss of confidence it must instill in all its reported measures, we are not nearly so constrained. The trend toward increased outsourcing *and* offshoring and the concomitant neoliberal policies of the government can go far to explain the specific pattern of corruption in the R&D statistics. But to explore that, we must dig deeper behind the numbers.

The NSF does not collect the underlying component data that enter into its reported measures; for that it relies on other arms of the U.S. government. Much of the private sector expenditure data, as well as data on international trade in services, come from the Bureau of Economic Analysis (BEA) in the Department of Commerce. Most people can be forgiven for not being aware that BEA data have been regarded within government and policy circles as particularly unreliable over the course of the last decade or so, so much so that the BEA was upbraided by the Government Accountability Office (2004) as having failed in its task of tracking offshoring and outsourcing of services. The breakdown in monitoring has many antecedents, but one of the more important identified in the GAO report was the neoliberal legislation passed during the Clinton era dubbed the "Paperwork Reduction Act of 1995." Although the BEA only had previously enjoyed a formal mandate for corporations to respond to its surveys concerning foreign affiliates when their assets topped $100 million or for external transactions greater than $1 million, the

---

Figure 3.2. Difference in U.S. performer- and agency-reported federal R&D from 1980 to 2006. *Source:* National Science Board 2008a, 4–29.

Paperwork Act (with some backpedaling at BEA) made it possible for even large firms to evade government initiatives to standardize data collection. The ability of multinational corporations to shift definitions of activity at will, combined with the neoliberal drive to get the government off their backs, rendered most official data on offshoring and outsourcing to be minimally credible. And this corruption consequently affects all R&D accounts, not just those initiated by federal funding.

In the few cases where one might compare BEA data with that collected from the other side of the divide—say, by an organization in a foreign country attempting to gauge inflows of purchases of services—the U.S. version of the data has not even been in the ballpark. One egregious example was the comparison between BEA data on software imports into the United States from India from 1998 to 2004 (Aspray, Mayadas, and Varda 2006, 2–9) and similar data provided by the Indian government. Incredibly, the BEA reported software imports falling to a level of $76 million, while data reported by NASSCOM, the Indian software trade association, had this level rising from $1 billion to $4.7 billion in the same period. Maybe the Indians had fallen prey to boosterism, but the numbers were so far apart as to have come from different planets. Even *Business Week* (2005) reported how BEA staff economists were forced to prettify the data and massage the statistics whenever the topic came close to offshoring services.

Similar sorts of problems beset important offshoring data collected by other U.S. agencies, such as the Mass Layoff Statistics at the Bureau of Labor Statistics (Bronfenbrenner and Luce 2004); Aspray et al. (2006, 2–11). Data collection has therefore been weak and ineffectual. Yet even that minimal level of intrusion into corporate prerogatives was an anathema in the current globalization era, for, as we have seen in Chapter 1, the BEA was directed to cease collecting *all* data on R&D expenditures by corporations in 2008 (McCormack 2008). Has the National Academy of Sciences or the NSB raised a voice in protest? Or, as in the case of global warming, are we rather enjoined to "acknowledge the controversy" and simply lapse into intellectual paralysis?[41]

Just as with recourse to academic capacity, global outsourcing tends to be concentrated in a few industries, such as pharmaceuticals, electrical machinery, automotive, computer software, and telecommunications equipment. Nevertheless, surveys within these industries reveal a sharp increase in research carried out beyond the home country's boundaries from the 1960s to the 1990s (Kuemmerle 1999). A survey by the Economist Intelligence Unit (2004) reveals the globalization of R&D gathering pace over the 1990s, with over half the respondents indicating that they would expand their overseas R&D investment in the next three years. A study for the European Commission found that 70 percent of companies included in the survey admitted having increased offshoring of R&D in the five years preceding 2005 (OECD 2008, 20). A survey of corporate R&D officers in 2006 revealed that American-based firms do not plan to site near-term research investments in the home country, but instead cite China as the most popular venue (Thursby and Thursby, 2006a). When queried about the major considerations governing their decision, the most popular responses cited strong protection of IP, lower costs, and the tapping of indigenous research capacities. It is access to lower-wage labor in the context of an academic infrastructure, disengaged from any corporate obligations to provide ongoing structural support for local educational infrastructure, that explains the shift in research funding to countries like China, India, Brazil, and the Czech Republic (Economist Intelligence Unit 2004). Offshore outsourcing to developing countries demands disengagement from earlier narratives of the ways in which capitalist firms and economic development depend on scientific advance.

Another way to cut costs is to absolve the firm from nationalist appeals to help support home scientific infrastructure, accompanied by improved opportunities to further reduce or avoid corporate taxation. "Just as the internationalization of manufacturing had important cost advantages, the internationalization of R&D is also motivated by cost cutting and outsourcing of R&D to countries with low costs" (OECD 2008, 44). Indeed, one might go so far as to suggest that, once knowledge has been commodified and detached from its cultural and academic contexts, the latter follows the former with a degree of inexorability.

Approaching the commercialization of science from this angle profoundly revises the usual narrative of the privatization of modern academic science as a straightforward case of cash-strapped universities following the money, albeit with a few nagging qualms concerning the propriety of telling corporations only what they want to hear.[42] Rather, I suggest here that many of the novel institutions of globalized privatized research were first pioneered *outside* the academic sector per se, especially as adjuncts to the modification and reengineering of the modern corporation, and only then foisted on universities. Consequently, the universities have been forced to react to these benchmark citadels of the new globalization regime in their own internal restructuring of scientific research. Government revisions of policies with regard to IP or educational subsidy may have constituted incentives for changes in the universities, but they could not have unilaterally imposed the structure of the new regime. While legislation such as the Bayh-Dole Act was enabling, it should not be confused with the *cause* of the privatization of science, which was instead attributable to the larger shift in the nexus of science management and funding. Indeed, since most universities still lose money in their technology transfer operations, it could hardly have happened otherwise.

Indeed, one of the great unspoken presuppositions of modern commentators on the commercialization of science is that the individual scientist or the community at large is still capable of choosing "how much" public open science one wants to preserve, while leaving the remainder to be covered by the private sector. Contrary to this presumption found in most models of the economics of science that one can rationally choose a menu in any combination from columns A and B, we have observed that once the institutional structures of the globalized privatization regime have been put in place, the very character and nature of public science is irreversibly transformed.

Science Regimes in a More Global Perspective

Thus far, I have offered a predominantly "American" perspective on the economics of science in the twentieth century. This should not preclude parallel consideration of science funding and organization in other countries nor suggest the dispensability of making allowances for their own special historical circumstances. (Think about how this narrative might have differed if America had suffered widespread indigenous devastation in World War II, for instance.) The immediate question confronting us in this chapter is whether the three-stage schema used to describe American experience has any ability to illuminate the economic history of science in other countries in the twentieth century. The only serious answer is that it is eminently an empirical issue, and solid evidence must await serious synthetic surveys of the economic history of science in the experience of Germany, Japan, France, Great Britain, and elsewhere. But in lieu of that, it may be possible to make a few tentative generalizations based on the existing literatures in the history of science. The first and most sweeping generalization is that the characterization of the erudition regime probably does not extend all that well to distant climes. Without wallowing in the appeal to "American exceptionalism," that old standby of U.S. historians, it does seem that there were just too many institutional peculiarities that diverged from turn-of-the-last-century Europe to expect a similar dynamic of incubation of laboratories in corporate and university settings. The mere fact that continental Europe was nowhere near as hostile to government as a science manager as were American contemporaries practically guarantees that. But foreign corporations and long-established universities bore little resemblance to American forms as well. Nevertheless, World War II essentially wiped the slate clean, and it could be argued that Americans subsequently imposed a Cold War–style science regime on many of the losers in that conflict (Krige 2006). Thus the hypothesis explored in this last section is that the last two regimes—the Cold War and globalized privatization—actually can do yeoman service in organizing the histories of certain select other countries, but for entirely different reasons. In the case of the Cold War, it was imposed upon other cultures ill-disposed to receive it; in the case of globalized privatization, it occurs because of a larger dynamic of convergence of science systems brought about by the very mechanisms of globalization.

When it comes to the laboratory, the European version of the early twentieth-century regime of science organization has been reconnoitered by Fox and Guagnini (1999); in some national instances (such as Germany) the academic lab predated the industrial lab, while in others they were simultaneously constructed. German labs set the tone for the European research organization, but Germany's neighbors were sometimes unwilling to imitate them. One specific illustration is NatLab, the Philips Physics Laboratory in the Netherlands, which was established in 1914 (Boersma 2002). Perhaps unusually, the early history of NatLab resembled that of many American in-house labs. In addition, NatLab significantly shaped the technical physics degree developed at the Technical University of Delft in the Netherlands.

Turning to a European incarnation of the Cold War regime, it is prudent to acknowledge that capitalist firms were very different animals on the Continent (Djelic 1998) than in the States. French firms were the smallest on average, with joint stock companies experiencing relative backwardness. British firms were still mostly family-run affairs. German firms were much larger and better financed, primarily through bank loans. Even still, in both 1914 and 1930, the two hundred largest German firms were uniformly smaller than their corresponding American cousins (ibid., 52). They offset this disadvantage to some extent through the formation of cartels such as IG Farben, and these cartels tended to define the nature of industrial research. The cartelization of industry begat public/private research entities such as the Kaiser Wilhelm Institutes (J. Johnson 1990). Upon German defeat in 1945, part of the "new economic world order" imposed by the Allies was decartelization and the imposition of American ideas of antitrust law, even though this was utterly foreign to German legal traditions. The lineal predecessor to the EU, the European Coal and Steel Community, was a major vehicle for the imposition of these reforms. The reformation of European corporate structures brought their research arms closer (though not identical) to the American-style R&D units. The commercial laboratory could not assume precisely the same shape, however, if only because the Continental higher education sector retained much of its prewar configuration in each national context. Nevertheless, the relative separation of academic and applied research was taken for granted in Europe as well as in the United States, and it was enforced by having the intellectual property generated by academic faculty devolve to them personally, rather than to the university.

Postwar European state building and the nurture of indigenous research capacity joined hands in ways very similar to those found in the United States, for the relevant reason that Americans had a large role in shaping postwar science capacities, as recently argued by Krige (2006, 256): "American assistance for the postwar reconstruction of science in Europe was not something that happened parallel to, and independently of, the construction of empire: it was part and parcel of the same project and cannot be fully understood without it." In both defeated Germany and Japan, Americans intervened in incredibly specific ways to recast both the educational infrastructure and industrial research capacities of their former adversaries (Beyler and Low 2003). After the German science base was selectively plundered by the Americans (in Project Paperclip) and the Russians, the Allied Research and Development Board took charge of the remaining German science capacity, suppressing much applied industrial research in the name of preventing rearmament (Krige 2006, 49). The Allied Control Council Law No. 25 imposed the basic/applied research split on German firms and universities, in much the same ways tax and contract regulations had imposed it in the American context. "Basic" research was heavily promoted, at least in part to preclude Soviet interest in its follow-up (ibid., 33). Both academic and industrial labs were enjoined to remain open and transparent for years to follow. German patents and trade secrets were to be made freely available to the Allies: "Europeans willingly cooperated in the reconstruction of their scientific capacity: they had little choice" (ibid., 13). Although these impositions may not have been successful in every respect, and were eventually dismantled with devolution of firms back to the indigenous population, the effectively homogenized economic context tended to bring European systems closer in line with postwar American practices of R&D.

The central role of military concerns in the postwar European science base is illustrated by the founding of CERN, the European Organization for Nuclear Research, now the world's largest particle physics laboratory. It was created at the height of the Cold War, with an agreement adopted in 1952; publicly it was justified as an effort to restore European physics to its former grandeur, to reverse the putative brain drain of the United States, and, coincidentally, to consolidate postwar European integration (Pestre and Krige 1992). In fact, it was promoted by the Allies as a way to have European physics capacity centrally located in a "glass house" in Geneva, to improve intelligence gathering and rapidly detect any military applications, and most important, to "contain" the German physics community within a supranational European laboratory (Krige 2006, Chap. 3). The scientific Panopticon would persist well into the Cold War after the immediate postwar years in the guise of the science directorate of NATO.

As for the advent of the globalized privatization regime, the case for parallel developments throughout the world has been the occasion for prodigious commentary and controversy.[43] The logic of the spread of multinational corporations would seem to suggest the possibility of something like convergence of prior diverse "national systems of innovation" to a more relatively uniform advanced transnational model of the commercialization of science, especially since barriers between differing economic systems collapsed after the fall of the Berlin Wall. European corporations have since experienced their own twilight of the Chandlerian firm;[44] NATO and indigenous member militaries have progressively withdrawn from previous levels of an active role as science manager. The outsourcing of corporate R&D has become a global institutional phenomenon.[45] Concurrently, there has been a trend toward convergence in university systems over the last two decades.

In the advent of the commercialized modern regime throughout the developed world, the withdrawal of the state from its Cold War role as science manager and funder, while not quite so precipitous as in the United States, has been progressing apace. Table 3.2 illustrates the uniform contraction of state funding of R&D in the university sector, and even the part remaining that is publicly funded has been converted from direct institutional subsidy to mechanisms that imitate market processes through competitive grants and research contracts. Although some variance persists—France, for example, has a history of resisting privatization of research, whereas corporate

Table 3.2. Percent funding of higher education R&D originating in the government sector

	1981	1992	2003
Canada	79	71	63
Denmark	96	88	84
Germany	98	92	85
Japan	61	52	51
Spain	100	89	70
United Kingdom	81	70	65
France	98	93	90
Sweden	93	84	71

Source: Vincent-Lancrin 2006, Table 4.

subventions have a longer legacy of being well established in Japan—the overall trend is clear.

In response to a common (but mistaken) perception that the United States had been outpacing Europe in science and technology along many dimensions,[46] the European Union has been on the forefront of fostering the reengineering of institutions of research within European member nations. Worried about the so-called European paradox (an observed discrepancy between the role of Europe in global scientific production and its place in the production of patented inventions), the EU stepped up its involvement in science and technology policy (Larédo and Mustar 2001). Since economic development has become a dominant issue of the EU agenda and the pursuit of competitiveness is a major *raison d'être* of the Union, a shift may be observed from more traditional Cold War science and technology policy toward what has been internally dubbed innovation policy (Borrás 2003). The push for "harmonization" has operated along many scientific fronts nearly simultaneously, from pharmaceutical testing and regulation (Abraham 2007a, 2007b), to high-tech trade policy, to the foundation of a European Research Council to oversee the grants process, to the top-down establishment of a European Institute of Technology, patterned upon MIT.[47] In the process, government, or national, laboratories such as the CNRS in France and the Max Planck Gesellschaft in Germany were thought to be losing ground to universities, and consequently they were encouraged to promote or renew their links with corporate research. These changes should be situated within the context of pervasive budgetary retrenchments and privatization programs of neoliberal governments worldwide, a particularly European fascination with the construct of the knowledge economy, and the hasty imitation of Bayh-Dole–style legislation throughout the EU as well as Japan (Yonezawa 2003; Lynskey 2006) and China.

On the pedagogical front, governments have been eager to move higher education in Europe toward a more "transparent" system and to create a standardized higher education that is more attractive to the rest of the world. Acquiescing to the standard neoliberal jargon, the avowed aim of the European Union has been to recast itself as the most competitive and dynamic knowledge-based economy in the world. This movement began with the voluntary Bologna Declaration of 1999 but gained impetus with the Lisbon European Council of 2000, which sought to bridge the perceived gap with the United States and Japan by coordinating member research activities and laying the foundation for a common science and technology policy across the European Union (Rodrigues 2002, 2003, 2009). The task started out as university standardization to permit free movement of academic labor within the EU: a standardized three-year undergraduate and two-year master's degree regimen, a European Credit Transfer protocol, and a "reform" of the PhD degree. As the Lisbon Process became established, it also grew more ambitious: goals set to raise the numbers of graduates in the natural sciences and mathematics by 15 percent and, more germane, a 2005 directive to raise the level of industry funding of academic research (Marginson and Van der Wende 2007) and naturally, reformers promote favorable mentions of "deregulation" of higher education. The temporary setback of the ratification of the EU Constitution may have frustrated some of these ambitions but surely has not altogether stifled them over the long term. The euro crisis of 2010 may be another matter. The goal was to create a European Research Area, which had been referred to officially as "an internal knowledge market," and it was conceived as an R&D equivalent of the "common market" for goods and services by 2010. In the process, the EU proposed to increase its overall expenditure on research to 3 percent of GDP by 2010, although there were no explicit mechanisms to bring this about. Instead, what ensued was a "one-size-fits-all" set of policies for the commoditization of knowledge, even while the Washington consensus was being roundly abjured.

The globalized privatization regime has fostered a radically transformed science landscape in Europe, consequent upon shifting CGE alliances. Calls for a renewed social contract of science, for value for money, and similar neoliberal themes originated in the United Kingdom, and spread to the Continent soon thereafter. Early reforms in the United Kingdom include the Research Assessment Exercise (RAE) and the Teaching Quality Assessment (Hargreaves Heap 2002). The former was introduced in 1986 as a mechanism of control to enable U.K. higher education funding bodies to distribute public funds for research selectively on the basis of neoliberal criteria, by making universities accountable for their use of public money. It has been conducted roughly every four years, and since 1992 about £5 billion of research funds, which constitutes the bulk of the research component of the so-called block grant, has been allocated through the RAE. The effect of the RAE has been to concentrate funds at a few elite institutions, as well as to redirect research activity in a more commercial direction. This was finally openly acknowledged in the recast "Research Excellence Framework," which now stipulates that roughly 25 percent of the research rating should be governed by "impact criteria," which include the creation of new business, attracting business R&D investment, commercialization of new products, and improving health outcomes (Collini 2009).

Evidence of the desire to bring about convergence to a transnational model of the commercialization of science abounds in the EU. For instance, after two decades that witnessed an exodus of top students and scholars as well as a decrease in government support per student by 15 percent, Germany announced plans to form a group of ten American-style elite universities and award almost $30 million a year for five years to increase their competitiveness and quality (R. Bernstein 2004; Hochstettler 2004). Germany's attempts at American-style reform further involve the establishment of alumni organizations to raise money, greater selectivity in admission of students, and payment of faculty salary based on performance, running counter to Germany's long-held egalitarian ideal.

It is even more striking to observe the effects of the encroaching regime of globalized privatization on the country that had initially been used to justify the Bayh-Dole reforms in the United States in the 1980s. In the period of its perceived technical superiority, Japan displayed none of the trappings of the neoliberal marketplace of ideas: universities could not own patents, IP was handled informally between firms and professors, and direct industry funding of university research was small. It was no accident that state control of science funding and education were reminiscent of the Cold War pattern: they had been imposed by the American Occupation authority. However, when the Japanese economy stalled in the 1990s, the very same system of science management that had been praised for its postwar economic success was then equally indicted as an explanation of its stagnation. Invidious comparisons with neoliberal regimes abroad abounded, leading Japan to adopt many globalization devices, such as a 1998 law encouraging the opening of technology licensing offices at universities, a 1999 Japanese version of the Bayh-Dole Act, and a loosening of restrictions on civil servants participating in new commercial start-ups (Yonezawa 2003).

Further experiments in privatization of research in Japan have ventured well beyond their supposed American inspiration (Brender 2004; Miyake 2004; Lynskey 2006). In a significant departure in April 2004, national universities were themselves privatized into independent administrative agencies, tantamount to being transformed into independent public corporations. University-industry relationships, which had been previously based on a dense web of informal personal ties, were being rooted out and forced to be recast as more "market-mediated" structures. This has essentially ignored the fact that patents and IP had been used by Japanese corporations in very different ways than in America; for instance, antitrust had never played the role in Japan that described above in the U.S. regimes. Nevertheless, neoliberal notions of an institution-free marketplace of ideas and global competition seemed to overwhelm and throttle all such considerations.

The neoliberal approach to universities has since spread throughout Asia. Many Chinese universities have rushed to nurture start-up firms under pressure from the country's open-market policy, straining the national budget. The Chinese situation holds special significance for the offshore outsourcing of science from the United States, and, while referenced briefly in subsequent chapters, this easily deserves a separate book of its own. By 2004, China had managed to build up the largest higher education system in the world, at least in terms of numbers of students. Yet the Chinese situation constitutes a historical anomaly in that some commentators have noted that the blurring of the public/private divide in education has been induced much earlier than in the history of other national university systems (Tyfield 2009). Unusually for the history of a rising economic power, China has not profoundly shifted proportionate funding toward the higher education sector and state research enterprises as it grew richer, but it has rather encouraged the corporate sector to take the lead in the building of new scientific capacity. In this respect, the policy shift away from the state sector and toward commercialized science has mirrored the contemporary experience of the United States, as revealed in Figure 3.3.

The rising Chinese presence in world science is not a simple straightforward reflection of a state-led burgeoning university sector, which would just have "naturally" occurred in early stages of accelerated economic development. Rather, China has become the testing ground for a neoliberal science base, built from scratch.

Indeed, to play the neoliberal game, Shanghai Jiao Tong University began producing world research university rankings, realizing that the laggard can

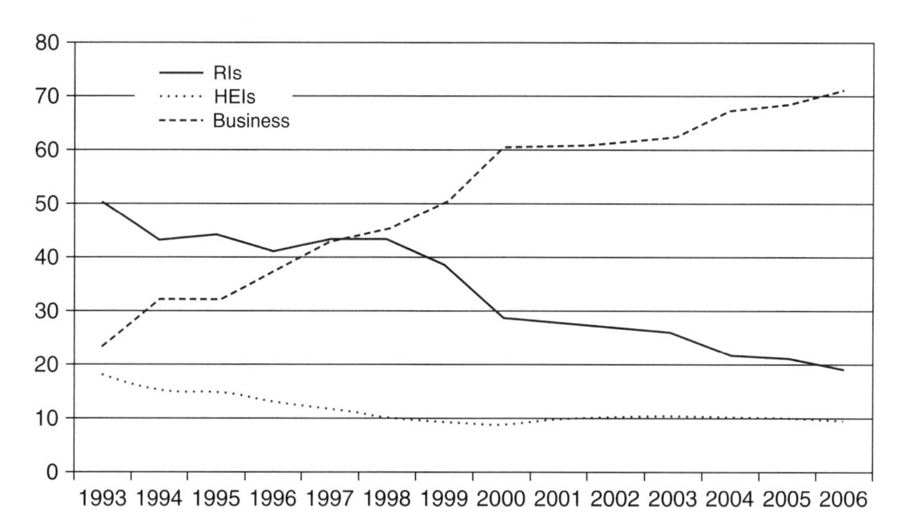

Figure 3.3. The transformation of Chinese R&D from URI-led to business-led (percentage total R&D appropriation by institutional type). Legend: RI–State research institutes; HEI–Higher education institutions; left axis: percentage total expenditure. *Sources:* Ministry of Science and Industry of the PRC, Statistics Data Book, http://www.most.gov.cn/eng/statistics/2006/index.htm; Tyfield 2009.

often make a splash by imposing a poorly justified set of league tables predicated upon meaningless criteria, just as *U.S. News and World Report* had done by fabricating university rankings within the United States. In the audit society, people readily confuse unidimensional rankings with "competitiveness." Just because U.S. universities initially came out on top in the Shanghai rankings does not constitute evidence of their cogency. Education reform in Japan, now emulated to some extent in successful economies like China, is presented as a way to give universities more autonomy, while being manipulated to reduce levels of state financial support at the institutional level. Changing governance structure plunges universities into an uncharted and ill-conceived era of competition, where a government tries to have it both ways in cutting subsidies but retaining control. These are global symptoms of what has been called the modern corporate audit culture (Apple 2005).

In short, there appears to be a concerted global effort to emulate the putative advantages of the U.S. privatization regime in advanced science and technology, while at the same time curtailing the amount of government support and reducing the accompanying bureaucracy. Universities outside the United States have responded to these incentives by seeking to raise money from the private sector, conducting more privatized and contract research, hiring more casual and temporary labor, and competing for a worldwide pool of footloose, paying students (Marginson 2007; Marginson and Van der Wende 2007). Indeed, in the latest foray of rendering the university to be more like the corporation, some universities have sought to offshore parts of their own education business, setting up wholly owned subsidiaries in various prosperous outposts, or trying with little success to establish a foothold in China, the new Wild East (Lewin 2008). A branch campus in Abu Dhabi has become the latest bauble in the university crown of excellence for institutions like New York University. In the process, these universities frequently have encountered cultural and economic barriers due to the fact that non-American universities have performed a different societal role than those in the United States because they exist in a somewhat different CGE environment. Nonetheless, if this trend persists, its effects on the existing U.S. system of scientific research are bound to be far-reaching.

I bring this chapter to a close by suggesting one more distinguishing characteristic of the modern regime: In a more elaborate history, it will prove necessary to expand our previous CGE analysis to take into account a distinctive fourth class of actors in the modern world system of science organization. Science has not only been promoted by firms, governments, and universities, but increasingly in the twentieth century it has also been organized and funded by international agencies that have propagated commercialization and standardization of research practices and institutions. These comprise scientific non- and intergovernmental organizations (scientific NGOs and scientific IGOs) as well as nongovernmental civic society organizations (civic NGOs). For our subsequent purposes, the various agencies may be divided into three classes: (1) the World Trade Organization, which has spread and enforced standardized rules of IP and trade in services under the guise of providing a stable platform for international trade (Drahos and Braithwaite 2002; Sell 2003); (2) the United Nations, which through UNESCO and WIPO has promoted targeted aspects of international science policy; and (3) a raft of international nongovernmental organizations (INGOs), from the World Bank (Goldman 2005) to the International Foundation for Science and the International Council for Science, which have played a crucial role in the spread of the globalized privatization regime.[48]

An earlier Mertonian approach tended to treat science as subsisting beyond or outside of politics, but nothing reveals the obsolescence of this belief better than an inventory of the means by which some pivotal scientific institutions have been spread by these international organizations. Some entities, of course, are merely the international arms of national professional organizations of scientists, and as such they conform to prior Mertonian images of the self-organization of science. But increasingly after World War II, these have been augmented by politically oriented INGOs that combine both scientists and laypersons into activist groups seeking to spread a model of "best-practice science" in the third world in the name of economic and political development.

The activities of these INGOs go some distance in explaining how it has become more readily possible that corporations can begin to take advantage of a globalized standardized research capacity in such a wide range of cross-cultural settings, particularly when it comes to offshore outsourcing. The pervasive similarities of science policies in almost all of the developed world and now, increasingly, those parts of the developing world where corporate R&D is moving in the near future are due in large part to the elaborate work of INGOs in propagating a relatively generic culture of commercialized research within local national education systems and government bureaus of science policy (Drori et al. 2003). The implications of this push to standardization extend far beyond the simple spread of something like a disembodied "scientific worldview" or the export of the Internet to previously untapped areas. For instance, standardization of scientific institutions and the displacement of local knowledge have turned out to be necessary prerequisites for the globalization of for-profit higher education (Morey 2004) as well as harbingers of the outsourcing of much routine scientific labor to low-wage countries.

The high-wage university sector in countries like the United States is consequently the next major economic area slated for another serious round of downsizing, cost cutting, and outsourcing from the advanced metropolitan countries to low-wage areas, and I would hope to convince the reader (and of course Viridiana) that this will have unprecedented effects on the content of scientific research just over the horizon.

4

Lovin' Intellectual Property and Livin' with the MTA

Retracting Research Tools

Down in the Boondocks

Throughout the twentieth century, many observers fretted about the prospect of patents stifling the course of scientific research, but lately, the situation has become more recondite and tangled. Indeed, of all the topics covered in this book, the effects of intellectual property (henceforth, IP) on science, corporate innovation, the university, and everything else under the sun have become a rampant pox upon the economics of science, infecting every argument with every possible permutation of evaluation, ranging from fevered images of a nirvana with no IP whatsoever, to an Orwellian nightmare where every utterance gets revised whenever convenient by TruthCorp and comes electronically accompanied by its own price tag. Why have economists apparently gone bipolar on IP? The problem seems to be that once the neoliberal conception of the market as an ideal conveyor belt of information had become second nature, leached into the very circulatory systems of the corporation, the university, and the state, then it no longer makes sense to attempt to decompose "the effect of patents on variable Z" into simple isolated correlation exercises or comparative statics calculations. "Responsibility" for the salutary (or deleterious) effects of any particular bit of IP can no longer stay neatly confined to the stipulated boundaries of the particular legal instrument or narrowly conceived commodity. This fact alone will sanction us to bypass a lot of the existing economic writing on IP.

If this book adds anything to the endless academic commentary on IP, it will be a heightened appreciation of the fact that it was the *process* of science as a whole that mutated as a consequence of the last three decades of tinkering with IP on a global scale; that in turn reverberated back on the legal instruments themselves and eventually on the institutions devoted to escalated levels of commercial activity. In this chapter, you shall discover that to keep your eyes fixed solely on the patents—as we have been repeatedly exhorted to do by numerous economists, university technology transfer officers, university provosts, and apologists for the Bayh-Dole Act—is to neglect the great majority of the ways in which scientific research becomes fettered and trammeled in the modern regime. To really keep your eyes on the prize, you must first feel the burden of the guerdon.

The notion that the U.S. patent system is broken, and that things have gone badly awry with regard to neoliberal ripping tales of market signals drawing amazing feats of imagination and innovation out of the woodwork, is now fairly well understood among those concerned with such issues.[1] The self-defeating notion that modern science necessarily must adjust and conform to this travesty of IP will be one of our major arguments against the neoliberal narrative. In the first section of this chapter, I will mostly recapitulate one of the best centrist evaluations of what went wrong, namely, from Jaffe and Lerner (2004). When I happen to depart from their orthodox interpretation, it will be mostly in the direction of the ideas of Mario Biagioli (2006b), the Harvard science studies scholar and historian, who reminds us that the original politics behind the patent system was never all that concerned with economic growth or ideal marketplaces of ideas. Some empirical consequences of these considerations will then resurface in Chapter 6. But even before that, it will become apparent that the orthodox focus on patents in the literature of the effect of IP on science is significantly deceptive.

To round out this overview of IP in science we will need to explore an obscure and seemingly insignificant legal document that turns out to be the portal to a world of frustration, delay, and obstruction of research, the dark doppelganger to the bright patents that commodify the fruits of science. The instrument is called a "material transfer agreement" (MTA), although we shall soon learn to regard the MTA as merely a subset of a class of legal documents that travel under a variety of rubrics, from "software evaluation license" to "bailments" to "material collaboration agreement" to "evaluation agreement"; the important lesson will be to not get hung up on the "material" part, but rather to regard them all as species of contracts between parties engaged in the same or similar scientific research, which aims to legally encumber that research in some fashion. MTAs are a central symptom of the current regime of globalized privatization, and they thus warrant the attention accorded them in the rest of the chapter.

How are such contracts born? To a first approximation, suppose our companion on this journey, Viridiana Jones, would like to do some research that involves the use and manipulation of research input ♣, which she has heard Scientist Dan Boone has in his possession. The input ♣ is not commercially available, precisely because science at the cutting edge is not standardized but rather riddled with uncertainty and laced with tacit knowledge intimately bound up with Boone's local activities in his lab; hence there is simply no other way for Viridiana to pursue that particular line of inquiry without Boone's help. However, when Viridiana approaches Boone, someone (perhaps Boone, or maybe the tech transfer officer at Boone's institution, Boondocks University) insists that the provision of ♣ is not only costly in time and money in Boondocks, but that there are all manner of *legal* considerations that it would be imprudent of Boone's institution to overlook. Here the Boondocks clan start off mentioning liability—it would violate simple justice to be held liable for anything that might happen when Viridiana deployed ♣ and (heaven forfend) something went wrong. In effect, Boone's representative informs Viridiana that there is no implied warranty with regard to the use of ♣ in any way whatsoever, including that scientific paper that enticed Viridiana to want to research it in the first place. But then prudence shades rapidly into due diligence: Boondocks needs to protect itself, so certain obligations could be imposed upon Viridiana as a condition for access to ♣. For instance, presuming that Viridiana should (wittingly or not) want to avoid harm to or compromise of Boone's ongoing research, Boondocks might demand that she delay publication until the new research is first reviewed by Boone. Boondocks, in the guise of its technology transfer office, might also want to clarify what sorts of property rights Viridiana might eventually claim are growing out of her use of ♣. This really begins the slide down the slippery slope, because this entails Boondocks asserting control over Viridiana's future research, insofar as it involves any commercial aspect. In the extreme case, Boondocks might want to claim proprietary rights over anything valuable that might arise out of her research, which are known in the trade as "reach-through" rights. Or perhaps Boone imagines that he might himself seek to patent some finding closely related to ♣ in the near future, in which case he worries providing ♣ to Viridiana might be deemed prior art to his claims, thus barring his patent application. It would therefore only be due diligence to get Viridiana to agree to confidentiality at minimum, or perhaps a few restrictions on how ♣ might be used, or stipulations as to how a licensing scheme might be structured between Boondocks and Viridiana's institution. And should ♣ turn out to be a living entity, they should get clear on how to treat the derivatives or offspring in Viridiana's lab. So—whew! This is getting pretty darned complicated;—Consequently, Boone suggests that his people should call her people, and therefore Boondocks' legal counsel will contact his counterpart at Viridiana's institution, convene a powwow, and negotiate a *contract* that would cover all these considerations: usually, it is delegated to someone who deals with MTAs. (For convenience, henceforth we shall refer to all such contracts as "MTAs," whatever their specifics.) All Viridiana has to do is . . . put her research on hold until this negotiation is brought to its conclusion, then sign the contract (that is, if she can live with the provisions it contains).

These sorts of agreements have been kicking around in the corporate sector for a long time; indeed, that is where they originated. What is new about the interaction between Viridiana and Boone over ♣ is that, first, they were essentially absent in most university settings before 1980; and second, the MTAs proposed by corporate providers have grown to an unwieldy size and reach in terms of the aspects covered in the period since 1980; and third and most significantly, *universities* have begun to demand MTAs from other universities since the 1990s. Something strange and deformed has thus sprung up, like mutant fungi after a rainstorm, and it is starting to derange the ivy and lawns of academe. What is striking about this development is that nobody really likes it: not Viridiana, sitting there fuming over the delay in getting hold of ♣; not Boone, who mostly only cares about the extent to which he discourages or encourages Viridiana's research (for, who knows, someday she might get elected President of the ♣ Society, or perhaps help him escape from the Boondocks); not many university tech transfer officers, who tend to delegate responsibility for monitoring MTAs to the least experienced and newest members of their staff;[2] not the patent lawyers, because that's not where they get to make the real money; and not the university provost, who has never given his TTO a bonus because they have greatly increased the MTA output of their office. Not even a mother loves her MTAs. So how is it that everyone is now issuing them? That is one of the questions addressed in this chapter.

Some tendentious distinctions are frequently made concerning MTAs. For instance, a technology transfer officer (TTO) at MIT once lectured me: "If you patent it, it is IP; if it is a material, then it's not IP, but tangible property." Definitions are the soul of law, but not, we venture, the gist of the economics of science. She was committing what we might think of as the fallacy of misplaced concreteness. Two considerations are relevant here for seeing the error. First, these instruments are *contracts* between researchers nominally engaged in a joint project of the furtherance of science, enjoining the parties to cease and desist from certain stipulated activities. Second, while these contracts are clearly different from intellectual "property" such as patents, they are equally clearly accessories to the patent system, and more to the point, crucial tools in the commercialization of knowledge. While it may be true, as Richard Posner says, "if the only people who have access to your property happen to be the people with whom you have a contract, you can regulate their access by means of contract and forget about property law" (2002, 6), in the special instance that the system is in the midst of implementing a whole new class of intellectual property, one cannot ignore that the shape of the contracts are conditional upon that innovation. Imagining patents and MTAs as existing in two wholly disjoint universes, as she clearly intended to suggest, does nothing to illuminate the roles and functions of either instrument.

When it comes to the economics of science, we shall discover the tangible character of research tools is not really a substantive consideration (although undoubtedly critical for some legal purposes); ultimately the parties to the agreement are still attempting to control ideas and their fabrication by means of control of things and quasi things like software and business procedures. There are some who think that they can wave away problems of the modern shape-shifting character of MTAs through semantic flailing: "software evaluation license," "bailment," "tangible property license," "material collaboration agreement," or "evaluation agreement." Those are all instances of the displacement of a fixed IP instrument (the frequently noted "one-size-fits-all" problem of a patent system) by a more flexible contractual agreement. We shall explore the possibility in this chapter that the "reform" advocates of a boilerplate standardized MTA at the National Institutes of Health (NIH) and elsewhere in the mid-1990s made this same mistake—they thought the focus should be on a fixed procedure or "thing" that could be standardized, another one-size-fits-all proposition, when in fact the history of MTAs is primarily a saga of the evolution of new flexible forms of contractual control aimed at upstream conduct of research.

It has become commonplace to treat patents as the alpha and the omega in the structure of incentives within a marketplace of ideas; but, in fact, patents are only one possible means to that end, as is well understood in the corporate community, as well as by many paragons of a neoliberal economics of science.[3] Trade secrets, disclosure agreements, mergers and takeovers, employee exchanges, internal security measures, and much more are all ways to either foster or discourage the generation and transmission of knowledge. TTOs frequently misconstrue patents as ends, rather than means. It is relatively rare to find patents treated solely as stand-alone revenue generators in modern corporate strategy; more frequently, they are just one subset of a larger business plan. Different industries and different sciences tend to deal with these problems in their own idiosyncratic manners; as university science becomes increasingly privatized, the university has had to confront this diversity in its quest to sell its wares. A lesson that has only begun to be learned is that patents are not the alpha and omega of the process, and furthermore, they have turned out to be an excessively cumbersome vehicle for the control of commercialization in fast-moving research fronts in science. The purpose of this chapter is to try and situate MTAs within the larger context of corporate strategy, scientific competition, technology transfer, and the academic deployal of intellectual property, as a prelude to understanding one of the more stark examples of the feedback of unintended consequences of commercialization in the modern regime upon the health and well-being of contemporary sciences.

The Problems with Patents

Do patents help or hurt science? It is amazing that after at least two centuries of experience with patents in the developed economies, in both law and economics it appears the question is no closer to a satisfactory resolution than when it was first posed back in the nineteenth century. Classical liberals, including many illustrious neoclassical economists (Johns 2006; Watt 2000) had basically opposed patents (and copyright) primarily because they were an unjustified grant of government monopoly that could only serve, in their view, to obstruct further inquiry and the spread of useful technologies, and to line the pockets of a chosen few who were unlikely themselves to offset the harm done. Opposition to IP was ordinarily deemed a corollary of opposition to monopoly in general, back then a hallmark of being a sound economist. No one had seriously suggested that patents played a major or indispensable role in past economic development, say, as actually "sparking" the industrial revolution or American rise to economic dominance. The shift of attitude among pro-market thinkers, from a default position of relative hostility to patents to one where patents were praised as a most wonderful inducement to the growth of knowledge and encouragement of innovation coincided, not accidentally, with the elaboration and stabilization of neoliberalism.[4] It is important to pay close heed to the intimate connection between the recasting of the market as an information processor, and the growth of the conviction that knowledge should be concertedly commodified and rendered more friendly to market allocation. The fact that we shall encounter a raft of enthusiasts for the introduction of IP into the academic sphere of science in this chapter would more plausibly be attributed to the general spread of neoliberal ideas, and not, we shall insist, because we found new discoveries languishing stymied and tangled up in ivy.

Viridiana despaired at the cacophony of voices concerning the necessity or superfluity of the need for legal protection of ideas at the beginning of Chapter 2. Our current touchstone on the modern problem with patents, Jaffe and Lerner, are "centrists" in this regard because they simply presume without argument that "to make innovation rewarding, the government must give or grant something valuable to people or firms that produce important innovations" (2004, 38). Their dictum contains the important weasel-word "innovations" because as they know, and the legal literature insists, neither patents nor copyright were intended to protect *ideas;* they (until recently—more below) instead granted rights over certain formats of the *exploitation of ideas.* Neoliberals, for reasons already covered, tend to slide indifferently between these two classes of phenomena. This matters for science because almost all IP law in the modern world displays very little concern for where ideas come from and blithe insouciance about how one nurtures the development of these ideas, even though a spurious *noblesse oblige* toward the repressed creativity of the little guy has become standard PR for the globalized privatization regime.[5]

A patent, far from being a license to invent, is rather a state-granted right to sue others if they make, use, sell, offer, import, or (even!) offer to import the entity covered by the patent. Considerations of international commerce have only grown in importance for patents since the watershed foundation of the World Trade Organization (WTO) in 1995. The fact that this has nothing whatsoever to do with the actual process of discovery is demonstrated by the fact that independent innovation of the same "thing" is no defense against litigation. Because it is only a right to sue, there are no "patent cops": The full weight of discovery and punishment rests on the patent holder, and this must be pursued through special courts in the United States, something that lends the entire field a very skewed legal structure. The "thing" protected has to meet certain tests to initially be granted this right by the U.S. Patent Office (USPTO), nominally categorized under the rubrics of allowable subject matter, utility, novelty, and nonobviousness. While we shall not go into the gory details of how these legal tests are deployed, it will be important to take note that the tests themselves have been substantially revised (and many would say gutted) in the last few decades.

Jaffe and Lerner have mooted essentially four indictments of the modern U.S. patent system, which have been conceded to a greater or lesser extent by most participants. The first is an unprecedented and unwarranted expansion of the range of "things" that qualify to be patented, partly through revisions of the tests at the USPTO, and partly through court cases that have expanded the range of patent grants. Some of the latter have had profound implications for scientific research, such as the Supreme Court ruling in *Diamond v. Chakrabarty* (1980), which affirmed the patentability of biological organisms under the excuse that they were in fact "human-made."[6] Since the Cohen-Boyer patent on the production of "biologically functional chimeras" was seen in retrospect as firing the starting gun for the breakout of the biotechnology movement (see Chapter 5), and dated from the same year, the lowering of the bar for patentability of organisms and the transformation from the Cold War into the privatized regime of science are intimately linked. But the weakening of what qualifies as patentable was not restricted to living organisms: equally significant for science were easing of restrictions of patenting computer code, mathematical algorithms and business practices, to such an extent that the patent system has come dangerously close to allowing the patenting of ideas themselves, *particularly when those ideas arise in scientific research.* In a development that postdated Jaffe and Lerner's book, but render it all the more poignant, in June 2006 the Supreme Court rendered a decision in the case of *Laboratory Corporation v. Metabolite Laboratories*[7] "which allowed this patent on a biological fact to remain in effect" (Andrews et al. 2006, 1395). The aspect of this case, which sends red flags shooting in every direction, is that the original research finding that a high level of homocysteine (an amino acid) is correlated with a vitamin deficiency and the development of a test for its presence in the body was originally conducted at Columbia University and the University of Colorado. As has now become commonplace, the investigators formed a start-up firm in Colorado, Metabolite Labs, and filed for a patent to capitalize on their discovery. However, in a bit of hubris, the patent application asserted that the petitioners should be allowed to patent the basic physiological fact, so that they could claim a royalty whenever *any* test for homocysteine was sold. Now, other tests for homocysteine already existed and were used to diagnose a variety of medical disorders. In a sequence of events too baroque to recount in any detail, Laboratory Corporation of America (LabCorp) had licensed the right to the test from Metabolite, but it then published a scientific article discussing how high homocysteine levels might indicate a deficiency to be treated by a vitamin regimen, and Metabolite sued for breach of contract and patent infringement. The U.S. Court of Appeals for the Federal Circuit astoundingly ruled that publishing the fact infringed the patent and, not satisfied with garden-variety outrageousness, also ruled that doctors would infringe the patent merely by contemplating the physiological relationship. The Supreme Court originally allowed a review of the case but then dismissed it on essentially technical grounds, so the lower court ruling still stands. Just to be perfectly clear about this, it was *university scientists* who doggedly went the extra mile to assert the extra-strong patent claim; then incongruously, a large for-profit company, LabCorp, then published the information in the open literature; and finally, *university scientists* through their start-up firm, Metabolite, brought the suit that produced this poisonous legal precedent. Who were the "good guys" here, seeking to protect the virtue of science? Academic scientists have not been the hapless innocent bystanders in the process of extending patents to facts, contrary to the folktales found in such outlets as the *Chronicle of Higher Education, American Scientist,* and *Science.*

Jaffe and Lerner have not devoted very much attention to deciding whom might be deemed culpable for this grand patent deformation, but they take the position that the capacious augmentation of determining what falls under the purview of the patent system has not occurred in a principled or logical manner. This is linked to their *second* indictment: that the USPTO has been badly weakened over the years so that it could no longer stand as a bulwark against the unremitting pressure to patent everything under the sun. As a major tenet of neoliberal doctrine deems government functions should be privatized, Congress decided to treat the USPTO as a cash cow, not only dictating that "the primary mission of the Patent Business is to help customers get patents" (Jaffe and Lerner 2004, 137), but then, as the Office charged applicants bigger and bigger fees, it played accounting tricks to expropriate the "surplus" and allocate it to other uses. Once the Patent Office was treated as a money-spinner, all the incentives were skewed for examiners to allow patents, rather than reject them. From a neoliberal perspective, it is better to let the Market, not some underpaid bureaucrat, decide whether knowledge was new and valuable. It certainly made it easier to sneak dubious applications through when patent applications per examiner rose from the mid-1980s onwards (ibid., 131). The amazingly high approval rate of around 70 percent of all applications per year has only started to come down in the last few years; but no one should breathe a sigh of relief that crippling weaknesses of the patent review process are finally getting fixed.[8] The quality of patents granted continued to fall throughout the first decade of the new millennium, as we shall discover in Chapter 6. Jon W. Dudas, the undersecretary of commerce for intellectual property, was quoted as claiming in 2008: "We are getting more and more unpatentable ideas, worse and worse quality applications" (Pear 2008).

The hyperexponential growth in patent applications is a key variable in understanding the utter incapacity of the USPTO to exercise some discretionary control over the patent system, and this is explained in Jaffe and Lerner's *third* and *fourth* indictments of American patents. As economists, they pledge allegiance to the principle that patent applicants are rational economic men: the primary reason why patent applications show no sign of tapering off in their exponential growth is that simple cost-benefit calculations have been rendering patents more and more lucrative as investment vehicles over the last few decades. It's not that there's been proportionately more science performed, and it's not rocket science: More lucrative patents translate into more patent applications. The increase in value can partly be traced to the outrageous extension of the length of time of patent protection in the United States. Not only was the time extended from seventeen to twenty years from filing in 1994, but a myriad of special concessions to all sorts of patent holders, particularly in the pharmaceutical sector, have permitted an even longer horizon in specific cases. A patent that lasts longer is worth more to the holder, thus encouraging more filing. But even more significant was the creation of a special Court of Appeals for the Federal Circuit (CAFC) in 1982 to deal with patent infringement cases and in international trade disputes (a combination that is not as incongruous as it might at first seem, as we shall shortly discover). Due to its special composition, this court has tilted the entire system in favor of patent holders and against the defense in infringement suits. Jaffe and Lerner found that patent suit litigation exploded soon after the institution of the CAFC (2004, 14), and more to the point, the share of patents found infringed on appeal jumped substantially in the late 1980s (ibid., 105). The average levels of damages and penalties hit the stratosphere (110); not to be outdone, the CAFC began its own crusade to loosen restrictions on the tests that had to be met for valid patents (115). The whole process had become so profitable that the 1990s saw the rise of the "patent troll," an individual or firm who accumulates patent portfolios solely for the purpose of pursuing infringement suits, without having any intention to actually "work" any of the patents or produce anything. Neoliberals actually promote this phenomenon as one more instantiation of their precept that any apparent "market failure" can be rectified by instituting another meta-market: In this case, venture capital flows in to start-up firms to serve as patent retailers and even introduce derivatives (Lohr 2009a).

More than one legal scholar has gazed upon the current system with vertigo and wondered what madness would allow this travesty to persist, if not to actually degenerate over the last two decades.[9] One answer has to be the widespread neoliberal conviction that the marketplace of ideas is the supreme mechanism of intellectual advancement, and that if some property rights are good, more are better. As Metlay (2006, 590) points out, before 1980 the "public interest" was interpreted in the courts as providing a check to untrammeled private enterprise, but thereafter it was translated as thwarting inefficient government intervention and wily foreigners. And through it all, more lucrative settlements of infringement cases meant more patent applications of dodgier provenance.

This vicious spiral of the fortification of the patent system hit science right around the same time as the passage of the Bayh-Dole Act, which encouraged universities to claim IP in discoveries made by their faculty in the course of research funded by federal grants. Some analysts such as Mowery and colleagues (2004) have asserted Bayh-Dole didn't matter, because universities would have escalated their patenting activity even in its absence. From my perspective, this is just another example of the blinkered analysis characteristic of neoliberal economists: *of course,* some American universities pioneered patent ownership even when federal rules were less than clear, and, *of course,* once the cost-benefit calculus tipped in favor of patent holders, more universities would be tempted to follow their lead. Bayh-Dole was neither the sole nor even the primary reason why science was becoming increasingly commercialized; it was just one component in a whole range of roughly simultaneous "reforms" being engineered into corporations, the government, and the universities—all calculated to instigate the marketplace of ideas throughout the entire culture. But Mowery and his confreres proceed to ignore the ways that Bayh-Dole mandated the neoliberal restructuring of the university, treating it merely as offering "incentives" to commercialize research. For instance, Bayh-Dole irreversibly shifted the faculty-university relationship by requiring employees to inform their employers of all inventions, as part of the reporting requirement for federal grants; this brought the universities closer in line with IP law for corporations, which had deprived employees of rights since the 1920s. It also tipped the balance in favor of licensing rather than assignment of patent rights, which one might argue could have made more sense for a university sector not straining to imitate corporate behavior.

This lesson that you have no choice over whether to act like a business once you start playing the patent game for keeps has been enshrined in a 2009 Court of Appeals ruling against Stanford University (Lederman 2009). Stanford administrators like to portray themselves as deploying a relatively "permissive" patent policy with respect to its faculty, as a means of luring them to sunny California. However, when Stanford faculty sought to enforce its patent rights by bringing an infringement case against Roche Pharmaceuticals, the court dismissed the lawsuit because the firm, which collaborated with faculty in developing a test to measure the concentration of HIV in blood plasma, had asserted immediate proprietary rights over *any* work that was done by any employee. Because Stanford permits faculty latitude to decide how they would assign their rights, the corporate IP vested first, and Stanford lost that patent (and maybe many more in coming months). Patently, if you want to keep your IP, then all freedom of the individual faculty scientist must be revoked. So much for the market defending your birthright freedom to choose.

Once the visions of sugar plums temporarily ceased dancing in the heads of provosts, and they started informing their faculty scientists that patents were a tenurable index of high-quality research, then universities themselves began to learn in earnest about the consequences of their love affair with IP. Throughout, most scientists have believed there was a "research exemption" from patent infringement: As long as they were engaged in scientific research, they would not become the targets for litigation over infringement remedies. Interestingly enough, this was a myth, since there appears to have been no developed body of legislation or case law in the United States spelling out that notion. As long as universities did not pursue IP very strenuously, the system more or less worked as though such an exemption was in force. Then along came the federal court ruling in *Madey v. Duke University.*[10] That story itself is a lesson in the consequences of privatized science, and in a scientific field, microwave physics, that is far less pervasively privatized than biology.

John Madey, while a professor at Stanford, invented a microwave electron gun and free electron laser. Incongruously, given Stanford's reputation and activist technology transfer office, it declined to patent these devices and devolved the rights to Madey. He obtained the two patents, and he subsequently joined the faculty of Duke University. Duke, as part of the enticement package, moved his devices from Stanford into a "Free Electron Lab," where Madey proceeded to use and further develop these inventions, with the support of government funding. For reasons we can skip here, Madey and the Duke administration grew to be at odds over the activities of the Free Electron Lab, and Duke appointed a second director and an oversight committee. Madey regarded this as a hostile move and objected; Duke then replaced Madey as director and petitioned the Office of Naval Research to replace Madey as principal investigator on the relevant grant. Understandably, Madey then resigned his Duke position. So far, this was just another sad, vicious tale in a long sequence of academic dustups and scholastic vendettas in the history of the research university. Maybe it would have remained that way, except for those two patents . . .

Madey, piqued, then sued Duke in federal district court for patent infringement, because those lasers were still in place at the Free Electron Lab and were still being employed for research purposes. The district court ruled in favor of Duke, but the CAFC reversed the lower court ruling, stating, "Our precedent does not immunize any conduct that is in keeping with the alleged infringer's legitimate business, regardless of commercial implications. For example, major research universities such as Duke often sanction and fund research projects with arguably no commercial application whatsoever. However, these projects unmistakably further the institution's legitimate business objectives." In other words, if Duke wants to act like a for-profit firm, accumulating patents and enjoying pecuniary benefit from encumbering its discoveries, then *even if the individual research project was conducted exclusively for noncommercial purposes,* Duke can no longer hide behind some imaginary façade of public-spirited research exemption when making use of patented materials. Crudely, what's sauce for the goose . . . This had far-reaching implications for what are now often called "research tools," that is, research inputs covered by some form of IP protection. Combine this with the progressively sloppy extension of patent protection into the realm of ideas, and it begins to dawn just how much this one decision can potentially come to totally reorient future practices of science. Contrary to the economists David and Dasgupta (1994), the CAFC has dictated that universities can no longer arbitrarily divide their research up into a conveniently protected public sphere and a competitive private sphere: There is no such thing as getting a little bit pregnant, or turning a little bit for-profit.

Even though it is far too early to make any solid pronouncements, it has already become common for neoliberal apologists to assert that *Madey v. Duke University* has had little real impact on science (Blumenstyk 2007a), partly because many (and maybe most) scientists still don't bother to monitor the IP status of their research inputs, and furthermore, no one else has been subject to a high-profile infringement case since *Madey*. Some speculate this is because universities still can't be bothered to enforce discipline at the lab bench: in one counterexample, Iowa State conducted an audit into the IP ownership of research tools in *one laboratory*, which involved contacting seventy-one different external entities and an expense of $24,000 to conduct checks and query patent owners (Boettinger and Bennett 2006, 321). There is plenty of evidence that universities have started to regularly sue corporations for infringement (Blumenstyk 2007b; Lederman 2009), only the strategy has now started coming to grief. But the seeming calm may also be an inadvertent artifact of the fact that more and more university infringement cases have had their settlements sealed, as part of a prudent business strategy. *Madey* was unusual because, in the final analysis, it wasn't about the money, and hence became public.

Whatever the reasons for minimizing the impact of *Madey*, even strong critics of the patent system like Jaffe and Lerner oppose a strong research exemption for university scientists (2004, 67), because they automatically subscribe to the neoliberal premise that markets are still the best form of organization of intellectual inquiry. It is hard to be an orthodox economist these days and deny this credo.

The utter incoherence of the U.S. patent system with regard to research exemptions was cemented by an even more recent judicial intervention. On June 13, 2005, the U.S. Supreme Court decided a crucial case in *Merck KGaA v. Integra Lifesciences I, Ltd.*[11] involving the scope of the research use exemption, which entered U.S. patent law through the Drug Price Competition and Patent Term Restoration Act of 1984, sometimes referred to as the "Hatch-Waxman Act." It turns out to be a different exemption from that denied in *Madey v. Duke University*. This particular act contains a research exemption that permits persons other than the patent holder to "make, use, offer to sell, or sell within the United States or import into the United States a patented invention (other than a new animal drug or veterinary biological product) . . . solely for uses reasonably related to the development and submission of information under a Federal law which regulates the manufacture, use, or sale of drugs." This is another example of how special pleading by pharmaceutical companies has further distorted the patent system.

In its opinion issued on June 13, 2005, the Supreme Court found that the Hatch-Waxman research exemption "extends to all uses of patented inventions that are reasonably related to the development and submission of *any* information under the [Food, Drug and Cosmetics Act]." Further, "this necessarily includes preclinical studies of patented compounds that are appropriate for submission to the FDA in the regulatory process." That is, after having the CAFC denying research universities the research exemption in *Madey*, the Supreme Court granted a liberal exemption in just one extremely narrow area of science having to do with preclinical drug development. There are few more cynical examples of the power of money in the modern marketplace of ideas than this: "Creating the ironic result of providing broader research exemptions to the pharmaceutical industry than to basic scientific research" (Lieberwitz 2007, 67).

As if the situation could not be manipulated further to the benefit of a handful of parochial interests, along came the 9/11 disaster and close on its heels various appeals to the war on terror to further jigger IP. The Project Bioshield Act of 2004, having recourse to the excuse of rapid response to bioterrorism, created all manner of further special dispensations to the pharmaceutical and biotechnology industries.[12] Among those relevant to our current concerns, it conjured a way for firms to get around FDA approval if the executive branch deemed it an emergency measure, permitted the NIH to sidestep peer review and other quality monitoring of research grants and contracts in order to "accelerate R&D," and sanctioned payment in advance for stockpiles of designated treatments, all the while maintaining the strong patents firms would be permitted to obtain through government funding. And to top it off, neoliberal legislators sold the bill as "creating a market" where none before existed.

The Anatomy of MTAs

By all accounts, the legal instrument of the MTA originated in the corporate sector, but the actual birthdate seems lost in the mists of time.[13] The reason for the missing birth certificate is that, in the corporate context, it existed merely as one minor instrument in a whole armory of contractual agreements and licenses that corporations had developed over time to deal with issues of nondisclosure, the management of portfolios of patents and trade secrets, the defense against fraud and liability for damages under breach of contract, and so forth. There was nothing especially noteworthy about it before 1980; it was just one small component in the prudent exercise of due diligence that became the standard of competence of corporate legal departments. In firms that had maintained a substantial capacity for in-house research, it defined an unexceptional part of the landscape traversed by any corporate scientist. No one would have been shocked in industry by a request to sign an MTA if they wanted to make use of some patented material or substance that originated in another firm's lab; collaborative research did exist, but it was treated gingerly by all parties involved. And if the other firm wanted to deny access—well, that was just part of the game. It was the price one sometimes paid to be employed in the private sector.

As with so many other instances in this book, the turning point came around 1980. Some TTOs at major universities have told me they first saw MTAs showing up in university labs in the early 1980s. Most observers speculate this had something to do with the passage of the Bayh-Dole Act, in that universities became more attuned to the possibility that they might lay claim to IP in rivalry with private firms, particularly in the area of biomedicine. Perhaps more to the point was the successful IPO of Genentech in 1980, which demonstrated for the very first time that a corporation with no discernible product revenue could nonetheless become a darling of Wall Street (Pisano 2006b, 87). With Genentech as avatar, a stream of biotech start-ups were created in the 1980s; most of them had to find immediate revenue streams in research tools, patents, and proprietary joint ventures with Big Pharma because they didn't actually make anything. All they had to sell was the science itself, since everything else was insubstantial promises and projections. This shift induced a much less open atmosphere when it came to sharing specialized research inputs, and universities experienced the chill partly in the form of a flurry of MTAs in response to requests submitted to corporate labs for organisms, drugs, and the like. Another relevant factor was the tendency for the USPTO to intemperately expand the range of entities for which they would award patents, as described in the previous section. These trends all came together to make it much more likely that research tools would be subject to all sorts of encumbrances after 1980. The uptick in MTAs appearing on university doorsteps was a symptom of these changes.

Most universities, even those who had enjoyed long-established technology transfer units, were ill-prepared to deal with the novel phenomenon. Many adopted what they thought was a pragmatic position: "We deal with the intellectual property; you handle your own lab's relationships with outside labs." In many instances, there were no formal procedures for handling MTAs. Given that most tended to deal with biomedical materials, sometimes there would be a designated official within the medical school who might act as informal advisor on what to watch out for concerning incoming MTA agreements, without any counterpart at the university level. No one thought to monitor what their faculty members were doing with IP deals they had initiated with outsiders, so no thought was given at all to outgoing MTAs. Thus, in the early days, many MTAs never even registered on the administrative radar screen, and more often than not they were summarily signed by the researcher in question, without any oversight concerning the stipulated provisions. Something approaching this situation may still abide at many universities; various TTOs have confided to me that they still never hear about some substantial proportion of the MTAs that may have been signed by faculty at their home universities, even in the face of explicit regulations that require registration of all MTAs as part of the IP disclosure process.[14]

The IP That Dare Not Speak Its Own Name

To understand the full panoply of what is at stake, it becomes necessary to get more precise about the exact character of MTAs. The first and most important distinction to make is between incoming and outgoing MTAs. An incoming MTA is emitted from a lab that owns the research tool that you want to use, so they send an MTA for you to sign as a condition of receiving the material. An outgoing MTA is something you draw up for some external party who wants to use a research tool that you own the rights to and can control. The reason this matters is that many university scientists initially felt that they were not the ones hindering the advance of science by making unconscionable demands upon external labs who were interested in their research materials; rather, in their view it was primarily the corporate researchers who were making exorbitant demands in their incoming MTAs, rapacious requisites that required university faculty to exercise caution, and their university legal departments to exercise oversight regarding what they could be willing to sign. The upshot was that responsibility for incoming MTAs was often vested in one corner of the university structure, while responsibility for oversight of outgoing MTAs might be delegated somewhere else, if it was even a matter of conscious policy to monitor them at all. Of course, most university technology transfer offices are not even two decades old, so the fact that the exact nature of their responsibilities might still be in flux as late as the year 2000 would not be that unusual.[15] Even today, it is frequently the case that, even at major research universities with elaborate TTOs, responsibility for both incoming and outgoing MTAs is rarely consolidated together under one roof. Some officers I have encountered have proffered the opinion that they could see no reason to centralize university information on MTAs. In this respect, universities still lag far behind the chains of command and audit trails one finds in research-intensive corporations.

What is it that the university counsel and some university faculty were afraid of finding in those incoming MTAs? In a nutshell, the MTAs can contain a forbidding array of controls and restrictions over prospective research still to be carried out by anyone who signs them. Furthermore, as I shall discuss below, there is no such thing as a plain vanilla "standardized" MTA, so the actual contents of the agreement can range from a handful of boilerplate clauses to intricate legal documents of many pages. What immediately catches an academic's eye are the confidentiality clauses and various permutations of prior restraint upon publication or other disclosure of any findings that somehow make use of the material in question. The supplier of the research tool claims the right to enjoin how, when, and in what manner the scientific finding can be published, which strikes many as fundamentally contrary to the (admittedly old-fashioned) spirit of academic research.

Another variable I should highlight is the difference between MTAs conducted within the corporate sector, MTAs negotiated between a university and a corporation, and most significant, MTAs signed between universities. While we should not want to pretend that there ever existed a harmonious Mertonian heaven where scientists unselfishly and unhesitatingly provided their rivals with all the materials they needed to replicate and surpass their benefactors' research,[16] it certainly never was the case prior to the spread of MTAs that academic scientists had the contractual right to *sue* other scientists for damages if they published their findings in a fashion that offended the party of the first part. Furthermore, it is unprecedented that universities would resort to contracts over research tools to control research happening in other universities. But if nondisclosure clauses were all there was to be found in MTAs, then their impact would have been much attenuated relative to experience in recent days. The other rapacious class of codicils in MTAs is the so-called reach-through clause, a claim on any IP that might arise in the future research of the recipient of the research tool. Many regard this as an unacceptable price, ransoming the future of your research to someone whose only contribution was the provision of a single material accessory input into the process. These IP-related clauses have become quite baroque of late, including options clauses for licenses on future research materials, grant-backs for newly discovered uses for the existing material, splitting of future royalties (or costs), royalty-free access to the organization's patent portfolio, broad claims over "derivative" materials (such as offspring of organisms, related cell lines, collateral secretions), patent prosecution controls, indemnification against any liabilities that might arise from use of the research material, and time limits on the use of the material. Another perverse and disturbing clause that has shown up in some MTAs is a provision that states that the parties must keep *the very existence of the MTA itself confidential*, usually with the excuse that if the agreement were common knowledge it would implicitly signal the IP strategy of one or both of the parties (Henry, Cho, Weaver, and Merz 2003).

There is a widespread misperception that there is something peculiar to the specific science of modern biology and biomedicine that has summoned the elaborate system of protections now inscribed in MTAs, something about the need to handle slippery live "things"; but what is significant is that, as the university has become progressively commercialized, the circle of research tools that have become encumbered by MTAs in the university context has widened ever further. Not only are they attached to drugs and "patient-derived materials" (tissue samples, tumors, and the like), but they also have become increasingly prevalent for software, radiology pulse sequences for MRIs, dog food, banks of test questionnaires, psychological assessment protocols, computer chips, absorptive particles in gas masks, plastic polymers, and all manner of machinery used in research.[17] In a recent survey of members of the AAAS, 65 percent of the MTAs self-reported by respondents who were attempting to get research tools from others were in the biological sciences, and 35 percent were for nonbiomedical materials (Hansen et al, 2006, 18). It would seem that we have stumbled on a cumulative trend here, which is rather more pervasive than it is usually portrayed; in the interests of providing supporting evidence, I report the fruits of two years of repeated requests of data from university TTOs in Table 4.1.

I want to begin by admitting this may seem a paltry amount to show for two years spent in data gathering, but I would like the reader to contemplate that coming to an understanding of the reasons why universities are reluctant to provide this kind of information constitutes the beginning of wisdom concerning the effects of fortified IP upon academic science. There was a time in the late 1990s when many university TTOs proudly endorsed the research ethos of the institutions of which they were a part, for example, openly publishing elaborate statistics describing their various internal and external activities bearing upon IP and making them available to the public and on the Web. One by one, however, perhaps at the behest of their lawyers, or perhaps because of some tense moments at faculty meetings, or perhaps after having some academic provost question the allocation of effort within their office, they came to the conclusion that this openness was not altogether conducive to their overall strategic place in the university and had suspended that practice in favor of producing glossy annual reports that would make any corporation proud, both in terms of their production values but also in being devoid of any solid information whatsoever.[18] We should be clear about this: not only has a certain modicum of secrecy come to be taken for granted *within* the research process in many of the natural sciences in the twenty-first cen-

Table 4.1. Total TTO reported MTAs by university, incoming/outgoing

Year	Harvard	Duke	UCLA	Emory	UMass	Stanford**
1997			181/			
1998			201/			
1999			242/			
2000			290*/			
2001	186/52		340*/		164/	
2002	173/71		325*/	158/69	163/	
2003	160/109	355/148	371*/	221/65	189/	360(550)/
2004	203/118	370/297	516/	233/75	225/	400(600)/
2005		384/299	657/	287/132	268/	400(624)/
2006	562†		730†	236/100		450(650)/
2007	531†		722†	287/123		
2008	709†		990†	416/120		

Notes: Reporting period differs—due to academic year, business year, or calendar year. Some sources report only incoming MTAs: hence there is only a number to the left of a solidus in the table.

* Since reporting period varies over the course of these years, the complete series was converted to academic year by interpolation.

** Stanford figures are round numbers for MTAs, with "total agreements," which may include amendments to prior MTAs in parentheses. The TTO claims it does not track outgoing MTAs. Upon provision of these figures in August 2007, Stanford had a stock of 2,700 fully executed incoming MTAs in force.

† incoming + outgoing MTAs combined

tury, but that secrecy has now come to blanket university information *about* the very process of commercialization of research itself. Commercial reticence comes to erect a veil not only around the putative "products" to be sold but also around the business practices and strategies of the commercial unit: in this case, the business arm of the university.

But that is only the beginning of the explanation for the obstacles to finding out what is happening to MTAs in the university sector. In general, because universities have not made a deliberate conscious effort to monitor and control MTAs, but rather have tended to backtrack into dealing with them piecemeal and on an ad hoc basis, in many instances TTOs themselves still possess only an unclear and incomplete picture of the situation at their own universities. It seems no two universities deal with MTAs in the same precise fashion (and practice in non-American institutions superimposes another layer of complexity; Rodriguez 2008), which in itself makes a mockery of the pretense that there are "standardized" boilerplate documents that everyone understands. Many universities still allow their medical schools to handle some subset of the agreements without notification or coordination with their university-wide TTO. In others, the office of university counsel is delegated to handle some classes of incoming MTAs, and often they don't bother to keep statistical records. In the unlikely circumstance that there persist separate offices of sponsored research and technology transfer, then MTAs might be shunted off to one or the other. It is not unusual for many universities not even to require faculty disclosure of or administrative signature on outgoing MTAs. (This last fact alone explains why so many entries for "outgoing MTAs" are missing in Table 4.1.) In the contemporary university, responsibilities for MTAs can often be treated like hot potatoes, passed off to someone else as nimbly as possible; no one unit thus can claim oversight for the institutional aggregate or the Big Picture.

MTAs are the unloved orphans of the research university for the basic reason that they do not, in and of themselves, make money. Statistics of inventions disclosed, patents filed, patents granted, and royalties collected are the sorts of indicators that get TTOs raises and more resources. "The experience of most universities is that very few MTAs result in intellectual property of value (or any IP at all for that matter!)" (Ku and Henderson 2007, 721). Even though technology officers will proffer a lecture on due diligence and liability indemnification at the drop of a hat, the simple truth is that they do not like to discuss MTAs as a *necessary* intermediate input into their touted success indicators of accumulated IP. It is undeniable that their faculty hate MTAs, and some of the most common complaints to the TTO are, Why are you holding up my research, why won't you just sign the damned thing, and why are you strangling my life's work in red tape? For precisely this reason dealing with MTAs is reputedly the lowliest task for any TTO, often consigned to the employee with the least seniority, office clout, and least background in IP. Indeed, in many offices, dealing with MTAs is talked about as a kind of apprenticeship position, a pit stop on the way to bigger and more important assignments. This, of course, effectively ensures that little thought or effort is devoted to monitoring, or indeed to following up, knowledge about the impact of MTAs within the university. In the existing scheme of things, hassles with MTAs can only go wrong, leading to litigation or interminable hard-nosed negotiating tactics; a good MTA is one that, from the viewpoint of the receiving end, has been so automatic that it doesn't matter. No one in existing TTOs has any incentive to seriously consider the emitting end, as we shall see shortly. MTAs are such a nuisance that many TTOs can't even be bothered to file and warehouse them in such a way that their provisions could be readily retrieved. I have had some TTO officers in a reflective mood concede that they may have file cabinets full of ticking time bombs; but the tone of their musings has been that ignorance is bliss. No one has informed me that all their MTAs are neatly archived in an easy-to-retrieve electronic database somewhere—although there is a contemporary movement to supposedly automate MTAs, also considered below.

This conjuncture has fostered an interesting atmosphere in which it is in the interests of TTOs to cast a blind eye toward many of the MTAs that pass through their universities. It also explains why many officers will privately admit in interviews that they see but a fraction of the total MTAs signed or emitted by their faculty. As it is, a TTO office sees on average two or more new MTAs showing up per business day and they are obliged to acknowledge them. A studied neglect helps diminish an unwanted and unwelcome workload but also keeps a lid on a phenomenon that has threatened to balloon all out of proportion into something that has serious long-term consequences for the viability of the commercialization of the university. Confronting our dearth of data, it is therefore important to appreciate that the numbers in Table 4.1 are almost certainly underestimates, with TTOs being confronted with annual compounded growth rates of incoming MTAs of somewhere between 6–15 percent, with no end in sight. Because these figures do not include other comparable contractual instruments (nondisclosure agreements, evaluation agreements, tangible property licenses), the level of legal encumbrance of research tools and inputs is pronouncedly understated by the data.

The understatement and relative neglect are partly due to simple rational allocation of manpower, but there may be another motivation. The real embarrassment of this trend for university TTOs is the concomitant rise in *outgoing* MTAs, and in particular, MTAs going out to other universities. It is one thing to blame the rise in MTAs on rapacious corporations and their crafty legal departments, but it is quite another to acknowledge that the university sector has been doing more and more of this to itself. Bluntly, those hewing to business models force universities to gouge other universities. If this specific trend were more publicized and better known, then much of the rhetoric about the utter blamelessness of the TTO in the perception of hold-ups of the timely pursuit of research would collapse. It very much contradicts the standard image of technology transfer as the university doing what it always did before 1980, with now the TTO merely speeding up and facilitating the "transfer" of findings to an external commercial organization and grabbing some of the benefits. If the university itself becomes more commercial, then it starts acting more like a firm, which means using its IP offensively as well as defensively. In other words, the commodification and control of research tools and research inputs have been necessary and perhaps inevitable consequences of the Bayh-Dole Act and the neoliberal transformation of the university. There is no such thing as exerting fortified control over downstream "outputs" of scientific research from the university without the blowback of seeing that control creep back upstream, especially when the corporate biotech sector (consisting primarily of university start-ups and spin-offs staffed with university faculty and grad students) has been pushing trends in the same direction. The growth of outgoing MTAs in the modern university is thus the dirty little secret of the new century, which is why those entries are so glaringly absent in Table 4.1. It is the IP that dare not speak its own name.

I anticipate that the reader will object that I have not demonstrated that research is actually hampered or obstructed; all I have done is count contracts (when the TTOs would let me). Perhaps the science faculties are really not the villains in this tale, or perhaps, with experience, most of the MTAs have been structured as simple letter agreements or otherwise in such a manner that they really are almost harmless. In either case, counting contracts would not have had the dire implications that I have attempted to draw out here. A more extreme (but unfortunately prevalent) reaction would be that MTAs don't really affect the modern process of research much at all, and if you ask the scientists, they will tell you that their lab protocols have been much more resilient and flexible than I portray them here, so all the warnings about the drawbacks of MTAs are just crying wolf. I shall deal with the first set of objections in the remainder of this section, and the latter set in the final section.

Founding an Empire on a Mouse

Sometimes the attempt to track down the effects of strengthened IP on modern science can feel like getting lost in a maze there are just way too many distractions and you are always being sent to someone else's office for further details. If neoliberals don't manage to change the subject with their visions of sugarplums, then a hubbub concerning tangential related aspects of science tends to hypnotize attention instead. That seemed to be the case in the contretemps recounted here, that of the Harvard Oncomouse. The aspect of this incident that got extensive press attention when events were playing out was that this constituted the first patent granted on an animal rather than on a plant or a bacterium; most of the controversy swirling around the murid metaphysics had to do with moral and procedural objections to the idea of patenting living organisms.[19] The vexed question of the moral legitimacy of the expropriation of life is not one we would wish to dismiss or otherwise denigrate; however, that did frequently divert attention from what I will argue here was an even more significant change in the scientific landscape. This case involved the first U.S. patent to be granted on an animal, but what garnered much less attention at the time was that this turned out to be one of the earliest patents on what can only be classed as a research tool, because there was no other commercial use for mice who had been genetically altered to have a specific predisposition to certain cancer tumors. It therefore stands as one of the very first attempts by a *university* to make serious money off a research tool.

The lowly mouse has long been the standardized organism of choice for much of biomedicine and biological research. Alongside the tobacco mosaic virus, the nematode worm *C. elegans,* and the drosophila fruit fly, the mouse has been and continues to be the synecdoche for the animal kingdom in biological research. Something carefully standardized and reproduced on such an expanded scale for such a wide array of researchers could not operate on a simple "gift economy" basis, and so an arrangement approaching a market for purebred mice had been in operation in the mouse community since 1933. One main producer had been the Jackson Laboratories of Bar Harbor, Maine (known in the trade as JAX), and its story has been nicely told by Karen Rader (2004); indeed, she provides material for an institutional economic history of science of the sort we advocate in this volume. In particular, Rader relates the story as a bit of a tragedy, with her main protagonist, Clarence Little, finding over time that the more he succeeded establishing his inbred mice as an indispensable research tool, the less he and JAX were able to be players in the basic genetics community as "pure" researchers in the 1930s and 1940s. But it would be a travesty to paint the vicissitudes of history as permitting either only factory mouse production or only biological research, as if they were mutually exclusive options. What happened instead was that JAX became built up as a necessary research infrastructure for a large community of researchers around the inbred mouse. Without this infrastructure, American universities would never have been enabled to become major players in many areas of biomedical research. Along the pattern laid out in Chapter 3, it is significant that the nonprofit JAX was funded by very different expedients in the different science regimes: in the erudition regime, it was supported out of the largess of the Hudson Motor Company and the Rockefeller Foundation; during the Cold War, it was the beneficiary of direct government subsidy (particularly but not exclusively by the National Cancer Institute); and from 1980 onward, it was utterly transformed by the privatization of transgenic mice.

Once again, the lesson to be gleaned here is that some subset of research inputs have long been "priced" in a species of market, but that nowhere near constituted a thorough privatization of the research enterprise. Given what will be said shortly, it is necessary to note that there was always a commercial segment to mouse production spanning this entire time period; it is just that the full-scale neoliberal commodification of the research tool in universities only occurred during the modern regime. Rader's history unfortunately leaves out the role of the other major mouse producer in the same time frame, Charles River Laboratories.[20] Charles River Breeding was founded in Boston in 1947 as a for-profit firm that supplied rats and mice raised in a germ-free environment largely to pharmaceutical and chemical firms. The passage of the 1960 Hazardous Substances Act and the expansion of the National Cancer Institute's huge program in chemotherapy drug screening were both good for its business, and in 1968 Charles River was able to successfully go public. In the 1970s it had subsidiaries in Canada, Britain, France, and Italy, and it branched out into supplying disease-free hamsters, guinea pigs, rabbits, and rhesus monkeys. By 1980 the company was solidly profitable, with revenues of $35 million and a net income of $3.85 million; in 1984 Bausch & Lomb acquired it. This demonstrates it was indeed possible to make a modicum of money off some generic research inputs being used in certain spheres of biomedicine, but the university sector was neither a lucrative nor growing market segment. To a great extent, the market was not dependent upon strong IP for commercialization of laboratory mice and other mammals. Thus, the commercial and nonprofit production of customized research tools like mice apparently lived in peaceful coexistence, at least until 1980. It was at this juncture that the "designer mouse" came to dominate murine research tools (Maher 2002). It was soon thereafter that the Harvard Oncomouse thoroughly upended the lab cart.

JAX had been conducted up until the 1980s as a nonprofit service, supplying inbred mice specially customized for cancer research at low cost, as well as performing various other services, such as publishing a "Mouse Newsletter" for the user community (it takes superhuman fortitude not to call them "Mouseketeers"). Rader discusses the circumstances that permitted a JAX mouse to become the "standard" mouse for many medical specialties, which included its policy of not imposing much in the way of IP restrictions or control over the mice. The situation changed irreversibly in 1980 when Jon Gordon and Frank Ruddle devised a technique for introducing fragments of viral DNA into a mouse pronucleus in newly fertilized eggs, and they managed to get the mature mouse to produce offspring with the same plasmid DNA in their somatic cells. To many in the mouse community, it was apparent that this capability would open up new horizons for the ways that mice would be used in the labs of the future, but how it played out turned out to be beyond the scope of parochial imagination. One researcher who perceived an unprecedented opportunity was Philip Leder, who had just moved to Harvard from the NIH in 1981.

Leder and Timothy Stewart, a postdoc, conceived of an experiment to produce a transgenic mouse where the expression of the foreign DNA fragment (here, the *myc* gene) would be pronounced in the female mouse's mammary tissue; the intention was further use as a lab model for breast cancer—hence the designation Oncomouse. Leder's project was funded by both DuPont and the NIH while he was at Harvard, a fact shortly to assume some significance. By 1983, Leder had managed to fuse mammary tumor virus DNA and the mouse genome, and by 1984 he had demonstrated that a proportion of the transgenic mice had developed breast cancer; female offspring with the gene had also developed breast malignancies. The fused gene by itself was not a sufficient cause of breast cancer, but this opened up the possibility of using the Oncomouse either for carcinogen testing, or conversely, on testing therapies seeking to counteract a predisposition toward contracting breast cancer.

In a sponsored funding contract that was arranged prior to the DuPont funding, it had been agreed that Harvard would claim and own any patents arising from Leder's work but that DuPont would be granted an exclusive license on such patents. Since Leder's "invention" happened close on the heels of the *Diamond v. Charkrabarty* Supreme Court case, which deemed organisms to be patentable, Leder believed that he was obligated to report the possibility to the Office of Technology Licensing (then housed in the Harvard Medical School). External counsel subsequently decided that the methods used or the plasmids involved in the process were not themselves sufficient, and so drafted an extremely broad patent claim on the mouse itself and filed it in June 1984 on behalf of Harvard. The patent, granted in April 1988, covered not just the specific mouse line with the transgenic *myc* gene, but *any* "transgenic non-human mammal all of whose germ cells and somatic cells contain a recombinant activated oncogene sequence."[21]

The general claim of property over *all mammals* reeks of a Frankenstein-style hubris (or, more correctly, *Wall Street*-style brazenness); but for the time being, the Thirteenth Amendment still seems to bar slavery and, therefore, ownership of IP in humans. The Harvard TTO was wildly overreaching here; the fact that the USPTO ratified the travesty was just the first of many infractions that caused an uproar both inside and outside the mouse community. Leder had not been the first to implement the transgenic procedure, and he was not the only researcher transferring alien genes into mice. Nonetheless, the fact that Harvard had itself become committed to imposing broad IP on research tools, and not just appeasing DuPont, was revealed in the subsequent filing history. Other countries were nowhere as quick to grant IP on the Oncomouse. The European Patent Office denied the application in July 1989 but Harvard, undaunted, devoted considerable efforts and resources, got the ruling overturned on appeal, and was granted the patent in October 1991. When the Canadian Patent Office reported it would grant a patent on the process, but not on the mouse itself, Harvard was not satisfied and challenged it all the way to the Canadian Supreme Court. Bucking the general trend, that court denied a patent on the Oncomouse (Jasanoff 2005, 211–213). As if all this were not enough newly minted IP, Harvard (in conjunction with DuPont) also managed to impose a proprietary trademark on the mice: hence Oncomouse™.

Given that there was always some commercial element to the distribution and provision of mice, most of the mouse community probably thought things would be run more or less as they had been before through JAX—after all, the Harvard mouse work had been done in part using public NIH money—but they were brutally disabused of such notions after a short hiatus. Because JAX refused to deal in mice with extensive IP encumbrances, DuPont and Harvard opted to run their Oncomouse provision services through Charles River Labs.[22] Initially, Charles River set the price quite high, more than ten times what JAX had been charging. But fairly quickly, DuPont realized something (which constitutes the theme of this murine part of the chapter)—namely, that the point of owning patents is rarely the direct revenue generated from the covered commodity itself but rather what you can strategically do with your IP to control your target clientele and your competitors. Thus, by the mid-1990s, DuPont set out to transform the existing market for biomedical research tools into a new field for privatizing science, and its primary instrumentality to achieve this aim was the insertion of the MTA into the Oncomouse supply chain. Charles River/Bausch, as an outsider to the mouse community, was happy to cooperate.

When the mouse community began to be asked to sign these novel MTAs, they were served up another cup of bitter gall. For instance, the MTAs prohibited the previous practice of breeding and sharing descendant mice from the initial Oncomouse. Doggedly, DuPont sought to make this provision stick even against scientists who had not obtained their mice from Charles River but who had produced their mice through transgenic transfers in their own labs. DuPont sought to impose annual disclosure requirements on the scientists using Oncomice in their own research, claiming a right to monitor (and supposedly intervene) in research projects located far from the original breast cancer research. And most stunningly, DuPont initiated reach-through clauses, attempting to exert IP claims on future downstream research that had made use of the Oncomouse. Even researchers who had developed similar Oncomice in their own labs around the time of Leder but chose not to patent (professing their old-fashioned commitment to "open science") were hit with the same restrictions. Researchers were livid with DuPont (but curiously enough, not with Leder or Harvard) for attempting to "leverage its proprietary position in upstream research tools into a broad veto right over downstream research and product development" (Blaug et al. 2004, 762). Other biotech firms, with little more than promises and research tools to sell, knew a good thing when they saw it, and they soon followed suit; for instance, GenPharm International began to issue MTAs to place similar restrictions upon its novel "knockout mice."[23] In the meantime, Leder never suffered reproach for this outbreak of primitive accumulation: he was awarded the National Medal of Science in 1989 by President George H. W. Bush.

These practices not only hurt other mouse researchers (other than those at Harvard, that is) but also harmed the scientific process in a number of unforeseen ways: for instance, they slowed the emergence of a standardized Oncomouse, in exactly the reverse of the manner JAX had earlier encouraged the convergence upon a standardized lab mouse. Evasion happened if you could somehow keep your head down and lab protocols vague. If you were at a high-profile university where cancer research was an active area, then somehow you had to get your administration to sign a repugnant MTA agreement with DuPont. Here again, we observe it was not only the carrot that got universities to institute something like a TTO if they didn't already have one: it was also the stick, and here the stick looked like an MTA.

The reaction of the mouse community to these depredations was somewhat ambivalent. On the one hand, many of the scientists were themselves actively engaged in pursuit of IP by the mid-1990s, and many more were not afraid of a little competitive elbowing and shoving, so few felt pure enough to stump for a return to some (imaginary) Mertonian nirvana where mice simply teleported between labs of their own accord. But on the other hand, biologists could see the structural effects, such as freezing JAX out of its role as institutional mediator, thereby rendering vector comparisons between labs more difficult. In the short term, many scientists simply elected to flout DuPont's intellectual property and either breed their own mice or violate the MTA in their exchanges. Their rationale (and this is still very relevant today) was that DuPont or Charles River would never actually take a university researcher to court—what could they expect, since treble damages on a pittance is still a pittance? However, it was unseemly that there was all this public NIH cancer money sluicing around in a neighborhood that was starting to get a reputation for "illicit" activity (Jaffe 2004).

The major research universities could have banded together to challenge the patent, because it was dubious from many perspectives, and thus they could have made an early brave stand against the patenting of research tools. However, here is where the neoliberal regime really began to kick in: university TTOs were not inclined to attack Harvard's IP just when they were all scrambling themselves to jump into the licensing lottery. The main resistance to control over the Oncomouse, such as it was, came from Harold Varmus, who had won a Nobel Prize for identifying a mammalian oncogene; as director of the NIH he imposed formal negotiations with DuPont over the license. After four years of lawyer-on-lawyer activity, DuPont and the NIH signed a Memorandum of Understanding in 1999,[24] stating that "public health service" scientists or NIH "recipient institutions" could use Oncomouse patent rights free of charge, provided the research was not intended for any commercial purpose or conducted by a for-profit institution. In the press, the agreement was praised to the skies as a victory for free and open science; in practice, though, it was riven with further ambiguities and loopholes and therefore was hardly any kind of triumph at all.

The first critical gray area was not knowing whether modern research universities were acting like for-profit institutions. The 2002 Supreme Court decision in *Madey v. Duke University* seemed to stipulate they were; and DuPont took the position that any university research sponsored by commercial entities or leading to commercial products should fall outside the memorandum. The University of California rejected that interpretation, but nothing was really settled, at least until there is some relevant litigation on the subject (Blaug et al. 2004). The second and more significant gray area was the fact that the NIH memorandum gave in on the most important point: still permitting DuPont to require an MTA with every Oncomouse used. Citing the memo, DuPont went on the warpath, dunning numerous major research universities breeding and sharing any mice with cancer-prone germ-line transgenic insertions with demands to sign restrictive MTAs (Marshall 2002). At this juncture, JAX finally capitulated and, to remain a player in the mouse research community, found it had to supply mice with all manner of attached IP restrictions, now including the Harvard/DuPont Oncomouse™. To remain relevant, JAX was forced to become just like Charles River, to the extent of agreeing to let Charles River distribute JAX mice under the new more stringent IP regime in August 2001. And as for the NIH, all it really ended up doing was protecting its own parochial interests, and not mounting some general defense of the free provision of research tools to all and sundry, contrary to how it likes to portray itself in its promotional materials.

Why didn't the press (including *Science* and *Nature*) or the National Academies understand this donnybrook for what it really was? First, of course, this struggle had played out gradually over time, so there was no manifest tipping point where one could say, yes, that's where the whole process went off the rails. Various redoubts of resistance were encountered along the way, but it took time for them to learn that resistance was futile. Second, much of the information about events was and still remains proprietary: there is no history, as yet, of what DuPont really thought it was trying to accomplish by imperiously asserting IP rights on the Oncomouse. Harvard, for its own part, has never explained why it didn't simply cancel the exclusive license once the mouse community had roared.

Yet the stronger message should be that if you take your cue from neoliberal apologists and keep your eyes focused merely on the patents, you miss most of the real action in this era of globalized privatization, especially the places where IP has transformed the process of science, which in this instance happened with the MTAs. The Oncomouse was the pretext for the introduction of an entirely new way to structure and control research through contractual obligations and restrictions on research tools. What the NIH has done is effectively endorse and validate this system, even while it was pleading otherwise. True, it would be nice in the future to avoid anything quite so unseemly as one firm like DuPont throwing its weight around in so many different research arenas, but the lesson they have taken from this is that commercial relations should be handled with tact and discretion behind the scenes, in veiled contracts like the MTAs, and not in something so painfully apparent like a single high-profile patent.

Perhaps the most dramatic effect is that TTOs can now feel free to claim that the Oncomouse was just a fleeting learning experience, the transitory teething troubles of the tyros, and now everyone has ratcheted down to more responsible levels of rhetoric, and biomedical scientists have learned how to come to terms with the new commercialized regime. Not surprisingly, the Harvard tech transfer office is especially desirous of making that case:

> We try and say to companies that we follow the NIH guidelines for research tools. So in general, we do not do exclusive licenses for tools anymore. The early licenses were not like this and so Harvard had no choice in the way they set up their sponsored research with DuPont. Now we have changed so that we keep the rights to use any research tools and disseminate them through a non-exclusive [license] and also start to have academic exemptions. If it is exclusive we try to put in language in that our scientists at MGH have the right to continue to do research, but at times there may be push back. Not every TTO can do this, especially smaller institutions. We always try to make sure there are exemptions, due diligence terms to try and avoid cases like the Oncomouse.[25]

Of course, avoiding the Oncomouse contretemps should never be confused with avoiding the intentional commercial encumbrance of research tools. And always raise your guard when someone with a law degree starts mentioning "due diligence."

Any Flavor Other Than Vanilla

As broached above, it is more than a little misleading to discuss "the MTA," since there really is no such coherent thing. Contracts are very flexible instruments, and the provisions vary widely; if anything, they have become more baroque and rococo over the last two decades. As complaints have proliferated about MTAs, a number of commentators and policy makers have taken this to signify that the problem somehow is rooted in this multiplicity, and *not* in any larger structural attempt to impose a particular kind of unprecedented market upon research tools, as argued above. The solution, from that perspective, is simply to impose a relatively harmless plain vanilla MTA on universities, stripped down to a modernist bare minimum, which should ameliorate the worst of the problems, leaving scientists free to choose. One frequently hears TTOs repeat this diagnosis as well. The recent history of MTAs and the unavailing quest to standardize them argue against this interpretation.

The first foray into standardization came from within the AUTM, the professional organization of university tech transfer officers, which has *never* taken the position that universities should not commercialize research tools. A special interest group was constituted to begin discussions with the NIH about MTAs in the early 1990s.[26] Arguing that a streamlined MTA should facilitate research congress between universities and corporations, an internal committee produced something called the Uniform Biological Materials Transfer Agreement, or UBMTA, in 1995.[27] In an attempt to circumvent government regulation, AUTM proposed that nonprofit organizations who were enlisted as signatory to their charter could simply register the plain vanilla agreement with an AUTM clearinghouse by signing a simple implementation letter. The large number of signatories to the charter initially seemed to bode well for the spread of the UBMTA, but there clearly arose a gulf between preaching minimalism and openness and proprietary practice, since it became evident that very few universities actually used it with any frequency.[28] Nonetheless, most university TTOs provide a version of the UMBTA on their Web sites as a sort of starting point for faculty who contemplate entering into some sort of MTA with counterparts at other universities.

Harold Varmus, as director of the NIH, and in part due to the Oncomouse controversy, was not satisfied to leave the issue solely to the auspices of the AUTM. In the spring of 1997, he commissioned a Working Group on Research Tools and signaled he was not assuming the "business as usual" stance of the AUTM by appointing Rebecca Eisenberg as chair. Eisenberg's overview of the results of the inquiry (2001) was the first serious published survey of the nature and problems of MTAs as they impinge upon university science, and it still remains unsurpassed down to the present. She began by pointing out, as we have done, that no one seemed satisfied with current practice and had complained that the progress of research was being threatened; but, if these are smart people, why couldn't they just get together and rectify the situation? Putting this in the language of the neoclassical economists, "If transactions costs are consuming the gains from exchange, why haven't the communities that confront this problem figured out mechanisms for reducing these costs?" (2001, 226). Perhaps the problem with phrasing the issue this way was that it conceded too much to the neoliberal framework at the outset, beginning with Ronald Coase's explanation of institutions through "transactions costs" and extending to implicit acceptance of markets as the predominant means to organize scientific inquiry and resolve conflicts.

One early warning that this was not merely a problem of temporary grit in the gears of efficient markets was the following finding: one "user's research tool might be a provider's end product . . . many of the people that spoke with the Working Group were eager to establish that the term 'research tool' means something other than their own institution's crown jewels . . . When one institution's research tool is another firm's end product, it is difficult to agree upon a universe of materials that should be exchanged on standardized terms" (2001, pp. 228–229). If the parties can't even agree on the commodity definition, then the stable ontology presumed in every single economic model just goes right out the window. But the interviews also revealed a more jaundiced assessment: "Companies complain that universities do not understand business and suffer from a cultural schizophrenia about whether they are businesses or academic institutions" (238). In other words, universities were seen by firms as cynically trying to have it both ways, getting special dispensations from IP when it suited them, but then soaking the commercial sector for all it was worth when it appeared they had the upper hand. Firms bemoaned the tendency of universities to slip all sorts of unwarranted conditions into the MTAs they sent to for-profit entities (245), the diametrical opposite of the contention of universities that corporations' demands were the problem. Outsiders opined that faculty had been given too much autonomy in these matters, acting like they would not have to knuckle under to the commercial imperatives of their own universities; after all, weren't they just employees, no different from the staff scientist in the industrial lab (242)? University administrators, or maybe TTOs, would have to learn to get tough with these prima donnas.

The working group found that in order to appease all the stakeholders, it had to phrase its recommendations in terms of "balance" of the interests of the competing parties. Implicitly, they rejected any notion that the health of the research process itself might require staking out a preserve wherein the needs of researchers would override the claims of any patent holders, or perhaps that a well-defined research exemption was a necessary component of any flourishing system of IP. The NIH in particular had already rejected any whiff of a principle that the needs of patients might preempt those of patent holders, so perhaps it was a foregone conclusion that the same would apply to any nonprofit entities as well. In any event, the NIH responded to the report by issuing a draft of principles and guidelines applied to research tools in May 1999. The "compromise" therein proposed was that research tools (implicated in NIH-funded research) should in general be transferred under plain vanilla MTAs. The guidelines pointed to certain clauses as involving flavors other than vanilla—"commercialization through option rights, royalty reach-through, or product reach-through rights back to the provider are inappropriate"—but there was no actual attempt to prohibit any such clauses across the board. Even this was too much for the biotech firms and their faculty CEOs, however, and the guidelines were subsequently weakened in a revision to the guidelines published in December 1999. In effect, the NIH was driven back to the AUTM position: instead of addressing the structural problems for which MTAs were the symptom, the NIH opted for the legalistic solution of supposedly jawboning all participants to make use of a standardized, plain vanilla MTA. Even then, the AUTM was critical of a further revision of the guidelines released in March 2004.

All the while, universities had been willing and avid participants in the proliferation of clauses and innovation of encumbrances in the spread of MTAs. "We have seen repeated violations of the 1995 UBMTA . . . The primary violators appear to be the technology transfer officers who draft the agreements . . . NIH pressure on universities to disseminate unpatented materials widely simply leads those universities to patent [research tools] more extensively" (Rai 2004, 123–124). Here we observe a vicious circle where TTOs persist in proclaiming their allegiance to "open science" by making research tools readily available, all the while simultaneously extending the reach of IP to encompass a much wider array of research inputs than ever before in the history of modern laboratory research. When confronted with the possible hypocrisy of their actions, they plead that they have been forced into the uncomfortable situation by the activities of their peer institutions.

It is disheartening to observe a parade of doughty reformers in recent years thinking they will somehow succeed where the NIH and the AUTM failed, all taking as their credo the precept that once the contractual instruments are sufficiently standardized, then everyone will play nice, and the obstructions to the free flow of research tools will simply dissolve. It seems no one is more vulnerable to being taken in by the crudest technological determinism than technology transfer specialists. For instance, Science Commons, an offshoot of the Creative Commons project, seems to believe that if plain vanilla won't work, than a limited repertoire of standardized component clauses (Neapolitan?) available in an online database linked to the Public Library of Science will break the logjam. Addgene.org, a nonprofit plasmid bank that has been set up to facilitate the transfer of DNA plasmids between universities and nonprofit labs, attempts to impose a UBMTA on each transfer, except, of course, in the many instances in which the gene is covered by some existing patent or other IP restriction. Another start-up along the same lines, www.materialtransfer.org, was intended to become the Napster of research tools (but seems to have given up the ghost during the time this book was undergoing the editing process), offering peer-to-peer functionality for universities that join the project, allowing a scientist to share research tools (but not animals or human tissues) with a scientist at another cooperating institution, mainly by signing one of a menu of predrafted MTAs, which might differ for each institution.

Each of these proposed "reforms" displays an amazing paucity of analysis in tendering explanations just why their "technological fix" should be expected to ameliorate the primary problem, which is the delay or frustration of research programs due to the lack of access or availability of research materials and tools. Speeding up the signing process for MTAs is no panacea for anything; it is merely a Band-Aid plastered over a pustulating sore. Automation merely addresses the problems of internal monitoring and record-keeping. The fundamental issue left unaddressed is the proliferation of MTA formats, and the progressively expanding tendency to use them to control the research of others. I have been told by Joyce Brinton (formerly of Harvard) that back in the 1990s everyone knew that universities would never successfully standardize MTAs once they took hold in academic settings; at best, documents like the UBMTA would be considered focal points to jump-start the process of negotiation over research collaborations and IP agreements. This is indeed exactly what has happened; the core reason for the increasing resort to contracts is to transcend the one-size-fits-all character of patents through legal customization in the new regime of globalized privatization. The repeated crusades to re-impose standardization, here of MTAs, merely have the effect of distracting attention from the real motives behind the commercialization of scientific research.

Who Says Science Has (So Far) Emerged Unscathed from Its Love Affair with Intellectual Property?

Thus far we have tracked the trajectory of MTAs through the groves of academe and identified the ways in which they have grown from modest indemnifications against harm into veritable Frankensteins of stitched-together chunks of IP control. Given the stark trends over the last two-plus decades, one would have expected that the academic scientists enduring the epidemic of MTAs would have risen up in revolt by now, storming the gates of their technology transfer offices, writing impassioned screeds to university administrations and newspapers and scientific journals, and banding together with their colleagues to rescue research from the plague of prohibitions and fetters. By contrast, some have suggested since the Bastille still stands intact, and that labs today are not seething cauldrons of rebellion, that all the brouhaha about MTAs that flares up from time to time is just a tempest in a teapot. Scientists are resilient and intelligent creatures, or so say the proponents of the New Order, and they have found ways to get around any impediments that temporarily may seem to have been erected by the spread of MTAs. Science can get along quite fine in the new regime of globalized privatization, since markets are the most exquisite form of information processors known to man.

Neither of these descriptions begins to get close to what has been actually happening on the ground of late in the sphere of scientific research; therefore, the purpose of this final section is to try and understand why an eerie calm seems to hang like a pall over the research landscape with regard to these issues.

To set the record straight, there have indeed been numerous public complaints registered about the proliferation of MTAs by scientists. There are lots of them to be found in earlier generalist journals, even before the Oncomouse controversy; they continue down to the present, but predominantly they are lodged in the biomedical literature.[29] Sometimes the complaints do show up in newspapers but usually only when someone actually gets taken to court, and then there is some quirky aspect about the dispute, like the incident when the American Type Culture Collection sued an art professor at SUNY-Buffalo for violating an MTA when he incorporated some *Bacillus atrophaeus* into his artwork.[30] In fact, university litigation over MTAs has itself been expanding appreciably over time, but beyond the odd controversy erupting out into the open, TTOs have proven unwilling to provide any information about the phenomenon, partly because most suits are settled out of court and under seal of confidentiality, but also because it belies the tenor of their public rhetoric about their disinclination to interfere in the research of external parties. The net effect has been to subdue public expressions of discontent over time.

Nevertheless, scientists have still spoken out, *but only when they were willing to entertain the notion that modern trends toward commercialization of research had deleterious fallout.* In other words, scientists only protest when they reject the neoliberal dogma that pervades our culture. Many such accounts exist in the interstices of the massive policy literature on science policy, such as the following excerpt:

> In my own laboratory, for example, we wanted to purchase some new gene-chip technology from a faculty colleague who had a small biotech company situated at another university. We were offered a reduced price for the materials if we would agree to give authorship to and share oversight of publication with the provider of the technology on any research that we published using his chip. Although it sounds simple, this agreement would have required us to share control of any IP generated by our research with someone whose only relation to the work was the sale of a piece of technical equipment. In another recent case, we hoped to obtain tissue samples from a colleague at another school to use in our research on stem cells . . . The MTA we were asked to sign to have access to the tissue required us to refrain from comparing the donated cells for safety and efficacy in case we found that the donated tissue was less effective or potentially harmful to the subjects in which they were placed. If strong financial interests had not been at stake, one would guess that the provider of the unique cells would have been delighted to know whether or not the tissue was useful. (Stein 2004, 8)

So criticism and protest emanating from practicing scientists have not been altogether scarce; rather, what has been noticeable is that circa 2005 any criticism had more or less vanished from the high-profile generalist journals like *Science* and *Nature*. There they were displaced by various studies, some commissioned by august bodies like the AAAS and the National Academy of Sciences, all conducted either by economists or by quantitative sociologists associated with business schools, which have generally reported that there is nothing whatsoever to worry about. This pattern has become so entrenched in recent years that a "review essay" on this topic has had the temerity to conclude that "the surge in university patents did not happen at the expense of their quality, nor of the quality of the research. Moreover, scientific excellence and technology transfer activities mutually reinforce" (Baldini 2006, 197).

What appears to have happened is that both generalist science journals and more unabashed TTO cheerleaders like the *Journal of Technology Transfer* have simply ceased reporting on complaints of natural scientists about the obstruction of research through IP, instead adopting the position that *social science* researchers have empirically demonstrated conclusively that there have been no deleterious effects on science of MTAs and patents. There is something noteworthy going on here, since otherwise august natural scientists rarely have the time of day for social science wisdom. The press then followed suit, reporting that "an area of particular concern for academic researchers—having access to technologies used to conduct research—may not be as constrained by intellectual property rights as many have feared, according to Stephen A. Hansen, director of the Science and Intellectual Property in the Public Interest project at the AAAS" (Blumenstyk 2007a). As another of these researchers told a European meeting of scholars concerned with the impact of IP on research tools, "Following *Madey v. Duke,* although researchers are now given instructions to improve patent awareness more frequently, no change in behavior has been perceived. The fact that research inputs are patented impedes research very little" (OECD 2006). One high-profile *Science* article declared, in a bit of a non sequitur, "Our results offer little empirical basis for claims that restricted access to IP is currently impeding biomedical research, but there is evidence that access to material research inputs is restricted more often" (Walsh, Cho, and Cohen 2005a, 2003). Elsewhere, those authors make an even stronger claim: "The results above suggest that patents rarely interfere with research, and even materials transfers are processed without incident" (Walsh, Cho, and Cohen 2005b). Another article in *Science* opined with even less actual documentation, "There is evidence that university licensing facilitates technology transfer with minimal effects on the research environment" (Thursby and Thursby 2003, 1052). Another similar review suggests, "There are concerns that the growth of patenting of foundational upstream inventions in cumulative technologies may constrain important follow-on research . . . The empirical basis for such concerns is, however, far from strong" (W. Cohen 2005, 68).

There are two things that every interested layperson needs to know about this literature. The first is that it is all uniformly predicated upon sample surveys, since these economists/sociologists have enjoyed no better access to hard quantitative data on MTAs incoming and outgoing at universities, their provisions, the nature and extent of consequent litigation, intervention in the research process by extramural players, and the direct impact on the particular trajectories of real individual researchers than I have been able to report in this chapter. Global data on MTAs are still scarce and often proprietary. Records of IP litigation between universities or faculty are often sealed. Given that the social science is all based on surveys and questionnaires conducted upon samples of scientists and/or TTOs, it turns out to be more than a little important to dig somewhat deeper into the circumstances of their conduct, as well as the exact character of the questions involved, in order to arrive at some preliminary evaluation of these broad sweeping absolutions pronounced over modern commercial science. The evidence, it seems, will not turn out as unequivocal as all these *ex cathedra* pronouncements suggest.

The second thing interested observers need to know is that when similar surveys are conducted, not by economists, but rather by biomedical professionals (and published in specialist biomedical journals), then the verdict is dramatically reversed: "Data withholding occurs in academic genetics and it affects essential scientific activities . . . almost half of all geneticists who had made a request of another academic for information, data, and materials had had that request denied . . . more than one-third of geneticists also believe that data withholding is becoming more common in their field" (Campbell et al. 2002, 473,478). "Nearly three-quarters of all respondents said they had at least one negotiation break down without agreement in the past year . . . All respondents reported using MTAs and NDAs for protection of nonpatented proprietary information and materials . . . Ten companies and seventeen universities said they used MTAs with both universities and companies" (Henry et al. 2003, 446). "Our findings suggest that a small minority of trainees (8%) reported denying other academic scientists' requests for information, data, or materials. At the same time, the quantitative impact of withholding was substantially larger, with approximately a quarter having requested but been denied direct access to either unpublished (21%) or published (23%) information, data, materials or programming while in training . . . The life sciences as a field, more so than chemical engineering or computer science, will likely have to wrestle with this issue among the next generation of scientists" (Vogeli et al. 2006, 133, 136). In one survey in Belgium, 60 percent of researchers reported abandoning projects because of IP on research tools (Rodriguez, Janssens, Debackere, and DeMoor 2007).

What could possibly account for the incongruous spectacle of economists teaming up with the AAAS and the National Academy of Science to give such sanguine accounts of the impact of MTAs and IP on science, especially as juxtaposed with the less visible parade of targeted exercises by biomedical researchers reaching diametrically opposite conclusions on a recurrent basis? Here, things do tend to get sticky. At the outset, it is hard to ignore that the particular researchers commissioned by the elite academies have themselves displayed long track records of propounding and defending the commercialization of scientific research and the privatization of the university more generally: Marie Thursby, Wesley Cohen, Richard Jensen, Robert Cook-Deegan, John Walsh, and Jason Owen-Smith. Indeed, the economists Thursby, Jensen, and Cohen have admitted admiring neoliberal doctrines, and the lapsed biologist Cook-Deegan writes for the neoliberal journal *New Atlantis*.[31] They are frequently funded by the openly neoliberal Ewing Marion Kauffman Foundation, an avowed cheerleader for the commercialization of science.[32] The extant universe of possible candidate economists and sociologists working on the topic surely has to be larger and far more diverse than this, but you would never know it from *Science*. Hence one question that presents itself is why the national academies, the supposed vanguard defenders of science, have proven so partial to tapping this group disinclined to detect distress in the scientific ranks for its self-scrutiny. But more to the point, when one takes the time to examine the nature of the data behind their prognostications, it appears that the survey evidence proffered by this group of economists has not really been all that different from that provided by the biomedical researchers: except, of course, in its interpretation.

How to Make the IP Impact on Science Disappear

For instance, let us begin with one of the first surveys commissioned by the National Research Council (Walsh, Arora, and Cohen 2003a), which concluded, "the anticommons has not been especially problematic . . . Almost none of our respondents reported commercially or scientifically promising projects being stopped because of access to IP rights or research tools" (331). But upon reading further, we find these bold generalizations were based upon a sample of seventy individuals, most of whom were "IP attorneys, business managers, and scientists from 10 pharmaceutical and 15 biotech firms . . . tech transfer officers from 6 universities, patent lawyers, and government and trade association personnel" (292). Only ten individuals, or one-seventh of the sample, could be classified as university scientists. Given the composition of the sample, the bulk of the respondents were simply not predisposed to detect any pathologies in the research process, and in any event, they would generally not have observed them, for reasons already broached in this chapter. And yet, this was published in a National Academies volume on *Patents in the Knowledge-Based Economy* (Cohen and Merrill 2003).

Cohen and John Walsh and Charlene Cho subsequently attempted to improve on their impressionistic survey with a much larger scale inquiry funded by the National Academy of Sciences (Walsh et al. 2005a, 2005b, 2005c). In this version, there was an attempt to compose a sample with one-third industry scientists and two-thirds academics; ultimately they evoked a 37 percent response rate with 414 responses from academics, which they characterized as "a modest response rate, [where] caution is warranted in making any claims of generality" (2005c, 9). Some of this might be attributed to the long questionnaire they mailed out, consisting of 159 items. Further, it appears that drug discovery was heavily represented in the sample, with 10 percent of the academics and 67 percent of the industry scientists self-reporting themselves engaged in that activity. Half the academic sample had self-reported ties to commercial firms of one sort or another, and 40 percent had applied for a patent at some point in their career. Interestingly, 8 percent of the academics already had a start-up firm based on their research, while another 11 percent were preparing a business plan as a prelude to starting a firm. Here we observe that the bulk of the respondents were well and truly integrated into the system of commercial science, even though they were classified as academics.

In asking their key questions on research inputs, Cohen and colleagues gathered only 381 answers from academics (33 refusing to respond); of these only 8 percent (=32) admitted to requiring knowledge (NB—*not "materials"*) from some other researcher covered by a patent. This extremely low number, trumpeted in Cohen's other publications (Walsh et al. 2005a; Cohen and Walsh 2007), becomes a little more comprehensible when one notes that only 5 percent of their sample admitted checking for patents on the research inputs that they requested, even though about a quarter also admit to being informed by their local TTO that they were obligated to do so (2005c, 47). It may also have had something to do with the strange way the question was asked—most people before the stunning Supreme Court ruling in *LabCorp v. Metabolite*[33] in 2006 believed that "knowledge" or "facts of nature" were not patentable, so the question may have struck many as meaningless when the survey was actually conducted. Nevertheless, treating those prudent thirty-two souls as a basis for analysis, twenty-four of them reported contacting the IP owner for permission, while one changed his project to avoid infringing on the IP. Quoting the authors, "For obvious reasons, the number of those reporting that they proceeded without contacting the IP owner may underestimate the true figure . . . of 381 academic scientists . . . none were stopped by the existence of third party patents. Even modifications or delays are rare, each affecting around 1% of our sample. Even relative to the small number of respondents (32) who were aware of a patent related to their research, the figures are modest, with 13% modifying their project, 16% having a delay of more than one month, and none stopping a project" (2005c, 17).

This is the source of Cohen's oft-repeated message that "only 1% of academic researchers report having to delay a project, and none abandoned a project due to others' patents, suggesting that neither anti-commons nor restrictions on access were seriously limiting academic research" (Cohen and

Walsh 2007, 12). Far from demonstrating any such thing, it only reiterates once more the salience of the maxim that how you pose the question in a survey setting makes all the difference in what you receive as an answer. Elsewhere in their same report, we discover that Cohen's test questionnaire made a strong distinction between patents on "knowledge" and requests for research tools denied for all other reasons. Somewhere around 20 percent of all these "other requests" were denied in their sample, which, when standardized for rough classes of genomics researchers (2005c, 20), is actually *higher* than that reported a few years earlier in the medical literature (Campbell et al. 2002). Here, "over a two year period, 35% of academic consumers have had a project delayed by more than one month as a result of failing to receive a requested research input from academic suppliers. Fourteen per cent of academic consumers have had to abandon at least one project over the last two years" (2005c, 22). Most germane to our present concerns, 42 percent of these "other" requests required signing an MTA, and 26 percent of those MTAs required negotiation of more than a month in duration; 12 percent ended in failure, being denied. In a number that particularly jumped off the page, 38 percent of the MTAs signed included reach-through clauses, and 30 percent imposed publication restrictions on academics. This track record begins to sound much more like the world of pain and frustration we have been concerned to describe in this chapter; and yet, Cohen et al. still maintain, "Are the requests for MTAs interfering with the transfer of materials? This does not appear to be the case" (2005c, 24). Cohen et al. also run a number of regression analyses on their data to make further causal correlation statements, but given the small number of degrees of freedom involved, we shall simply ignore them.

The other major mid-decade survey by the AAAS is a little harder to dissect and analyze but turns out to have similar small-numbers problems when it comes to those issues of the dissemination of research tools. In just one example of the challenges of interpretation, the authors admit in a footnote to including TTO officers in their sample, where they are classified as "academics." Out of a much larger initial sample of 1,111 "scientists" distributed across the gamut of research fields (by contrast to the Cohen sample, which was confined to genomics and biomedicine), only roughly 200 respondents admitted acquiring a patented research tool from an external lab (Hansen, Brewster, Ascher, and Kisielewski 2006, 16). This is a much higher proportion than the (roughly) corresponding proportion reported for the Cohen sample (ibid., 21fn14). It seems that the AAAS survey did not explore as-yet unpatented research inputs in its questionnaire, and therefore (by design?) would miss out on a much larger universe of MTA-encumbered exchange. Nevertheless, of those here admitting acquiring a *patented* research tool,

35 percent (=67) arranged the transfer through an MTA. Of all those who opted to respond to the question as to whether their research had been adversely effected by seeking to obtain patented tools, 40 percent (=72) reported that it had. Among the difficulties reported by these seventy-two, 58 percent said their research had been delayed, 50 percent said they had to change topics, and 28 percent said they had to abandon the line of research altogether (ibid., 22). And, despite the fact these proportions are not unsubstantial, *if you normalize them to the "right" base,* the authors chose to declare in their upbeat executive summary that "it appears that academia has been less affected than industry by more restrictive licensing practices in the acquisition and distribution of research tools" (2006 p. 8).

The bottom line is that these social scientists, if and when they venture to ask the "right" questions, find proportions of scientists inconvenienced or worse by the whole panoply of IP surrounding research tools requested from external sources that are in the same ballpark, or even of larger magnitude than those reported by the biomedical scientists cited above. And yet, to a fault, they consistently gloss their findings as revealing that IP has had no deleterious consequences on science. It takes great fortitude not to accuse them (and possibly their sponsors) of strong pro-commercialization bias. These social scientists come up with prognostications that are diametrically opposed to the biomedical scientists who also conduct surveys. Their own colleagues in the business schools who have approached the issue from the vantage point of the biotech firm have suggested that commercialized research tools have hampered the scientific success of that sector (Pisano 2006b). Using the Cohen survey as a stalking horse, we can briefly lay out the tricks of the trade that permit the spinning of dross into gold.

First and foremost, both the Cohen and AAAS studies insist on a sharp empirical separation between "patented research inputs" and the whole range of contractual quasi IP that gets attached to research tools. This reprises the same sharp separation that TTOs were insisting upon at the outset of this chapter; and for most purposes, it is fallacious. Once one comes to understand the recent rise of the MTA in academic contexts, then it becomes possible to appreciate the extent to which imposition of contractual obligations on those with whom you share your valuable research know-how and materials is part and parcel of the new commercialized science regime. Contracts are the weapon of choice to frustrate or encourage others doing research in your area. To restrict any questions of harm to the narrow rubric of formally patented research tools when attempting to gauge the level of disruption wrought by IP is like an insurance company restricting its estimates of covered catastrophes after Hurricane Katrina solely to demonstrable instances of wind damage.[34] Yet that is primarily how Wesley Cohen arrives at his "1 percent" solution, as well as the bulk of those who attempt to assert that in the case of patents on the human genome, "neither anticommons nor restrictions on access were seriously limiting academic research" (Caulfield et al. 2006, 1092). The more correct approach would be to combine all common legal means of discouraging research into a single portmanteau category in order to ask: what constitutes the totality of effects of fortified IP on scientific research? Indeed, that is generally the way the biomedical scientists pose the question (under their rubric of "withholding") in their sample surveys. If we normalize to this numerator, and make the denominator the total universe of working bench scientists (but not tech transfer officers!) who make external requests for some kind of research input, then everyone agrees that somewhere between a quarter and a half of everyone encounters obstacles, from delay to outright denial.

Cohen's next line of defense is to assert that, even if the barometer of harm were defined in this fashion, then blame could no longer be laid at the doorstep of the commercialization of science, because (he claims) scientists have always withheld findings, data, tools, and expertise from other scientists. In a marketplace of ideas, it is merely a symptom of the level of competition in cutting-edge science. In this regard he cites the recent work of Mario Biagioli (2006a) on Galileo's various stratagems for turning his telescopes and maps into special instruments of credibility and upward mobility. Interestingly enough, he neglects the entire large literature in Social Studies of Knowledge (SSK) on the difficulties of replication of findings in the natural sciences, as well as the massive literature on controversy studies, since Cohen would never want to actually entertain the SSK thesis that the network of allies and opponents might affect the *content* of the knowledge arrived at, or the ways in which knowledge was stabilized. Indeed, he misconstrues the point of the SSK literature as purportedly demonstrating the economistic thesis that competition is endemic to the history of the natural sciences and that it has been a good thing for the health of science. No doubt there are and have been all manner of exclusionary measures deployed in the history of the sciences by researchers against their perceived rivals; but by focusing upon MTAs in a much more concerted fashion, we can begin to make some headway in inquiring precisely how the situation has changed (and perhaps worsened) since 1980. While it is unfortunate that we do not possess some solid benchmark measures of the extent of the withholding of research tools in the Cold War era (in part because security clearances performed much of that function, as noted in Chapter 3), even the few surveys cited above show the rate of withholding and obstruction rising over the period in which MTAs spread to the academic sphere. The relevant question is what has happened to contemporary structural changes in how scientists interact in treating the communal aspects of research tools, not some atemporal notion of a generic competition throughout the annals of science.

Cohen's third strategy is to argue that his data show that scientists have tended to ignore the patents on their research tools and that this can explain why, as he asserts, patents have had little deleterious effect upon the conduct of scientific research. It certainly seems to be the case that many scientists even at this late hour appear to be ignorant of the fact that the Supreme Court has struck down any notion of a "research exemption" to patents for universities (*Madey v. Duke University*) and that other courts have broadly extended the definition of what can be patented, but Cohen goes further in claiming that their studied indifference is a rational calculation. In a nutshell, treating scientists as neoclassical agents after the fashion of the public choice "new economics of science," Cohen imagines scientists could rationally calculate the odds of getting caught making use of someone else's patented organism or device and think the risk is small enough to justify their "rational infringement" of IP. From the converse side, Cohen also imagines patent owners rationally calculating that it is in their economic interest to allow a certain level of research infringement on their patents, since if they were successful, that would just make the IP all the more valuable. Furthermore, they might think a punitive lawsuit against academics would rarely itself generate much revenue. In Cohen's world, Ronald Coase's neoliberal thesis that property rights will simply be reallocated to their most efficient uses (in the absence of transactions costs) should be equally applied to the scientific estate. All the actors have done their homework on their cost/benefit calculations, and they have adjusted their behavior through rational infringement and rational forbearance to arrive at a situation where strong patents on research tools have putatively had almost negligible effect on the actual prosecution of research. In other words, the modern commercial regime "works" (or so he claims) because everyone knows just how far they can venture over the line of legality and get away with it.

This latter defense of IP in science might seem ironic, if it were not so very characteristic of the neoliberal approach to science (Cukier 2006; Nedeva and Boden 2006). In the neoliberal worldview, the allocation and rationing of information by the market cannot conceivably be surpassed by any other rival allocation mechanism; therefore anything that superficially appears as hitches or other hang-ups in the transfer of research tools must be redescribed as necessary and efficient outcomes, once one comes to understand the preferences of the agents and the constraints they encounter. The beauty of this analytical approach is that nothing short of absolute catastrophe would ever refute the neoliberal worldview.

Back to the Future

The ultimate message of this chapter is that there seems to be a growing problem with the transfer of research inputs in the sectors of modern science that have experienced the most advanced levels of commercialization, but numerous adventitious trends and currents have served to either obscure the phenomenon or else have explained it away. The reader may be tempted to interpret this chapter as dealing in conspiracy theories, but that would be an unfortunate mistake. Rather, I have tried to argue that select historical events, in conjunction with some larger intellectual trends, have come together in such a way that it has become nearly impossible to really understand what has been going on with MTAs in the last two decades in university settings. Some recent "empirical" work on the topic has only muddied what was an already murky situation.

Because the legal ins and outs of IP cause even the most perspicacious of intellectuals to zone out, it may be necessary to provide a Cliffs Notes condensation of the narrative of this chapter so far. The commercialization of university science took an inordinate amount of work to construct the "knowledge commodity" that would supposedly provide the structure for the neoliberal marketplace of ideas. The construction is still ongoing, but it is not so clear that it is working out all that well. Furthermore, it was not simply a matter of finding all sorts of prior commercialization opportunities simply as low-hanging fruit begging to be harvested by entrepreneurial universities. Indeed, in a fact not often admitted by TTOs but current among commentators on the modern university, once you take the full costs of TTOs into account, *very few universities make any money whatsoever, much less serious revenue, from their management of their IP assets.*[35] This curious situation forces us to contemplate the notion that the modern insistence upon the commercialization of science is more likely to have ideological, rather than simple behavioral, motivations. This is reinforced by the unwillingness to seriously confront the deleterious unanticipated side effects of the construction project.

Universities jumped on the patent bandwagon in part because patents were becoming easier to get in general, and the rights of holders had just been inordinately strengthened.[36] No university was getting rich by grasping at that brass ring. Yet by the same token, the patent system was simultaneously becoming more dysfunctional, and the taint of IP rot was transmitted to the university even as the AUTM and other promoters of the entrepreneurial reengineering of science were trumpeting the moral duty of universities to help society through ownership of their own research output. A major symptom of this corruption was made manifest through the MTA. Faculty regarded them as an irritant, and TTOs were not much more welcoming. However, as a subset of scientists were increasingly drawn into the commercial sphere, they saw it might be in their interest to themselves attach MTAs to research inputs requested by other academics; and thus began a tidal wave of MTAs that shows no sign of abating. Furthermore, all attempts to render MTAs relatively harmless by restricting their conditions to a few innocuous clauses have failed repeatedly, as many IP-related restrictions have been loaded into individual MTAs. One of the more important codicils dictates that the existence and content of MTAs themselves be treated as secret and proprietary.

The reactions to this deteriorating situation have been curious. First, the scientific community, and in particular the academic biomedical community, has stratified into a commercialized elite who essentially see nothing wrong with encumbrance of the research programs of those they consider interlopers, mainly because they already have all the access to research tools that they desire, and conversely, the bulk of less well-apportioned scientists, who have been painfully learning to adjust to a research environment studded with IP booby traps. Likewise, the TTO community has largely become inured to MTAs but relegates their negotiation to low-status employees and do not monitor them very closely. Hence research into the extent and implications of MTAs in universities has become blocked. Responding to the grumbling of the proletariat, as well as some studies funded through biomedical foundations, organizations supposedly charged with monitoring and lobbying for the health of science such as the National Academy of Sciences and the AAAS have commissioned studies of the phenomenon, but for reasons that still require further illumination,[37] they have sought out social scientists of neoliberal inclinations to conduct these surveys. Almost uniformly, the surveys conducted by these same social scientists conclude that patents on research tools and ideas have had little deleterious effect on the research process. Furthermore, if and when they consider MTAs, they invent special dispensations to explain away any harmful effect the MTAs might have had. Yet, careful review of their surveys reveals the levels of discontent they uncover among the relevant academic communities do not differ so very dramatically from the levels found in surveys by biomedical professionals and by analysts of the biotech firm sector, both of whom have pretty uniformly raised warning flags concerning the frustration of research traced to the commercialization of research inputs.

It is a very odd situation, and one that would seem to demand much more analysis than it has garnered to date. One potential way to approach it might be to compare and contrast it with the parallel transformation over the same time frame when copyright laws were fortified, and then downloading of music was criminalized, and old-line content companies fretted about how to offset the threats posed by the Napsters and Kazaas of the world. Only in this particular situation, the relevant analogue of old-line content providers are the universities, themselves undergoing wrenching reengineering to become more like corporations. The problem for them is how to get their scientists to act more like responsible employees, which means in this case learning to treat research tools more like the commodities they will have to become if universities are to engross the future revenue streams of knowledge production on something other than an irregular basis. Because it would be far too unwieldy and time-consuming to patent every single research tool, MTAs have proven the instrument of choice to control the commercial implications of cutting-edge research. This is why they will not vanish anytime soon.

You Can't Take TRIPs Home Again

We have spent an inordinate amount of time and attention on American legislation, case law, and legal peccadilloes, so it may appear to the reader that IP is just an American obsession and American malady, but when it comes to the economics of science, nothing is further from the truth. Exploration of the role of changing structures of IP in European, Japanese, and Chinese contexts would indeed be important chapters in a volume complementary to this one, but in lieu of that, it is imperative to focus on what is concertedly "global" about the regime of globalized privatization in science.

It may seem a little simpleminded to pose the question, "How does science spread from its country of origin to the rest of the world?" Science aims to be universal, so it should just spread naturally in what might be dubbed the molasses theory of knowledge. But that has little to do with the actual history of science, and even less to do with the economics of science. Ignoring the fascinating literature on how science was disseminated from the metropole to the colonies in the nineteenth and twentieth centuries, there are three answers to that question in the twenty-first century. The first answer is through training of foreigners in the metropole and then their return to their home countries. In a topic we have bypassed in this volume, many of the top natural science departments in the United States have been training increasing proportions of foreigners in total enrollments over the last few decades, thus in essence nurturing potential rivals.[38] The second answer is that it is spread by scientists themselves proselytizing abroad. This may happen through international conferences, visiting stints at foreign universities (encouraged by such programs as the Fulbright fellowships), or through an increasingly important channel, the establishment of dedicated science NGOs (Drori et al. 2003). The rise of science NGOs has yet to find its historian, as explained in Chapter 3, because they are truly a novel institutional player in the history of science, not formally reducible to the corporations, universities, and governments who have been the primary protagonists in the past.[39] The third answer, and the one that will occupy the bulk of our attention in the remainder of this chapter, is the spread of Western science as part and parcel of the spread of international trade and development.[40] Partly this has occurred due to the intentional programs of some key external agencies like the World Trade Organization, the World International Property Organization, and the World Bank; also in part it has become a mission of a few entrepreneurial universities seeking to emulate the operation of transnational corporations. A moment's contemplation will confirm that much of the historical drama of outsourcing of manufacturing, transnational flows of scientists, and the nurture and defense of the science base would never have happened without the prior spread of the science itself across national boundaries, both as a body of thought and as a complement of standard research and pedagogical practices.

One key epiphany of neoliberal economists in the post-1980 period has been that science cannot be construed as a marketplace of ideas unless some common market framework has been erected and maintained in the transnational context. This lesson may not have seemed quite so urgent in the Cold War era, but it certainly rose in salience once the outsourcing of manufacturing employment had gained momentum, as described in previous chapters. Once whole areas of the United States and Europe were devastated by the offshoring of manufacturing, there was bruited about the notion that the home country could nonetheless sustain its living standard by depending on its comparative advantages in finance, the service sector, and some IP-related industries like pharmaceuticals, software, entertainment, and other so-called high-tech products. This is the fiction of the "knowledge economy" discussed in previous chapters. Yet these Xanadus would remain nothing but pipe dreams unless the rest of the world, and in particular poorer countries, could be cajoled into acceptance of control of these sectors by the metropole and that would mean (among other things) respecting the level of IP protection in force in the metropole. Few if any countries would have seriously deemed that in their interest: Indeed, the economic history of previously industrialized countries is the history of shameless copying and stealing of technologies, knowledge, and experts as a way of evading the attempted lock on trade and development by the advanced nations. What makes the regime of globalized privatization dramatically different from everything that went before is the invention of effective means to force the rest of the world to accept the international trade and circulation of science and knowledge on the terms dictated by a very few developed countries—primarily by the United States, but with a little help from the European Union and Japan. Hence, when we say science is now "global," it refers to something completely different from the great bulk of experience prior to 1980.

This shift from the previous patchwork restraint of IP and knowledge by international conventions and bilateral treaties to the truly globalized standardization of control through the international trade system happened as the culmination of a conscious political intervention during the transformation from the General Agreement on Tariffs and Trade (GATT) to the World Trade Organization in the 1980s. It was accomplished mostly outside of the glare of public scrutiny (and absent participation by the university sector) by a small cabal of neoliberal economists and corporate representatives, and it resulted in the most significant linchpin of the contemporary international system of knowledge regulation, trade-related aspects of intellectual property rights (called TRIPs for short).[41] The effects of TRIPs on the organization of science have been nothing short of earthshaking, and the tremors continue to be felt down to the present. For instance, it was no accident that the bloodbath of in-house corporate R&D in America described in Chapter 3 commenced at just the point when TRIPs had become a reality early in the 1990s.

TRIPs was concluded as part of the text of the Final Act of Uruguay Round negotiations of the GATT (concluded on December 15, 1993) and came into operation on January 1, 1995. TRIPs constitutes a single unified framework for all members of the WTO to define and enforce intellectual property rights, a function previously imperfectly served in part by WIPO, forcing many nations to adopt multilateral agreements in an area they had previously avoided. In one fell swoop, IP agreements between roughly twenty to fifty states were extended to encompass 140 states. Furthermore, for the first time in history, international agreements on IP were fortified with real teeth, by subjecting them to the governance mechanism instituted to police international trade violations. In particular, Section 301 is the section of the U.S. Trade Act that is used by the U.S. trade representative to address unfair foreign trading practices, including unfair practices on IP rights. A 301 investigation may culminate in a bilateral agreement between the United States and the target state or, failing that, the imposition of trade sanctions by the United States on the target state. What at first might seem a highly technical and arcane set of legal provisions can now readily wreak havoc with the livelihood of large numbers of citizens of any country seeking to evade or otherwise get around TRIPs. It has proven an unprecedented sea change in the way knowledge would be commoditized.

TRIPs is unusually comprehensive and specific in casting its net widely to cover IP: everything from patents to copyright to trademarks to computer code to "appellations of origin" for wine, industrial designs, and trade secrets. While these may have some minor bearing on science, it is TRIPs Articles 27 through 34 on patents that have been most significant. Therein knowledge is treated as something discrete and individualized: There is no recognition of possible communal aspects of knowledge. This move was necessary because, as Sell (2003, 6) observes, "The framing of IP as being 'pro free trade' would not have been persuasive during earlier eras in which IP protection was seen, at best, as a necessary evil and at odds with free trade." These articles propagated the extended term of patent coverage to twenty years for all participants and dictated some definitions of patentable subject matters, such as genes, organisms, plant varieties, drugs (a major coup for Big Pharma),[42] and certain classes of biotechnological materials. Certain technical clauses in Article 31 essentially prohibit most compulsory licensing of innovations under public interest rubrics. Article 34 shifts the burden of proof in infringement cases from the owner of the patent to the defendant. Articles 51–60 render importation of any "pirated" generic pharmaceuticals illegal, even if a health emergency had been declared. Article 67 enjoins member states to develop the capacity to enforce TRIPs compliance, which in practice means the spread of the sorts of patent office practices prevalent in the United States and Europe. At minimum this implied possession of a rudimentary science base conforming to U.S. norms and practices. In other words, soon after the great fortification of IP in the United States, a similar set of cutting-edge practices were imposed on the entire world.

It has struck more than one observer as astounding that the world's poorer nations would voluntarily agree to such a draconian IP regime; I simply point the reader toward the relevant histories to seek to explain how this all came about. Nevertheless, it transpired that "twelve corporations made public law for the world,"[43] alongside some neoliberal economists and lawyers who spearheaded the effort. There was almost no one at Punta del Este to supply countervailing analysis or proposals. There were most certainly no scientists or university representatives keeping track of the consequences for world science: The venue was more than a little isolated, and in any event, the issues back then were not posed as having anything to do with the globalization of science. Another curio is that the Chinese were not yet on board with the agreement, as they had not yet joined the WTO. (This explains much of the subsequent opprobrium directed toward China's IP policies once it did join in 2001, as well as linking its disproportionate weight within the WTO to its ability to maintain an ambiguous stance toward Western notions of IP thereafter.)

Yet it would be misguided to regard TRIPs as the only means by which science was being recruited to be integrated within the new world order; the other arm of "reform" being the General Agreement on Trade in Services (GATS). As long ago as 1994 in a report entitled *Higher Education: The Lessons of Experience,* the World Bank was touting privatization of higher education and the scaling back of public funding for universities as a royal road to success in external trade. Another way that science would be rendered more fungible was to imagine *both* the teaching and research functions becoming subject to international market competition. It might seem that nothing could be more locally contingent and rooted in its immediate environment (language, cultural reference, socialization) than higher education, but that was no obstacle for neoliberals. To that end, in 2000 the U.S. trade representative announced a proposal to the WTO to reduce barriers to trade in international higher education. Just as in the Uruguay Round, no representatives of science or the universities were pressing for this reform, nor were any consulted before the United States decided to put this issue on the agenda (Bassett 2006). Without giving much thought to the consequences, two groups have rushed in to take advantage of these global opportunities: for-profit distance-education providers like the University of Phoenix, Kaplan College, and Jones International University, and a few reputable conventional universities seeking to leverage their "global brand" into transnational campuses in scattered affluent enclaves.[44] The contemporary spread of distance education has proven slow poison to the science base, since one effective way to cut costs is to prevent employees from engaging in research, as most of those entities do (Morey 2004). As Lynch (2006, 5) puts it, "The casualisation of academic and teaching staff is an inevitable correlate of for-profit education." More conventional universities aspiring to the status of multinational corporations are hardly behaving any better, because they mostly operate their satellites with itinerant "fly-in, fly-out" home faculty or more often, with low-paid part-time indigenous gypsy faculty.

Because these developments have been primarily monitored by lawyers and a few economists, it has mostly gone without comment how both TRIPs and the push to globalize higher education tend to devalue local and contextual traditions of research and displace them with a uniform American conception of commercial science: This ranges from explicit elevation of Anglophone publications in non-Anglophone countries to the status of privileged outlets, to the marginalization of traditions within specific sciences that have lost favor or never held any cachet in the United States. Hence we observe Western drug companies belittling indigenous psychopharmacology in Japan (Applbaum in Petryna, Lakoff, and Kleinman 2006); clinical researchers deeming certain diseases not prevalent in the metropole and not warranting serious research efforts in places like India (Chaudhuri 2005, Chap. 5); economic theory that does not conform to the American neoclassical school is summarily dumped from European and Asian business curricula (Goldman 2005); locally sensitive agricultural knowledge is rubbished by Western agronomists (Dutfield 2003). However much one might suspect that science in its ultimate realization is by its nature unified and transcendent of national borders,[45] the effect of the WTO in the here and now has been to render it blandly uniform and prematurely intolerant of multiple lines of inquiry. Indeed, the redoubled stress of the new globalized university on science is partly due to its commercialization and partly as a vehicle used to downplay and deny the local and contingent character of most education and much research.

Yet the most serious fallout from the globalization of intellectual property and university education has been the stabilization of the neoliberal approach to knowledge as a one-size-fits-all nostrum for everything that ails the world (Dutfield 2004a). As I noted at the outset, initiatives like TRIPs and the rendering of higher education as a trade issue put in place the framework that permits the offshoring and outsourcing of R&D, and consequently the decline of American scientific capabilities, as will be seen in subsequent chapters. The irony, if you can call it that, is that in the name of promoting a new knowledge economy the purveyors of globalization have made it possible for the first time in history to outsource the entire U.S. science base, should circumstances and existing prices and wages dictate. Far from some fevered delusion, I would argue that it has already traveled some distance down that path. The bloodbath of in-house corporate R&D in the 1990s proves to have been only the first act of this drama. Some agencies have tried to paint the phenomenon as promoting "partnerships" and "cooperation" (National Science Board 2008c) with foreign researchers, but the details bear witness to the incongruity of the euphemisms. It is unlikely that the loss of the science base was the conscious objective of those who built the international structures; rather, they were intentionally rigged so that the neoliberal ascendancy, once set in place, could never be reversed or neutralized by mere domestic political action. There is no path back to the Cold War regime of science organization; the bridges have been burned. Neoliberal reforms usually come cladded with nasty deterrents to thwart democratic reversals of the successful imposition of their ideal market.

Take, for instance, patent reform.[46] Suppose some segment of the American populace (or more likely, some segment of corporate America) decides that the fortification of patents has really gone too far and that the way to do something for science is to lessen the power of patent holders and tighten up on the classes of things that might be patented in the first place. If they actually got so far as getting something palliative passed through Congress, they would discover to their chagrin that most weakening of patents would be treated within the TRIPs framework as violating previous IP standardization, and this in turn could trigger trade sanctions. "Reform" is therefore automatically rendered ineffectual in advance (Kapczynski 2008). Or suppose that the WTO manages to get higher education to be regulated by international rules for trade in services. Then all manner of current academic practices might themselves be prohibited under international codes (Bassett 2006). For example, need-blind admission policies could be treated as an unfair trade practice, as could affirmative action; accreditation could be treated as arbitrary restraint of trade; admission of too many domestic students might be seen as the result of anticompetitive biases; public funding of higher education could be construed as illegal subsidy and monopoly restraint of trade: all of these could trigger trade sanctions. Both TRIPs and the Trade in Services agreements are engineered to push science and the university in the irreversible direction of further privatization, whatever the local preferences or creeping bouts of buyer's remorse.

Science and IP in the Long View

The phenomena discussed in this chapter tend to clash with a story that gets regaled time and again about IP, namely, that it is the outcome of a "bargain" struck with innovators and thinkers of the world. As the story goes, there is usually an appeal to some form of social contract[47] whereby we the public bestow on these chosen few sorts of abilities to arrogate, encumber, and restrict knowledge, and in exchange we get a number of supposed benefits, such as enhanced public disclosure of their inventions, heightened research effort and hastened transfer of results to those best situated to turn them into better technologies, and sharpened evaluation of the quality of their ideas.[48] The main drawback to this story, like most other fairy tales, is that the social contract was never actually drafted, signed, sealed, or delivered. Instead, the institutions with which we have currently lumbered grew up through a long sequence of accidents, antagonisms, and accommodations, most of which occurred without serious aforethought concerning their ramifications for the further development of IP, much less the consent of the governed. Science studies scholars such as Mario Biagioli (2006b) and a host of others[49] have been more informative about the crooked timber that framed the jerry-built structure of modern IP than have most economists, philosophers of science, and political theorists.

If we stick to patents for the nonce, then historically, they started out as gifts of monopoly privilege granted by monarchs to favored subjects. Shortly thereafter, they were sometimes sold by venal monarchs short of cash. Prior to the nineteenth century, they were generally predicated upon the presence within the nation-state of some working machine or process, and so specification of the abstract principles governing the protected endeavor was not a major consideration. They were about as far removed as you could get from any whiff of scholarship or atmosphere of the university. As we have said previously, patents were not about ideas, they were about power over implementation of technique; as such, they were clearly anticompetitive. Biagioli argues that this began to change at the beginning of the nineteenth century, although he does not offer much explanation as to why this happened. Nonetheless, his main point is that as patents were "democratized"—that is, thrown open potentially to anyone whom might apply, with inspection of existing grants in principle available to any citizen—then the very nature of what was being protected had to change. State bureaucracies could no longer monitor the actual working process, and if they were to subject applications to some sort of scrutiny for legitimacy, then the tests imposed had to become paper tests; they had to be quick and easy to deploy to boot. Biagioli suggests that utility and practice requirements for patents simply became unworkable and were slowly and progressively replaced by rather tendentious notions of "originality" and "novelty." Because this was precisely the period in which such ownership rights were being extended to corporate entities (something even he overlooks), then this also grew disconnected from any human personality trait, and instead "US patent law started to conceive of novelty in terms of the difference between a patent and another that proceeded it" (2006b, 1142). Patents systems became postmodern, in the sense they were predicated solely upon internal networks of textual referents. Hence, just when certain political powers were being transformed from special privileges into citizen rights, what was being protected by patents was being transformed from "things" to "ideas," in the sense that validation of patents became more "virtual": attempts to challenge patents had shifted from disputing issues of implementation to disputing issues of disclosure. Patents came to be not about what exists but what potentially could exist: a development with some deleterious consequences for modern science, as I shall argue in Chapters 5 and 6.[50] Something Biagioli may not realize is that this reduction of patents to sanctioned relationships between texts had the unanticipated effect of encouraging the unchecked expansion of coverage of the patent system: Things that had previously escaped patent control would leave no bureaucratic paper trail, and hence the spread of intellectual property would only need the complicity of the patent examiner (a lowly, poorly paid bureaucrat) and, intermittently, a court dealing with opaque technical distinctions. It was simply a given that supplicants would always desire a broader arena of patent protection, because the easiest route to patent ownership would forever be to venture just outside the perimeter of the sanctioned body of prior art.

There are a number of lessons to be drawn from Biagioli's account of the history of patent protection. First, there was never any primal bargain between "us" and the inventor, no "social contract" to trade some access to new knowledge in exchange for incentives to invent with intent or disclose the ephemeral vapors of their brains. This has been little more than a MacGuffin to get the neoliberal show on the road. The large-scale evolution of the patent system has been driven primarily by parochial political and bureaucratic considerations, and as such has been closely tied to reasons of state. Economists who insist otherwise bear the onus to demonstrate why the situation should be any different in the future. Second, it reveals that the neoliberals did not themselves conjure the primary notion that ideas should be treated as abstract objects subject to property law; that was an ingrained tendency within the *telos* of the changing patent system. What they did manage to do was locate the pressure points within the existing system and then intervene to recast the abstraction of ideas into the clarification of ideal commodities within the marketplace of ideas. They cleverly translated bureaucratic control over ideas into backstage control over *politics*.

Here, in a curious way, I find myself agreeing with the current most sophisticated neoliberal account of the trend toward fortified intellectual property under the regime of globalized privatization of science (Landes and Posner 2004). Landes and Posner had spent a previous book (2003) essentially seeking to explain the divergences from their own peculiar notions of economic "efficiency" of existing IP law by recourse to neoliberal "public choice" theory of James Buchanan and others, but by their own account, they had failed. If, according to public choice, IP invariably and always creates opportunities for "rent seeking," or jiggering the rules to rip off unwitting citizens, then why did the great fortification of IP only explode off the charts in the 1980s? To plug the embarrassing gap in their system, the duo resorts to politics and "ideological currents": "Given the historically and functionally close relation between markets and property rights, it was natural for free-market ideologists to favor an expansion of intellectual property rights . . . the political forces and ideological currents we describe, abetted by interest-group pressures that favor originators [*sic*] of intellectual property over copiers, may explain the increases" (Landes and Posner 2004, 23, 25). Nature and sympathy conspired to precipitate the virtual potential. This quote (as if

Posner were not a member of Mont Pelerin Society[51] and not one of the most visible neoliberal exegetes in America) suggests in a nutshell that economics cannot explain the great fortification of IP in the modern regime but that politics and neoliberal intervention can do an adequate job. I wholeheartedly concur.

5

Pharma's Market

New Horizons in Outsourcing in the Modern Globalized Regime

"Outsourcing is widespread in the business environment, but is only just beginning to be adopted by the scientific community. Biotech and pharmaceutical companies are leading the way . . ."

—PICHLER AND TURNER, 2007, 1093

Pharmadas of Entrepreneurs

Anytime Viridiana needs to clear her mind after reading what economists or science studies scholars or presidents of the American Association for the Advancement of Science [AAAS] have been saying about the commercialization of science, all she has to do is turn to one of my favorite journals, *Nature Biotechnology.* You might think such a journal would be chockful of mind-numbingly detailed accounts of microassays, chemical proteomics, genome sequences, multidimensional cell maps, transcription start sites, RNA interference, SNPs, and so on. You would not exactly be wrong, because those things are indeed archived there; however, Viridiana was shocked to discover that almost as much space is devoted to business models, puff pieces on the success of various biotech firms, legal disquisitions on recent developments in intellectual property (IP), how-to articles on dealing with the Chinese, editorials sneering at EU directives on genetically modified crops, think pieces on the role of greed in transforming academic medical research (Frangioni 2008), and much, much more—sometimes even grand generalizations concerning the modern economics of science, as in the quote that prefaces this chapter. Their data and analysis, often presented in four-color graphs more characteristic of a glossy magazine or corporate annual report than a scientific journal, are refreshing because *Nature Biotechnology* rarely adopts the sober academic caution of discussing what *might* happen, but confidently informs the reader of what has *already happened.* With the panache that characterized the breathless accounts of the triumphs of technological marvels of the postwar generation, it abounds with acronyms coined with wild abandon that only an MBA could love, and it trumpets the new institutional innovations of the globalized regime of privatized science as rational, complex, and inevitable. Take, for instance, the language in the following article:

> It is no longer possible to expect every technology to be readily available within a research institution, let alone a laboratory, yet access to such technology is often the difference between success and failure within today's competitive funding models. To fully embrace the emerging technologies, scientists are increasingly reliant on outsourcing to contract technology providers (CTPs). In this context, CTPs are companies or institutes that conduct partial or entire experiments on a commercial basis. CTPs take advantage of economies of scale to keep their costs of providing the new technologies relatively low. (Pichler and Turner 2007, 1093)

It just seems all so blatantly obvious to the authors : after you have already spun off the in-house R&D capacity of the really large corporation (even within Big Pharma), opened up the university to the privatization of its own research, and outsourced entire R&D units to lower-wage countries like China and India, and "monetized the page" with firms seeking to wring every last dime out of their ownership of the access to data fortified by copyrights, patents, and MTAs—well, then isn't the next logical step to "reengineer" the very scientific research program itself, to separate the various components of the process of inquiry and Taylorize them, and to outsource various modular parts of the activity to for-profit firms? If we are going to approach science in a scientific manner, would that not imply a rationalization of the process along market lines? And if you need a little help getting acclimatized to this brave new world, such as an outside management consultant (an MBA!) as proprietary guide to all the new firm start-ups just waiting to attract your business,[1] or even a "human resources" specialist to help deal with the inevitable disruption that ensues when "outsourcing represents transference of control from the scientist to the CTP" (Pichler and Turner 2007, 1094)—well, there is already someone there ready and waiting, from, say, Bain & Co (Lewin 2009) who is eager to be at your service, for a nominal fee.

The globalized privatization regime has already traveled much further down the road of revolutionizing the means of production of science than I think most people are aware. I cannot stress enough that this is not a "technological" phenomenon so much as it is an institutional one. Our intrepid authors for *Nature Biotechnology* realize that the economic rationalization of research even threatens to encroach upon the prior preserves of the philosopher of science: "Ultimately, science is more about the conceptualization of the experiment, its design, analysis and interpretation than it is actually conducting the experiment" (Pichler and Turner 2007, 1096). This vanguard believes that science is really just another manifestation of the generic division of labor, to such an extent that a small elite of captains of cognition can sit atop the entire knowledge economy pyramid and direct the serried ranks of worker bees in the mundane quotidian tasks of *doing* the research;[2] and if a few middle managers further down the pyramid manage to make a few bucks along the way, then who should object?

Once this mind-set becomes ingrained, the full pith and grit of the neoliberal imperative is brought to bear. Hierarchies are a temporary stopgap, the efficiency experts warn, but can never usurp the greatest information processor known to humanity: the Market. Once upon a time M-form hierarchical firms were confident they could steer the economy, but they were pitiful dinosaurs brought down by the chill winds of competition and the campaign to reengineer the corporation since the 1980s. The same is now happening to science. If you really believe that academic kingpins in their ivy cocoons can efficiently run the scientific enterprise, then think again. The final destination of market reform is to let commercial considerations modularize, standardize, and spin off almost every aspect of the process of scientific research, and consequently erase all boundaries between professional and wage labor. No human being, and especially no scientist, can comprehend the dispersed complexity of knowledge better than the market itself.

Viridiana reacts to this scenario with incredulity, preferring instead to cast a gimlet eye upon my own sanity; she insinuates my mental state is pitched somewhere between terminal paranoia and overactive imagination. But what can she say when one of the most sober trade journals of the chemical industry, *Chemical & Engineering News*, prints the following as if it were a quotidian fact: "Virtual companies have emerged [in China] as a new type of organization pushing drug discovery programs. Taking the biotech approach to the extreme, virtual firms consist of little more than a few brilliant scientists, a promising idea, and a research budget provided by venture capitalists. Virtual firms outsource the entire drug development process to third parties" (Tremblay 2008, 12)? What else is this, but a fantasy scenario read directly off the neoliberal knowledge economy script?

I am aware that many readers will regard this prospect of lean and mean science as bootless caricature; perhaps others may concede that something like it may possibly hold sway in certain limited precincts of biomedical research[3] but cavil that it has no bearing upon the great bulk of contemporary science. A part of me wishes they were right, but the task of this chapter is to explore just how far the means of production of science have been revolutionized in the past thirty years. It admittedly has been the case that many of the most radical of market reforms were pioneered within the biomedical sector, but far from being some narrow technological imperative, this illustrates the principle that certain specific sciences do indeed constitute the vanguard in any introduction of a novel regime of science organization and funding. In a similar manner, chemistry and electrical engineering had formed the vanguard of the erudition regime, and physics and operations research were in the forefront of the Cold War regime. Of course, those sciences, enjoying preferment, were perceived as being the most pathbreaking and progressive sciences of their respective eras, but the extent to which that was *caused* by any particular set of historical discoveries as opposed to the extent that their perceived white heat was the *result* of their elevation by the incipient regime is something we must decline to decide here. Nevertheless, the innovations in those blessed sciences merely tended to foreshadow for the laggards the shape of their own future, once the full consequences of the change in organization and funding had worked their way through the system. In this chapter I survey some of the contemporary developments in (mostly) biomedicine and pharmaceutical research in the new regime, of the sort readily encountered in the pages of *Nature Biotechnology*. It is a most eerie landscape, filled with rootless academics, pharmadas of entrepreneurs, ghosts, diseased bodies, silver bullets, golden parachutes and other hybrid vehicles, siren songs, and, of course, rocks—lots of rocks.

Wreck of the Themeter: The Biotech Model of Spin-off Science

There is a massive genre of academic commentary, especially popular among economists and boosters of the commercialization of the university, which portrays the rise of the biotechnology industry as the necessary consequence of the prior breakthroughs in biology in the mid-twentieth century, combined with the entrepreneurial proclivities of a few key protagonists.[4] The American pharmaceutical industry had begun to shift from branded over-the-counter nostrums to so-called ethical medicines predicated upon a science based regimen during World War II, jump-started by government crash programs in ramping up penicillin production (Athreye and Godley 2009). There is no doubt that the discovery of the structure of DNA in 1953 opened up an entire previously virgin landscape of possibilities for scientific research; that story was full of strange Cold War twists and turns, as documented by the best historian we have had of that period, Lily Kay (1993, 2000). And certainly there emerged certain aspects of "wet" laboratories that found no precedent in the conventional prior template of physics research. But the notion is preposterous that a certain topology of knowledge dictated that we would inexorably end up with Genentech, Amgen, Gilead, Genzyme, and buccaneer academics doubling as pirates of the Being CARI

(Creature Alteration Research Institutes). The fact that this literature almost always ignores the idiosyncratic prior structure of the pharmaceutical industry, its political activity, and the divergent ways it was structured in the United States versus in other countries is merely a symptom of how far we have strayed from a clear-eyed understanding of the modern regime of science organization. Indeed, the fact that many countries had express prohibitions against the patenting of drugs until fairly recently (Germany until 1968, Italy until 1978, and some developed countries even later [Dutfield 2002, 127]) should jar us out of the complacent notion that somehow it was the science that made us do it.

Thankfully, there have been a few analysts who have pointed to a different approach to understanding the nature of modern biotechnology, and they insist there has been something more going on beyond a straightforward biological imperative. The two I shall follow closely here are the economists Benjamin Coriat and Gary Pisano.[5] As Coriat has put it directly, "Changes in the knowledge base cannot explain the emergence of the biotech industry" (Coriat et al. 2003, 239); I shall concur. But that may seem to leave us in a worse quandary: if biology didn't begat biotech as we know it, then what did? We should already have become accustomed to phrasing the answer in macrostructural terms, as explained in Chapter 3. Many transformations within the corporate, government, and university spheres created the space for a potentially new format of scientific research. Some innovations were consciously wrought to smooth the road to a different kind of entrepreneurial science; but many were not, rendering the modern incarnation of the biotech firm as partially a residue of unintended consequences. The pharmaceutical industry circa 1980 was looking to radically transform the nature of its business, but there was little indication back then they would have succeeded to the extraordinary extent that they ultimately did, or that they would eventually conjure a whole knowledge economy to accomplish it.

The Biotech Model

It is no mean feat to separate out the post-1980 efforts to reengineer clinical testing (the topic of the next section) from the pharmaceutical industry's recent efforts to discover new drugs. Nonetheless, brevity dictates that I arbitrarily deal in this section with upstream drug discovery, and in the subsequent section with downstream drug testing. Forging our own way upstream, the history of the pharma/biology nexus can thus be crudely divided into three phases. In the nineteenth century, the pharmaceutical industry tended to prepare "extracts" of compounds from naturally occurring materials, generally suggested by previous folk cures; in the early twentieth century, synthetic chemistry began to be used for producing compounds that had not previously existed in nature but that possessed therapeutic properties. Synthetic chemistry has mostly been effective in producing "smaller" molecules (up to about 500 daltons); larger biologics, like proteins, would require different modes of construction. The third great transformation of the industry came later in the century with the integration of microbiology and information technology into research protocols for designing molecules, in many cases to block or enhance the operation of receptors or proteins. This phase of research could be automated on an unprecedented scale, with techniques of high throughput screening testing hundreds of compounds on genetically engineered sequences or protein targets. From the 1960s, combinatorial chemistry was promoted as a systematic means of assembling organic compounds. Recombinant DNA techniques were soon recognized as an alternative mode of synthesis, dating from the 1973 Cohen-Boyer discovery of gene transfer. The invention of the technique of producing monoclonal antibodies by Kohler and Milstein in 1975 pointed to a different route to interfere with disease-specific proteins. As the biology of metabolic pathways became increasingly differentiated, so, too, did the ability to detect and monitor them in clinical settings. This expanding research portfolio became predicated not only on the commercialized "products" of genetic manipulation, but also on the upstream "targets" and the array of research tools and materials for advancing biological research. Third-phase research was innovated by many scientific actors, not all located in the pharmaceutical firms.

The change in portfolio was mirrored by a change in the manner in which research functions were divided up and parceled out between the firms, government entities, and universities. Consequently, as discussed in the previous chapters, conventional abstract arguments over which parts of science should more correctly or appropriately be relegated to a public versus private sphere (Maurer 2002, Krimsky 2003) or to basic or applied contexts (Calvert 2004) have landed far from the mark. Since the 1980s the modern coalition of the pharmaceutical and biotechnology industries (in conjunction with contract research organizations) has effectively shifted the boundaries of what counts as public or private in biological research. As Diana Hicks (1995, 401) has perceptively observed, "Because there is no natural distinction, academic and industrial researchers construct the distinction between public and private knowledge in such a way as to provide themselves with maximum advantage."

"Ethical" pharmaceuticals have always been a "science-based industry"; insight comes rather from tracking precisely *how* the science has been put to use over time. The rise of the biotech model of scientific research comes into better focus when compared point by point with the preceding Cold War regime model. The postwar pharmaceutical firm and postwar federal drug regulation grew up together and mutually informed one another. First and foremost, the industry sought to demonstrate its "ethical" character by building up substantial in-house R&D capacity to satisfy regulators that it served the public interest by plowing back profits from the world's most market-oriented healthcare system. Publicly funded medical research at the National Institutes of Health (NIH) and in university settings provided "basic" insights into disease mechanisms and therapeutic candidates; pharma took these insights from the free and open scientific literature, combined them with their special competencies, and assumed responsibility for the "applied" end of things. Pharma did lean heavily on American patents to protect its market position and investments (by contrast with most other industries) during the Cold War; however, these were confined to the downstream side of the process, primarily to protect chemical entities as final products. Strong IP was publicly justified by concerns over quality control and protection of the patient. By and large, the entire R&D process (again, excluding clinical trials) was kept internal to the vertically integrated firm, except for the university/corporate interface, which was left reasonably porous, due to strong incentives on both sides of the boundary to keep identities distinct. It was a stable industry, with no new entrants from 1944 to 1976 (Pisano 2006b, 82). Knowledge would flow across the boundary both through recruitment of students and through participation in conferences and publication in the scientific literature. The modal shift from phase 2 (synthetic chemistry) to phase 3 (microbiology, combinatorial chemistry) was ushered in and supported under this ancien régime, so there was no simple one-to-one correspondence between scientific paradigms and types of structural organizations. There were few explicit complaints about medical/pharmaceutical knowledge being "bottled up" inside the universities or the NIH; most notably, some of the most commercially significant drug discoveries of the century were direct products of this system.[6]

Who or what is Big Pharma? Due to very substantial merger and acquisition activity since the 1980s, the roster has changed substantially over recent years, rendering it a bit of a moving target.[7] For a snapshot of the current situation, one can consult Table 5.1, which lists the top twenty pharmaceutical companies in the world, ranked by reported revenue.

Compared with the Cold War system, contemporary drug research doesn't look very much like it used to. Although a few representatives of Big Pharma have survived into the new millennium, the central players have outsourced much of their previous in-house R&D capacity; only in the last decade or so has Big Pharma experienced the bloodbath of in-house R&D that other sectors of the economy had endured in the early 1990s.[8] Instead, to the untu-

Table 5.1. Big Pharma

Pharma corporation	2007 annual revenue ($ billions)	2006 annual revenue ($ billions)
Pfizer	44.4	45
GlaxoSmithKline	38.5	37
Sanofi-Aventis	38.4	35.6
AstraZeneca	28.7	25.7
Novartis	25.4	23.5
Merck	26.5	23.4
Johnson & Johnson	24.8	23.2
Roche	21.9	19.3
Wyeth	17.1	15.6
Eli Lilly	17.6	14.8
Bristol-Myers Squibb	15.6	13.8
Abbott Labs	14.6	12.3
Schering-Plough	12.7	10.4
Boehringer-Ingelheim	11.1	10.4
Takeda	10.6	9.7
Bayer AG	12.3	8.5
Astellas	8.5	7.8
Daiichi-Sankyo	7.3	7.1
Eisai	6.2	5.5
Merck KgA	†	4.6

Source: Contract Pharma, http://www.contractpharma.com/articles/, 2007; CenterWatch 2008, 46. † = not in top 20

tored observer, the industry looks much less concentrated, with a vast veldt of small- to medium-sized biotech firms sprouting up between and around the few mighty pharma oaks. Several thousand biotech firms have been founded worldwide since the early 1990s. The biotech sector is made up mostly of small start-ups, usually spin-offs from academic university research units. The biotech firm typically starts out with some seed financing by venture capitalists; it might continue with various contractual joint ventures and cross-licensing with Big Pharma firms, but the real validation comes when the biotech "goes public" with an initial public offering (IPO) of stock. A few of the biotechs have eventually become full-fledged downstream drug producers in their own right, but that has not been the common trajectory. Instead, the bulk of biotech firms that make it past the IPO produce no downstream product whatsoever; rather, they are devoted entirely to the conduct of scientific research for profit. It is of paramount importance to stress just how unprecedented this state of affairs has been in the history of science. Never before had a multitudinous phalanx of for-profit firms, validated by the stock market, persisted for any length of time devoted solely to the conduct of scientific research.[9] I will refer to this unprecedented formation henceforth as the "biotech model."

This sea change has been nowhere more evident than with the elevation of so-called research tools to primary concern. Some of the earliest breakthroughs in genetic research involved processes or entities that enabled genetic manipulation: the Cohen-Boyer recombinant DNA technologies of Genentech; the polymerase chain reaction (PCR) controlled by Hoffmann-La Roche; and the Harvard Oncomouse distributed through Charles River—none of these were downstream products aimed at a consumer market. The first wave of biotechs did manage to bring some therapeutic molecules to market after some time had elapsed, and these are described in Table 5.2 and by Robbins-Roth (2000). However, there is something very curious about the track record of the first wave of firms relative to what came after, and I will discuss this further in the following section. Bracketing that for now, some of the earliest money made from biotechnology was to be found in the area of "research tools," rather than fully fledged therapies. Thus, many innovations of the early 1980s applied to entities that *might* have been shared with other scientists under the previous regime, but, given the transition that was occurring at that time, one must acknowledge it still would have been an open question as to what would have legitimately fallen under that rubric.[10] The biotechnology and pharmaceutical firms (and the CROs) did not want the question to be left open, however. Considerations of the bottom line for these new start-ups and spin-offs suggested extending commer-

Table 5.2. Early biotech firms

Company	Date incorporated	IPO date	First product to market	Date launched
Genzyme	1981	1986	Immiglucerase	1981
Centocor	1979	1982	Rabies diagnostic	1982
Genentech	1976	1980	Protropin (growth hormone)	1982
Biogen	1978	1983	Intron A (interferon-alpha)	1986 (w/ Schering-Plough)
Amgen	1980	1983	Epogen (erythropoietin)	1989
MedImmune	1987	1991	RespiGam	1996

Source: Fazeli 2005, 12.

cialization into the *process* of research, and not just controlling its products, well beyond anything that might have passed between scientists in the informal research economy before 1980. This trend has, if anything, only intensified in importance: An investigator who examined all U.S. patents issued from 1998 to 2001 relating to DNA sequencing estimated that as many as one-third were research tools rather than diagnostic, therapeutic, or other innovations (Scherer 2002). Biotechnology start-ups plumped for the narrowest possible construction of laboratory inputs that should be freely shared among scientists and sought to patent their research tools as part of the biotech business model.

Once the new model of science organization became established, biotechnology and Big Pharma became united in seeking to rein in the free dissemination of research tools. Universities tended to be caught up in the Great Engrossment, if only because they, too, were scrambling to jump on board what they had come to regard as the biotech gravy train, believing this meant playing the game in the same way that the biotechs and Big Pharma were already carrying on. When their faculty sought to spin off their research, university TTOs discovered they must possess some kind of assets to fortify a business plan, but almost no one had anything in the way of consumer products to show. Research tools were turning out to be the financial lifeblood of small biotech start-ups, but they also were pivotal for a strategy of patent-oriented research that emphasized secrecy, which had become the industry standard. For instance, the pharmaceutical firms had become wary of the possibility that academic researchers give freely provided research tools to competitors; that tool users would publish proprietary information and thus undermine future patent claims; or that they would reveal harmful side effects of a tool that doubled as a drug precursor and therefore create regulatory headaches. Their attempt to muzzle their licensees and/or benefit from their successes was a significant departure from previous uses of patents in the Cold War pharmaceutical industry, because it broadened the scope of patents to control upstream developments.

When one cuts through the starstruck jargon, what the new biotech model amounts to is the outsourcing of many of the upstream R&D functions that had previously been performed in-house by Big Pharma to this unruly sector of small start-ups and spin-offs from academic settings (and more recently from the NIH itself). Big Pharma has offloaded much early-stage funding of the research onto the universities and the venture capitalists (VCs), who pour resources into supposedly commercial projects with no other visible means of support. The biotechs themselves attempt to milk their research tools for cash up front, or, more likely, they negotiate temporary joint R&D projects with pharma for a little more cash infusion to offset giddy "burn rates" of finance;[11] this has proven a particularly cheap way for Big Pharma to monitor new R&D without having to foot much of the bill. If the research actually goes somewhere, the VCs and the star scientists cash out with an IPO and retire to Palm Springs; but this does not mean that the biotechs are actually showing a profit or are even viable. The stark truth is that most biotechs never produce a drug or other final product; they are just pursuing commercial science, *which almost never makes a profit.* This is frequently hidden by the fact that true nirvana for the biotech sector is to have a successful IPO followed by a lucrative buyout by a Big Pharma firm. In such cases Big Pharma is happy because it has bought a near-end stage drug-development sequence, and the firm itself disappears into the maw of an existing corporate structure.[12]

The big dirty secret of the biotech model is that, if one removes one or two outsized success stories (primarily Amgen and Biogen) from the mix, then the biotech sector as a whole has persistently lost money since the mid-1980s (Pisano 2006b, 115; Coriat et al. 2003, 238). "It used to be a joke to say that "biotechnology companies are pharmaceutical companies without sales" (Gottinger and Umali 2008, 584), but how amusing can it be when a major booster of the industry, *Nature Biotechnology,* wrote in its summary "State of the Biotech Sector—2007" that "the biotech sector as a whole narrowed its loss by 64% to $2.7 billion . . . This is the closest the industry has come to breaking even" (Anon 2008, 728)? And that was just before the Great Contraction, at the height of the bubble. Economists take note: *Commercialized outsourced pharmaceutical science in the aggregate* (that is, the "upstream" biotech model) *has not been a viable profitable system.* This fact is compounded by the further observation that the drug pipeline of truly novel therapies has been acknowledged to be "drying up" over the last decade or more.[13] It is truly curious that this combination of facts is not more widely known and appreciated; this development has got to be Exhibit 1 of any serious new economics of science. Much of the obscurity surrounding this state of affairs undoubtedly is related to the highly charged circumstance that it stands as profound mortification and direct refutation of the neoliberal agenda for the commercialization of science in general; furthermore, it makes universities and governments desperate to join the biotech gold rush by fostering their own corporate spin-offs look really foolhardy.[14] But there is also the complicating circumstance that the biotech sector has enjoyed a sequence of waves of infatuation in the stock market since the 1980s; that would be very hard to understand if the sector really were the walking wounded (or the living dead), as we have maintained. Economists have thus been loath to confront these facts; science policy scholars have not done much better.

If a phenomenon seems to all appearances to be a success when careful measurement reveals it is a failure, then one must widen the scope of the analysis in order to comprehend the sources of its staying power. The case to be made here is that the biotech model does indeed fulfill various needs and perform various functions, even though the economic process it most closely resembles is the time-honored Ponzi scheme—that is, it regularly incurs more liabilities than it accrues cash flow, with the trick being to cash out before the inevitable collapse ensues. Biotechs are therefore the Madoffs of the modern commercialization regime. The main requirements that the biotech model assuages are the strategic needs of Big Pharma to outsource most of its R&D process, while still depending on profitable "science-based" product development. In a very real sense, the biotech model owes its life support to Big Pharma, both through periodic infusions of joint-project financing deals and the small but real prospect of being bought out by a major. Sometimes the biotech might actually manage to charge for something Big Pharma needs for its own purposes, like specific research tools, or else access to IP controlling broad therapeutic areas, like gene patents; however, this never constitutes a major persistent revenue stream. Such dribs and drabs of cash flow could never justify the growth of the sector, or provide the economic basis for a new format of scientific research. The secondary functions it performs are to provide a steady stream of apparently legitimate high-tech IPOs for the entire panoply of financial managers: venture capitalists, funds managers, IPO specialists, merger and acquisitions specialists, and the like.[15] "A 'banner year' in biotech has come to mean a year in which financing is plenty and the IPO market is hot" (Pisano 2006b, 162). Few have noted that the timing of the birth of the biotech model was intimately related to changes in the finance law that limited the liability of venture capitalists and the pension funds that invest in them (Robbins-Roth 2000, 34).

The tertiary function performed by the biotech model is something that its boosters probably would never admit: It has served as the thin edge of the wedge that seeks to wean the university off public funding and transform some of its faculty into entrepreneurs. This explains why biotech entrepreneurs and neoliberal theorists tend to enjoy vibrant elective affinities.[16] Part of the magic of its charmed existence has been the historical accident that none of these constituencies were actually directly responsible for the genesis of the biotech model; that instead can be attributed to a confluence of events already covered in this book: the rationalization of corporate structures, the Bayh-Dole Act, and the related raft of legislation, which encouraged universities to claim intellectual property in publicly funded research; the *Chakrabarty* ruling by the Supreme Court; the vast fortification of IP generally, both nationally and internationally through TRIPs; various political developments that serve to guarantee that the existing industry structure of Big Pharma will not suffer from attempts to nationalize medical care; the general crusade to undo vertically integrated corporations in the name of rendering them more "lean and mean"; and the push to outsource corporate R&D to lower-cost performers. Yet all that, however formidable, was not sufficient. One thing Benjamin Coriat has reminded us is that innovations in the financial sector were critical to the genesis and success of the biotech model (again insisting that economics, and not the "science," bequeathed us this model). Indeed, the biotech model "was actively fostered, promoted and brought into being from the top down by a series of quite deliberate legislative and institutional decisions" (Cooper 2008b, 25).

The statement that a flawed biotech model of research was constructed and maintained by an awkward coalition of institutional alterations and flawed business models is one thing that many universities may find hard to accept; but in this book, I propose that the reader entertain one further implication of this realization, a fourth function of the biotech model. Not only did the biotech model embody a dubious economics of science, but it also resulted in inducing a peculiar set of biases to the sciences that it has thus transformed. For instance, as Jane Calvert (2008) has argued, the biotech model of commodification has tended to favor reductionist approaches in biology versus more overt holist approaches, such as systems biology and theories of emergent properties. The reason is that, in order to assert control over biological innovation and research tools, it becomes necessary for theory to stabilize discrete objects of ownership in accordance with the reigning rules of the property regime. Under the sign of commercialized science, those combinations of theory + empiricism that conform to the arbitrary definitions of the commodity will clearly be preferred to those that stress complexity and the interrelatedness of phenomena. One can observe this in the increasingly pharma-oriented health regimes in the late twentieth century.

Hence economic history matters. Frequently the Genentech IPO of 1980 is cited as a watershed in the history of the biotech model, and indeed, it was widely celebrated and discussed at the time as the first corporate share offering to be predicated entirely upon the performance of scientific research. However, Coriat points out that it was floated on the over-the-counter market: this was not a particularly attractive venue and certainly not one that would have sustained the subsequent commercial explosion of biotechs. Rather more decisive was the promulgation of special rules for NASDAQ in 1984 covering the listing upon the exchange of shares for corporations that *reported no profits before and after flotation*. This was a rather drastic change from standard practice, for instance, on the New York Stock Exchange, which required a track record of profitability as prerequisite. The NASDAQ justified what seemed an injudicious relaxation of discipline by stipulating that eligibility for listing could be met by the possession of substantial net assets, plus some rationale of promised high future growth potential. For the first time, "tangible assets" such as IP portfolios were included in the roster of assets that would qualify the listing to proceed. We can readily appreciate that these regulatory changes were tailor-made to incubate something very much like the biotech model, not to mention promote all manner of dubious company formations.[17] True to form, the actual biotech explosion happened just after the 1984 alteration in the financial system.

It has been noted elsewhere that a similar sequence of events had to happen in Europe to jump-start a biotech sector, and indeed the delay in altering financial regulation there almost exactly maps into a similar delay in the appearance of biotech start-ups, relative to the United States. Loeppky (2005a) points out that until the late 1990s no German firm could be listed on the Frankfurt exchange without seven consecutive years of demonstrated profit. The German government had attempted to subsidize a biotech incubator in that period, but to no avail. However, with the 1997 introduction of the "Nueur Markt" and its loosened regulations on profitability, in conjunction with changes in the tax law facilitating the sale of cross-holdings, the biotech sector took off in Germany; by 2000, Germany housed the most biotech firms in Europe. So we possess multiple historical instances demonstrating that a weakening of corporate governance regulations with regard to treatment of finance and profitability is a necessary precondition for the flowering of the biotech model.

Coriat then raises a very salient point. If much of the biotech model is not specifically chained to "biology" as such, isn't it somewhat more than likely that it would spread to other areas of science?

> If the biotech regime has been based upon the possibility of patenting and commercializing basic scientific knowledge, it is possible to envisage the creation of firms based upon patenting of knowledge in similar domains (math algorithms, for example), or generic applied knowledge, such as "business methods." What remains to be explored are the implications of the extension of patenting for [the current] technological regime, in different fields and sectors (and in countries other than the USA). (Coriat et al. 2003, 249)

Indeed, this question should be high on the agenda of a new economics of science, since it would demonstrate that the globalized privatization model had consequences well outside the domain of pharmaceutical research. There is very little literature on this topic; one piece of information comes from one attempt to conduct a survey of spin-off firms conforming to the biotech model from Oxford University (Smith and Ho 2006, 1562). Of the 114 firms that the authors managed to identify between 1998 and 2004, 29 were explicit biotechs, and another 16 were identified as pharmaceutical companies. Other sectors were identified as information technology (33), measuring instruments (6), optoelectronics (6), motor industry (3), and nanotechnology (3). The information is only partially indicative, because many of the more established firms in fact had dedicated product lines and so could not be considered as being solely devoted to research as a business model.

Yet nevertheless, there are murmurs of beginnings of the kind of thing Coriat imagines around the fringes of academe: for instance, the contract technology providers (CTPs) mentioned in the introduction to this chapter or, closer to home, the contract research organizations (CROs) described later in this chapter. These are usually firms inspired by the biotech model: commercialized scientific research in the absence of any product lines, heavily dependent upon early-stage venture capital and a later IPO launch, deriving from or displacing academic research, with mergers and acquisitions as the most common terminal state, pitched to facilitate the outsourcing of R&D from large corporations bent upon shedding their previous in-house capacity.

Something Clogging the Pipeline

So it seems that the biotech model comes standard equipped with a nasty predisposition to lose money; some have taken this to imply that "public equity [i.e., the stock market] was never designed to deal with the governance challenges of R&D entities" (Pisano 2006b, 143), such that this manifestation of the commercialization of science turned out to be a poor economic proposition. The neoliberal economist must surely be straining to retort, "But Hayek never claimed markets were designed!" Yet the point remains that it is a poor excuse for an industry to operate in the aggregate in the red over decades. I can imagine a more tough-minded neoliberal response, however, which would go something like this: No one ever claimed that the Market will provide sustenance to every single firm. Most firms fail in the long run, and market recompense rewards phenomena that the participants often cannot anticipate or comprehend. All that really matters are a few biotech firms that do manage to get rich and survive; entrepreneurs are thus encouraged to keep thrusting and pressing against the bounds of the known; all the rest is the ineffable operation of the market as superior information generator and processor. Life, for neoliberals, like science, is a gamble.

Uncomfortably for the believer in the marketplace of ideas, the biotech model constitutes a telling refutation of the neoliberal theory of science, venturing beyond the simple observation that most biotech firms operate in the red. Indeed, the very idea of the market as information processor is fatally compromised if it turns out that the progressive commercialization of science seems not to deliver the goods—that is, churns out fewer and fewer significant scientific discoveries and findings over time.[18] Usually, that is a case nearly impossible to make in a convincing fashion, because it is almost inconceivable that one would tot up "significant" discoveries as one would count cookies in a cookie jar. But here the situation with regard to pharmaceutical research is different, because the protocols for arriving at sanctioned, "successful" discrete end states are so stylized and codified—in other words, in the United States, the FDA formally declares at a finish line that you have found a new effective drug, or not—that quantification of the "output" of pharma research actually comes within reach. All controversy is, of course, not banished; no one believes the bureaucrats of the FDA possess this particular Philosopher's Stone. And the FDA has its own political weaknesses, which I must pass by here. But nonetheless, by provisionally accepting this criterion, something relatively solid can be said about the effect of the biotech model on science, and not just upon the economy.

As we have already intimated, the news here is not good. Just as with any other attempts to quantify the "outputs" of science, there will be reasonable and justified objections to heavy-handed quantification for its own sake. Nevertheless, the trade journals and the generalist press have been awash with lamentations that the pharmaceutical pipeline is drying up, and with some justification. This contention was elevated in the public consciousness by a few muckraking books that sought to attack the drug companies for their dominance of the political and scientific landscape.[19] The vulnerability they slammed is that the pharmaceutical sector tends to curry public favor by stressing its role in the discovery and development of new medicines and new therapies; that impression is precisely where their defense is the weakest. By any measure one chooses, the amounts of money channeled into biomedical research have never been larger: NIH funding *doubled* from 1999 ($13 billion) to 2004 ($26.9 billion); private cash infusions from venture capitalists and other sources funneled to biotechs have expanded massively in both Europe and the United States since 2000 (Fazeli 2005, 15); and industry spending also increased dramatically from $16 billion in 1993 to $40 billion in 2004, according to a GAO report. Nonetheless, almost every measure of valuable research output in pharmaceuticals has continued skidding on a downward trend, which dates back to the 1990s.[20] The negative correlation does tend to verge on the incredible: in fact, there has recently sprung up a cottage industry to try and explain it away (Agarwal and Searls 2009).

There are many possible ways to approach this topic, but one thing that must be said at the outset is that the FDA does not make it easy to bring these data together in such a way that the big picture can shine through in outline. Furthermore, one needs a fairly sophisticated background in the entire process of drug discovery and development to comprehend all the intervening variables that dictate which numbers do matter; if the reader wishes to pursue these questions in a systematic fashion, I would direct him or her to the notes. Instead, I will briefly provide the relevant time series of output proxies for new drugs, and provide some indication of what they may portend.

The *first* indicator to which attention needs to be drawn is the count of new drug applications (NDAs). To understand this (and much that follows), it is necessary to gain a rudimentary grasp of the modern drug development process in order to see how it shapes and restricts the way science is conceptualized and performed by the pharmaceutical industry. The process is summarized in Figure 5.1, taken from the FDA Web site: the NDA occurs toward the left of the figure. The drug development process in the United States has been effectively standardized through regulations promulgated by the FDA. This system dates from the 1938 Federal Food, Drug and Cosmetic Act (Marks 1997, Chap. 3; Rasmussen 2005). However, most observers agree that the real watershed was the Kefauver-Harris Amendments of 1962, which had been prompted by the Thalidomide controversy (Daemmrich 2004, 26–29). These amendments mandated that the FDA require drug companies to demonstrate the safety and efficacy of a drug before marketing it. The FDA was authorized to determine the standards and format of testing from the first animal trials through the final human clinical trials. Although immediately attacked by economists and industry as unwarranted government interference with innovation, delaying the marketing of new drugs, the FDA approach gradually became the gold standard of pharmaceutical approval.[21] Initially, because of the dominant size of the U.S. market, but later promoted as part of a process of "harmonization" of regulatory requirements across the European Union, Japan, and the United States, the FDA-mandated procedures now form the basis of corporate drug testing throughout the developed world (Abraham and Smith 2003; Abraham 2007a).

Briefly, the FDA-mandated process involves the following steps: the sponsor, in this case the pharmaceutical company, initiates the drug development process. This occurs regardless of whether the idea for the treatment originated in an academic, clinical, biotech, or corporate laboratory. The drug development process then comprises four stages: a preclinical (or animal) stage, a clinical stage, a regulatory delay, and a postclinical stage. During the preclinical stage, a new compound to effectively treat a disease is identified and tested on animals in order to ascertain pharmacological effectiveness, and potential for toxicity and carcinogenicity. The FDA has recommended a minimum of

12 months of tests on at least two species (typically mice and rats).[22] After filing an investigational new drug (IND) application with the FDA and receiving preliminary approval, the clinical stage consisting of four standardized phases begins, with each phase recruiting suitably informed patients. An institutional review board (IRB) oversees procedures and protocols of the clinical trial, and investigators at various academic institutions, hospitals, or (more recently) other sites administer the clinical trial. Phase I, which lasts about one year and involves a few dozen patients, typically aims to identify any deleterious effects on normal healthy patients. Phase II, which lasts several years and involves a few hundred patients targeted for the pathology of interest, determines if the drug has some therapeutic effect (either efficacious or deleterious) on the specific disease. Phase III, which lasts up to five years and involves thousands of patients, seeks to quantify degree of effectiveness, and can involve masked trials that compare the new drug with a placebo and/or existing rival treatments. If the drug proves promising in these trials, then the same firm files a new drug application (NDA) and then waits for FDA approval to market the drug. Further, Phase IV postclinical trials can be conducted after a drug is approved for marketing, perhaps due to concern over its longer-term efficacy, or perhaps because the FDA conceives of a need to monitor safety. In the pharmaceutical industry, the period between the initial Phase I trial and the submission of the NDA is often called the "developmental cycle time" (Getz and de Bruin 2000).

Therefore, the NDA is the first meaningful milepost in what one might consider to be the long and difficult slog to move a drug from tenuous candidate status to serious prospective therapy. Postponing until the next section the actual specifics of conduct of Phase I–IV trials, designated the "clinical" phases, getting a candidate to the NDA stage is one of the primary objectives of any research-oriented biotech or pharmaceutical corporation.

Here, the message of Table 5.3 starts to come into focus. If we date the maturation of the third stream of pharma technology and the rise of the biotech model from roughly 1980, then it becomes apparent that the initiation of a serious decline in the number of NDAs in the United States dates from roughly 1983. Recalling from Figure 5.1 that there subsists a substantial lag from the first glimmers of discovery through to the submission of the NDA, the coincidence is rather striking. Of course, there is no intention here to impute direct causality from one to the other; there are just too many intervening variables that dictate whether a therapeutic regimen exists and how a researcher might stumble upon it. One might speculate that the biotech model was simply too new to have such an effect on the entire sector so soon. There is also the question of the role of the FDA in slowing the tempo of the entire process over time; but here one must keep in mind that the developmental time

Table 5.3. Number of NDAs received by calendar year*

Calendar year	NDAs received	Calendar year	NDAs received
1970	87	1989	118
1971	256	1990	98
1972	272	1991	112
1973	149	1992	100
1974	129	1993	99
1975	137	1994	114
1976	127	1995	121
1977	124	1996	120
1978	121	1997	128
1979	182	1998	121
1980	162	1999	139
1981	129	2000	115
1982	202	2001	98
1983	269	2002	105
1984	217	2003	109
1985	148	2004*	115
1986	120	2005*	116
1987	142	2006*	123
1988	126		

Source: http://www.fda.gov/cder/rdmt/numofndareccy.htm. Beginning in 1994, receipts are counted based on user fee reporting requirements, which include all original receipts plus those NDAs submitted after refusal to file or withdrawal. The receipt figure does not include applications that are unacceptable because user fees were not paid.

* Includes the therapeutic biologic products transferred from CBER to CDER, effective October 1, 2003.

cycle part of Figure 5.1 is largely in the hands of the private companies; regulatory delay could only become a more serious obstacle after the NDA was submitted. Hence a prolonged fall in registered NDAs serves as one possible indicator of a decline in successful output of drug research, ignoring for the moment the possibility that the quality of each individual therapy is totally left out of this account.

Now, the FDA has been made painfully aware of this state of affairs, and it realizes that it looks bad for the poster child for privatized science to seem to be generating a decreasing number of candidate drugs. The FDA is also aware of the transformations that the third stream of biotechnology is wreaking upon drug discovery; for instance, genetic techniques permit the exploration of much larger therapeutic molecules than the older "small-molecule" paradigm, and this requires a somewhat different approach to regulation. In order to accommodate those trends, it has set up a separate track for biologic license applications. Indeed, many of the products coming out of the biotech

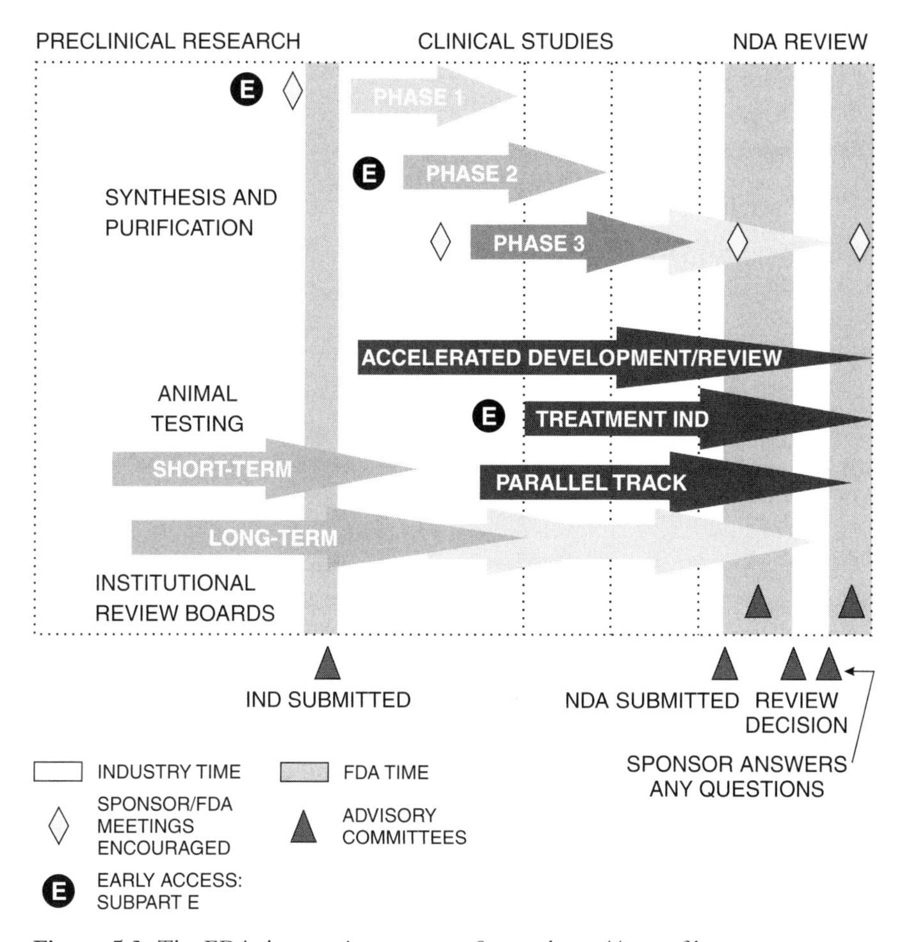

Figure 5.1. The FDA drug review process. *Source:* http://www.fda.gov.

model tend to fall into the biologic category. While it has been conducting this track since the 1990s, the FDA made the rather strange decision to consolidate all its reported biologic and small-molecule data together from 2004 onward. Without imputing motives, this has rendered most time series downloaded directly from its Web site highly misleading, certainly mitigating any appearance of decline in (for instance) the NDA figures in Table 5.3.[23] While we cannot separate out the effect of adding biologics for the NDA series, we can do so for statistics closer to the "finish line" of Figure 5.1.

The *second* indicator, and an even more contentious milepost for the FDA, is market approval after its review period. This is the sort of thing that gains press attention and moves those stock prices. Yet the FDA recognizes that a large proportion of drugs that it approves are in fact minor variations upon existing therapies and do not truly constitute any novel approaches to the conventional armory of drugs. It has been common to excoriate the industry for devoting so much of its valuable scientific effort to what are disparaged as "me-too drugs," profitable to market but effectively drags on the health system. It has tried to build this insight into the definition of what it calls a "new molecular entity" (NME). An NME designates a drug that contains no "active moiety"[24] that has been approved by FDA in any other application previously submitted. The FDA then makes a further judgment about the nature of the intervention, which is purported to take place with the NME, and further distinguishes between therapies that are truly novel and those that resemble prior known therapies. It designates the former as "priority" candidates and conducts its review with special alacrity, whereas the remainder it allocates to a "standard" designation. To qualify for priority review, the application must demonstrate evidence of increased effectiveness, reduced side effects and interactions, enhanced compliance, or possible use in a new subpopulation. Treatment for a serious or life-threatening disease is not a necessary prerequisite for the priority review determination. Only approved drugs receive these characterizations from the FDA; all data concerning rejected NDAs are treated as proprietary and secret.

The track record with regard to approved drugs has proven even more of an embarrassment for the pharmaceutical industry. The vast majority of all FDA market approvals do not even qualify as NMEs. The FDA reports 65 percent of all approved drugs from 1989 to 2000 were deemed noninnovative: that is, they did not qualify for characterization as NMEs. Of the remainder, 20 percent were deemed standard; so by their lights, only 15 percent of all ap-

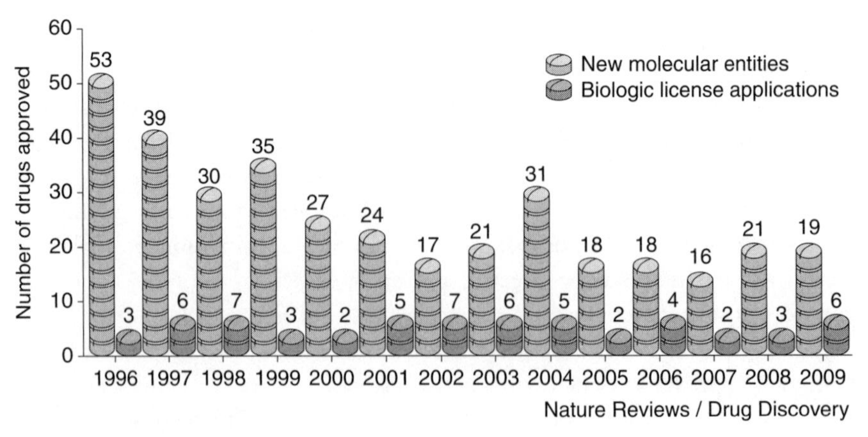

Figure 5.2. FDA-approved NMEs and biologics.

Table 5.4. FDA "priority" and "standard" NME drug approvals

Year	Priority NME	Standard NME
1993	13	12
1994	12	9
1995	10	19
1996	18	35
1997	9	30
1998	16	14
1999	19	16
2000	9	18
2001	7	17
2002	7	10
2003	9	12
2004*	21*	15*
2005*	15*	5*
2006*	10*	12*
2007*	8*	10*
2008*	9*	16*

Source: http://www.fda.gov/cder/rdmt/NMEapps93–06.htm; *from 2004, numbers include biologics, but from 1993 to 2003 they were excluded.

proved drugs were real innovations. Figure 5.2 and Table 5.4 pick up the story from there, paying closer attention to the possible impact of the spread of the biotech model of research on the track record of approval of new drugs. First, since the mid-1990s there have been a steady trickle of approvals of biologics, which mostly come from biotechs;[25] but there has been no substantial uptick in their production. Table 5.4 exhibits the same flaw as Table 5.3, in that biologics are lumped in together with NMEs by the FDA since 2004, masking what by all accounts is a marked fall in the number of priority NMEs being approved over the last decade.[26] Again, without imputing motives, the change in basis of aggregation occurring in 2004 looks a little too convenient to be an accident, given the sorts of attitudes toward science being promulgated at that juncture by the Bush administration. Second, the closer we move along the continuum toward "me-too drugs," for instance, by adding the standard category to NMEs, the more the decline appears mitigated. It becomes apparent one must be very wary, girded to be snowed by statistics and prepared to query the content of the categories of drugs "approved" whenever one encounters claims about the "drying up of the drug pipeline." Yet, at the end of the day, if we provisionally accept the FDA's judgment about which therapies are novel in a set of dimensions that it has worked out with the drug companies, then by consistent strict criteria, output has been falling in an era of unprecedented devotion of resources to precisely this pursuit. Divide priority NMEs by some appropriate dollar figure devoted to R&D, and the picture is undeniably grim.

As you can probably anticipate, the question of the role of the biotech model in producing this baleful debacle has come to be hotly disputed. One point of contention has been who or what should count as conforming to a biotech model (H. Miller 2007), because the majority of NMEs attributed to biotech are in fact small-molecule therapies developed according to conventional research protocols. "Many biotech firms producing small-molecule therapies only differ from pharmaceutical firms in their size and would be more accurately described as specialty pharmaceutical companies" (Hopkins et al. 2007, 576). Parceling out blame has thus proven elusive. Defenders of biotech have not been slow to emerge, particularly from neoliberal think tanks.[27] They seek to shift the onus, of course, onto plenty of other culprits. The industry seems more inclined to blame the government for dragging its feet or else imposing much more stringent conditions over time for the approval of drugs, especially after some recent extremely high-profile disasters, like the recall of Merck's Vioxx, Bayer's Baycol, and Novartis's Zelnorm, and the failure of Vytorin (Berenson 2008), the embarrassment over Avandia, as well as the great loss of confidence in SSRIs like Prozac.[28] The flaw with the neoliberal penchant for deflecting the blame onto the government is that measured times from NDA submission to a verdict in Figure 5.1 have been falling recently, at least in part due to "reforms" that have instituted the practice of the budget of the FDA being paid by fees and charges exacted from its "clients," the drug companies. But more importantly, consider a point almost never made: The FDA *cannot* be responsible for the bottleneck if the stream is drying up *all along the pipeline*. The *third* indicator of the drought is the fact that U.S. patents in specific biotech patent categories have been declining since 1998 (Adelman and de Angelis 2007). Although patents are a terrible indicator of scientific output, for reasons discussed in Chapter 4, there is something truly anomalous about an era where industry insiders admit that they are resorting to patents on a heightened basis because they feel driven to lock up IP due to competitive pressures, and yet they end up patenting less and less.

In sum, all drug discovery indicators point in the same direction; unhappily, that direction is downward. If it seems unlikely for apologists to pin the trend upon that old scapegoat the government, then what can possibly account for the seeming failure of the poster child for the new high-tech global knowledge economy?

There is another faction in legal science studies that argues that it is not the biotech model per se, but rather the activity of patenting itself that is resulting in what has been called "the problem of the anti-commons."[29] Authors in this tradition have come up with various reasons why patents might hinder the process of scientific research. While I argued in Chapter 4 that patents by themselves may not constitute the crux of the problem, some of the standard-bearers for this faction have pointed to the role of the expropriation and engrossment of research tools as a major trend in biotech research. There may be something in that complaint, because one recent exercise estimates that almost 50 percent of all biotech patents awarded in the period 1990–2004 were explicitly for control over "measuring and testing processes," the most stringent definition for research tools; if one throws in the 26 percent of patents explicitly for protein sequences and 9 percent for nucleotide sequences, then around 85 percent of all biotech patents actually cover some broad definition of research tools (Adelman and de Angelis 2007). In the previous section, I stressed the strategic role of resorting to research tools for short-term revenue enhancement in the biotech model. The work on MTAs in Chapter 4 would thus tend to reinforce the idea that it was the untenable business model of the biotech firm that drove it in the direction of attempting to engross economic control over research tools. Once that tendency started to take hold, then the obstruction of research has been deployed through contract and breach of contract suits, and not especially suits for patent infringement as such, with the exception of a few notable outliers like *Madey v. Duke University.*

The most convincing arguments that I have seen tend to deny that the decline in scientific output in pharmaceuticals can be attributed to some broad generic problems with patents or strengthened IP as such, but rather seek to nail the indictment to the door of the biotech model itself. In a very detailed survey of the actual scientific developments in the area, the economist Paul Nightingale and his colleagues concluded:

> Genomics technologies in drug discovery helped shift the bottleneck in drug innovation from the identification and creation of novel small-molecule drugs against known targets (chemistry) to the biological characterization and functional validation of large numbers of unknown drug targets (biology) at the molecular, cellular and system levels . . . While the decline in R&D productivity predates biotechnology, the application of biotechnology research tools, and in particular genomics, may be making the situation worse . . . Genomics technologies are generating large numbers of less well characterized new targets that are now being tested in experimental models that are increasingly removed from the intended patients (i.e., from patients, to animals, to cell cultures). Molecules aimed at these novel targets fail more often in the

> later, more expensive stages of clinical trials where efficacy is ascertained, increasing the cost of drug R&D. The financial risks of "first in class" drugs therefore tends to be higher, increasing the rewards for developing me-too drugs. Although the successful use of biotechnology in drug development may improve productivity in big pharmaceutical firms, there is little publicly available evidence that this is being achieved so far . . . At present it appears that new drugs against novel targets identified through genomics have a lower success rate during development than compounds against well-established targets. In this sense, at least in the short term, biotechnology has exacerbated the problems associated with drug development, rather than led to revolutionary improvements. (Hopkins et al. 2007, 571,574, 583)

If anything, these authors understate the extent to which the biotech model has proven a misleading paradigm for the pharmaceutical industry. Harkening back to Table 5.2, perhaps we can glean another reason why the subsequent evolution of the biotech model turned out to be so disappointing. The biotech model has been extolled and promoted on the backs of some of the earliest and most successful examples, like Genentech, Biogen, and Amgen. This has been more significant than it at first appears, particularly since Pisano (2006b) demonstrated that the entire sector has been unprofitable once one removes Amgen and Biogen from the record. From Table 5.2 we observe that the earliest success stories of these firms were indeed embracing large-molecule biologics, but essentially by using microbiology to artificially produce proteins or enzymes whose therapeutic uses were well understood, *because they had already been in use.* In other words, the earliest instantiations of the biotech model were not strictly used to discover and develop new drugs, but rather to turn technologies such as gene splicing to the production of known large molecules such as proteins and enzymes by other means than extracting them from animals or cadavers (Robbins-Roth 2000). Thus when Genentech struck the first R&D outsourcing agreement with Eli Lilly in 1978 to produce recombinant insulin (Pisano 2006a, 117), this was a form of outsourcing that could actually work. Even still, it seems that the current top categories of sales of biologics in the United States remain the original big hits: growth factors, monoclonal antibodies, hormones, and cytokines (Aggarwal 2007, 1098). Another way of putting this is that no matter how glamorous the new genetic topics might seem for investors, this early activity was more akin to applied engineering for manufacturing existing products rather than fundamental scientific search and discovery (Fazeli 2005). There was only a limited roster of known large-molecule therapies for humans, and once you had converted those to the biotech model, then the really expensive and unreliable commercialization of cutting-edge scientific research was left for those who came after. Nevertheless, every start-up and spin-off of whatever provenance was promoted by unscrupulous technology transfer officers and venture capitalists as bearing the mantle to become another potential Genentech. And, as we have seen, unless they were bought out by Big Pharma down the line or somehow managed to last long enough to bring a product to market, they just hemorrhaged red ink.

It is not even clear that Big Pharma has been totally satisfied with the motley mutant creature it has sired.[30] Certainly the biotech model has by now been thoroughly integrated into the modern drug discovery process. Yet, as Pisano (2006a) points out, most of the R&D alliances have been kept at arm's length and are fairly brief: the length of a typical contract is four years. This encourages the sort of just-in-time science that comes to grief with a higher rate of failure later on down the line, because drug development is inherently a long-term proposition. This, in turn, has resulted in numerous litigations between Big Pharma and collaborative biotechs—Genentech later sued Lilly, and Amgen fought a bitter legal battle with Johnson & Johnson—precisely because it is so difficult to identify the boundaries where one research project ends and a different one begins. Real science resists being reduced to commodity-sized chunks. And further, "Pharmaceutical firms see the biotechnology firms and universities that hold patents on upstream research inputs as so many tax collectors, threatening to dilute their anticipated profits" (Eisenberg 2003, 1117). Perhaps it is just a matter of time before universities, too, will grow more familiar with the downsides of the biotech model, and the infatuation with all things biotech will finally start to lose its golden allure.[31]

The Contract Research Organization and Just-in-Time Science

The biotech sector generally garners the lion's share of the attention from boosters of the new privatization regime of science: It's glamorous, with all the pizzazz and excitement of novel breakthroughs in fundamental biomedicine, the nobility of purportedly reducing human suffering, and the prospect of "getting the newest discoveries out into the public sphere without delay" (Shapin 2008b). But ironically, it has not been the ballyhooed biotech research sector where the "spin-off model" has proven most successful or, indeed, the most profitable. Rather, it has been the downstream side of drug discovery, the clinical phases coming after "IND Submitted" in Figure 5.1, that have turned out to be the real proving grounds of the brave new world of commercialization of research. There, a novel type of outsourcing has been happily embraced by Big Pharma, to the point of having almost totally displaced university research, only then subsequently to colonize it. In many ways, it provides a prevision of the economics of science in general in the future.

Consequently, I shall focus on the recent new new thing in the pharmaceutical sector, a purpose-built economic institution that exemplifies the strengths and weaknesses of the post-1980 era of commercialized research. That institution is called the contract research organization (CRO); for most intents and purposes it did not exist before 1980.[32] These CROs differ profoundly from earlier for-profit toxicology, bioassay, and pharmaceutical testing firms, which did long predate the CROs but which they have tended to displace. What began as small, specialized boutique firms offering narrowly targeted outsourcing services to pharmaceutical clients seeking help setting up clinical trials have come to dominate drug development and clinical trial management worldwide. And yet, there are no readily reliable data sources for the CRO. Astoundingly, although reformers regularly press for the open and transparent registration of clinical trials, there is no way to keep track of basic data concerning the main entities that carry them out. The FDA itself seems to have only the sketchiest knowledge concerning their operations.[33] Indeed, the only aggregate data tend to come from the industry itself, and therefore they must be treated with some caution.

Curiously enough, although CROs have been a subject of some anguish in the medical literature, thus far the literature on science policy and academic research on the economics of science has neglected their existence. Bustling biotechs may mesmerize the median venture capitalist and tantalize the TTOs, but it has been the CROs that have provided more consistent growth and transformation of pharmaceutical research in the current regime. A further neglected aspect of the recent development of CROs has been their gradual expansion into nearly every stage of the discovery, development, and marketing of new pharmaceuticals (Gad 2003). One survey pointed to preclinical research as one of the fastest growing areas of CRO services in the first decade of the new millennium (Milne and Paquette 2004). An industry source admits the existence of "a growing trend for CROs to cross-sell or package business, such as attaching phase I work to preceding preclinical work" (CenterWatch 2008, 60). Their activities have been known to range from initial screening of molecules for biocompatibility, in vitro screening, pharmacokinetic modeling, chemical synthesis and analysis, all phases of clinical testing, dosage formulation and pharmacy services, to all aspects of the regulatory process. In the 1990s CROs became much more preoccupied with the organization and provision of bioinformatics (McMeekin and Harvey 2002), to the extent of offering one-stop shops for archiving and control of clinical trials data. They also have become a major vehicle for exploration of the promise of pharmacogenetics (the genetic profiling and segregation of populations in clinical trials to better predict the efficacy of experimental drugs upon distinct subgroups).[34] CROs have sometimes been compared with modern accounting firms, which have also extended their services well beyond their initial remit of simple record keeping and audit. A further critical phenomenon of note is the rapid expansion of CROs in China, a topic I shall return to below.

Whatever the cause, analytical neglect by academics of CROs has been unfortunate, because their successes have catapulted them into the vanguard of a movement that insists that science conducted in a for-profit modality has had no deleterious effects upon the conduct of research. As a major spokesperson for the industry (and Quintiles executive) put it:

> Those of us who choose to pursue clinical science within the CRO industry reject the assumption that wisdom and ethical behavior are solely the province of the academy or the government. We reject also the presumption that the pursuit of profit along with the progress of science and medicine is inherently in conflict. In fact, in our experience the marketplace accurately reflects the public's hopes and expectations for science, and is a powerful guardian of behavior. It has little tolerance for shoddy performance or misapplied energies. It is a powerful mechanism for progress, for which no apologies are needed. (Davies 2001)

Estimates of growth of the CRO sector tend to be rather impressionistic but nevertheless quite impressive when one sets the zero origin at 1980. Some years ago, Davies (2002) presented estimates that the CRO market topped $1 billion in 1992 and had reached $7.9 billion worldwide by 2001. Shuchman (2007, 1365) puts it at $17.8 billion in 2007; CenterWatch projected it to grow to $25.9 billion by 2010. Industry-provided data in Figure 5.3 mostly confirm these estimates. For purposes of further comparison, Davies estimated that the market share of the top twenty CROs was half a billion dollars in 1992 and $4.6 billion in 2001, thus indicating a moderately concentrated industry structure. Concentration continues apace in the sector, with acquisitions of CROs by peers growing from five in 2001 to eighteen in 2004, thirty-one in 2006 and thirty-five in 2007 (CenterWatch 2008, 233). In 2004, one estimate suggests leading CROs managed nearly 23,000 Phase I–IV studies at 152,000 clinical sites worldwide.[35]

Another way to gauge the exuberant growth of CROs is to look at recent revenue growth of the four largest CROs in the United States in Table 5.5. Of these four CROs, Quintiles Transnational was incorporated in 1982 and Parexel International was founded in 1983; Covance was formed in 1987, as a unit of Corning.

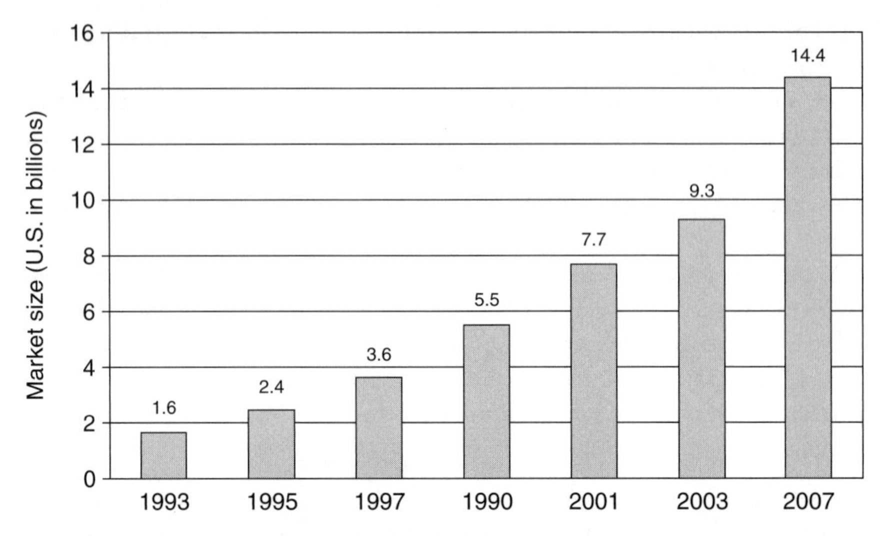

Figure 5.3. Estimates of growth of CRO sector. *Source:* Association of Contract Research Organizations Web site, http://www.acrohealth.org/.

Table 5.5. CRO firm revenue growth (in millions of dollars)

CRO revenues	2006	2005	2004	2003	2002	2001	2000
Quintiles	2,530	2,398	2,146	2,046	1,992	1,619	1,659
Covance	1,406	1,250	1,056	974	924	855	868
Parexel	760	671	**[NA]	619	564	387	378
Pharma Prod. Dev.	1,247	1,037	843	727	608	431	345

Source: Hoovers Online at http://cobrands.hoovers.com.

Perhaps a different yet more salient way to gauge the expansion and significance of the CRO is through comparison of the relative proportions of the pharmaceutical research industry R&D budget in clinical research that have been conducted through CROs with the amounts spent on their primary competitors, the academic health centers (AHCs).[36] One source suggested that the percentage of industry-sponsored clinical research captured by AHCs fell from about 80 percent in 1988 to 40 percent in 1998 (Davies 2001). Another source estimates the market share of AHCs dropped from 71 percent in 1991 to 36 percent in 2001 (CenterWatch, quoted in Parexel 2003, 130). A third source estimates that the share of outsourced pharmaceutical R&D going to AHCs was 23 percent in 2006 (CenterWatch 2008, 262). While there are unfortunately no official statistics on the absolute size of the CRO and AHC sectors, and none of these estimates can be treated as definitive, it would appear that an extraordinary displacement of AHCs by CROs is underway.[37] It seems the days are numbered for the previous Cold War model of academic medicine, at least when it comes to drug trials.

What is it that especially sets the CRO apart from anything that came before it? In the first two regimes of American science management, Big Pharma had gotten accustomed to outsourcing clinical trials to organizations like AHCs well before the inception of the CROs; there is a sense in which many of the component practices of the CRO were in fact innovated by universities. Outsourcing of this kind of R&D has had a long history in the industry: patently, it is *offshore* outsourcing that seems radically new. One might also object that commercialization of pharmaceutical research in the clinical sphere effectively predates the appearance of CROs. There seems to be evidence (particularly in the U.S. context) that some components of the market "innovations" we identify in what follows may have been present in some incipient form as far back as the 1930s, if not earlier.[38] Instead, my immediate intent is to provide an explicit census of how CROs have interacted with the reengineering of *modern* pharma laboratory, the post-Chandlerian corporation and clinical pharmaceutical research in the globalized privatization regime. By pulling together scattered evidence from journalistic accounts and the medical literature on what might be called a nascent "globalized mode of production" of drug research rooted in specific legal, social, and organizational structures, now centered in the CRO, I approach the topic by asking, what was the complementary innovation in the clinical sphere to the widely acknowledged innovation of the biotech model?

In short, how has clinical research in a CRO been altered by the imperative of commercialization? Was it in fact prompted by technology, as industry insiders claim, or, as in the case of the biotech model, much more by economic imperatives? We then can venture to broach the further question: how have these innovations then eventually found their way back to modern university science?

Who Wanted the Rise of the CRO?

In the postwar period, research into drug efficacy became a very formalized and ritualized process in America, due directly to government regulation. As the demands imposed on pharmaceutical development by the FDA have become more elaborate, they also came to be regarded by the industry as excessively onerous. Recruiting subjects, managing diverse trials in different settings, monitoring and recording data, subjecting data to statistical controls and higher-level analyses, and writing up the results for publication all absorb vast amounts of time and money. From the pecuniary perspective of the pharmaceutical corporation, the more time and money spent on FDA trials and procedures, the less time is available for reaping sales under patent protection. Because of a perceived need for speed, and because of the vast sums of money involved, there arose the impression of a conflict between conventional norms of (academic) science and the commercial imperatives operating in the drug development process. As one former FDA commissioner put it, "Many drug reviewers had become accustomed to working at an academic tempo, largely devoid of deadline pressure" (Kessler 2001, 40). Dilatory and dawdling scientists have historically been the bane of every neoliberal reformer, as we have observed repeatedly, but pressures also emanated from other quarters. In order to placate certain organized patient constituencies (especially AIDS activists) as well as address the concerns of the pharmaceutical industry, the FDA made numerous changes to its requirements for drug trials in the 1990s, under some circumstances dispensing with placebo comparisons, and implementing other rule changes that served to shorten the average time from NDA to drug approval. Indeed, industry research documents a sharp downward trend in approval times in the 1990s, with the percentage of new drugs receiving FDA approval within six months from NDA increasing from 4 percent in 1992 to 28 percent in 1999 (Kaitin and DiMasi 2000). Other sources point to declining clinical and approval times over the last decade (Hopkins et al. 2007, 575; CenterWatch 2008, 101). Because of various neoliberal deregulation initiatives, converting government activities into proto-market fee-for-service, somewhere between 12–17 percent of the FDA budget was accounted for by fees paid by pharmaceutical firms to expedite the regulatory process at the turn of the millennium (Abraham 2002, 1499).

Nevertheless, the pharmaceutical sector still criticized these attempts at quickening the pace in the 1990s for not going far enough, in part because, in the industry view, the problem resided as much in the academic clinical sphere wherein mandated drug testing had previously been conducted (Feinstein 2003). Interminable delays were not *just* the fault of the government. The new automated screening protocols were resulting in a tidal wave of new candidate compounds: One estimate suggests the number of drugs in U.S. Phase I clinical trials grew from 386 in 1990 to 1,512 in 2000 (Walsh et al. 2003a). Even though FDA time to approval from NDA was shrinking, the duration of the clinical developmental cycle was lengthening, at least until very recently.[39] In the corporate view, the remedy for this creeping prolongation of the clinical phase of drug testing was the promotion of a new breed of scientific researcher who was more comfortable with deadlines, and who focused more intently upon specific minutiae of the FDA guidelines and less on the complicated range of patient complaints; someone who appreciated the pragmatic importance of narrowly formulated questions as well as cost containment innovations, and less held in the thrall of academic advancement. Further, the need to recruit ever-larger patient populations seemed to require another kind of entity to coordinate clinical trials. Pharmaceutical corporations preferred to treat drug testing and research as fungible services, whether conducted in-house or contracted out, leading to a quantifiable output largely defined by FDA parameters, an output that could be monitored for its marginal contribution to the bottom line. It was not generic medical knowledge, but rather *data* that hastened approvals at the FDA. The pharmaceutical companies were casting about for a specially engineered research entity to impose real time cost-constraint, and a few far-sighted entrepreneurs came up with it in the 1980s: hence the standard "demand pull" account of the contract research organization.

A small body of literature attempts to explain why in the 1980s Big Pharma began to outsource its drug R&D and testing to external firms rather than to AHCs, initially in Phase II and Phase III trials.[40] This literature seeks to explain why CROs came to dominate the pharmaceutical sphere, displacing not just clinical drug testing in AHCs but also some in-house basic research. Mostly this literature focuses narrowly on financial problems in pharmaceutical corporations and the supposed benefits of network externalities, often neglecting larger issues such as the reconstruction of the research process, the changing nature of research questions,[41] concomitant revisions of intellectual property, and the consequences of globalization.

Conventional accounts (Petryna 2009) offer a number of reasons for the rise of CROs in the 1980s and 1990s, emphasizing the pharmaceutical industry's drive for increased efficiency and cost savings. Pharmaceutical R&D has grown substantially—as evidenced by gross costs climbing precipitously from $2 billion (U.S.) in 1980 to $30.3 billion (U.S.) in 2001 and $43 billion in 2006 (CenterWatch 2008, 68). This increased cost proved burdensome when stretched over longer horizons, as a day's delay in FDA approval has been estimated to penalize a firm with more than 1 million dollars in lost revenue (Abraham 2002, 1498). Consequently, the CRO filled a niche by offering targeted drug expertise, timely clinical trial completion, and eventually "end-to-end outsourcing support for all phases of clinical research" at a comparatively lower cost. Further, CROs offered the ability to start or cancel clinical trials on demand by smoothing the stop–go cycles of drug development and minimizing idle in-house research capacity.

Concurrently, the 1980s witnessed a gale of creative destruction in Big Pharma, with firms seeking to control emerging technology and establish global market share through buyouts and takeovers (Chandler 2005b). In the economic climate of the 1990s, the CRO industry rode the merger wave by acceding to the demands of the surviving pharmaceutical companies' desires to shed a portion of their labor force and cut back on expensive in-house laboratory capacity. Because proportionally fewer and fewer candidate molecules were panning out as successful new drugs, increased speed and ruthlessness in terminating unpromising trials would help contain costs; this would prove easier if the research was not conducted in-house or by quasi-independent academic contractors. Ironically, as the biotech model sought to outsource upstream research to quasi-academic entities, the downstream model sought to remove clinical research from academic settings. Therefore, to maintain managerial prerogatives and save costs, the pharmaceutical firms outsourced much of their routine development and clinical trials to CROs.

These conventional industry-sponsored accounts sometimes acknowledge that some aspects of globalization played a role in fostering a niche for the CROs. Pharmaceutical firms might lack expertise on foreign relations and regional regulatory differences in a globalized world, it is said, and thus a full-service provider to coordinate clinical research across national boundaries was needed. Some U.S. pharmaceutical firms engaged in regulatory arbitrage by pursuing a Europe-first strategy for drug approval in order to leverage a first-mover advantage for getting their drugs to market more rapidly. Moves to "harmonize" drug evaluation procedures across major markets, such as Japan and the European Union, only rendered this activity more attractive. Again, it made sense for these firms to offshore at least some of their clinical trials to other nations, rather than maintaining an elaborate far-flung global research and regulatory capacity. The CRO industry leapt in to supply the relevant cross-cultural expertise and transnational coordination in international clinical studies. CROs promised reduced time needed to find physicians and recruit patients for experimental regimens, and thus encouraged the clinical trial to proceed with relative celerity under diverse sets of regulatory circumstances.

One further conventional explanation of the rise of the CRO is that it was inherently better positioned than pharmaceutical companies or hospitals to take advantage of major technological changes in the way drugs were coming to be screened and tested. Some analysts suggest that pharmaceutical companies did not accord high priority to elaborate instrumentation specially geared toward evaluation of drug efficacy: "Pharmaceutical companies are not in the instrument business" (Lester and Connor 2003). Examples of such specialized technologies include custom-made information technologies and integration of genetic screening into the clinical process. An example of the former, PharmaLink, a CRO providing Internet-based clinical trial management, represented the vanguard of paperless clinical trial management via e-technology. This approach promised to ensure more accurate data collection, providing clients with instantaneous data access and expediting the clinical trial process. In the case of expediting the clinical trial process, the use of pharmacogenetic technologies for drug development would require, among other things, large, genetically screened patient pools and technologically complex clinical trials—an ideal task for a CRO. For example, PPGx, one such joint venture, "is one of the first attempts to build an integrated pharmacogenomics operation, including proactively genotyping healthy volunteers for Phase I." Both pharmacogenetics and pharmacogenomics are said to have opened up further opportunities for CROs to displace AHCs.

From our current perspective it is important to note, according to the canonical story, it was primarily external economic factors bearing down on pharmaceuticals that prompted the rise of the CRO industry, encouraging a new niche entity to augment efficiency and reduce cost. Big Pharma was therefore simply passively responding to market signals. However, the portmanteau term "cost" covers a multitude of sins: Analysts of CROs rarely explore the possibility that the neoliberal reconstruction of clinical research as a whole itself was the immediate *raison d'être* for the rise of the CRO and they do not clarify just why the innovation had to assume the format of a freestanding commercial entity rather than a restructured in-house research capacity. The conventional account rarely links the CRO to the concurrent invention of the biotech model or explores the ways they reinforced one another. Costs could have been reduced by conducting an invariant science under transformed economic circumstances or, alternatively, by changing the kind of science that was performed. In the case of the latter, an arm's-length relationship to the originating pharmaceutical firm would be necessary for the larger restructuring of the research process within a more thoroughly privatized framework. This turned out to be one of the most salient consequences of CROs, even if it may not have been paramount for the particular entrepreneurs who created the new institution.

CROs Transform the Very Nature of Drug Trials and Clinical Science

Representatives of CROs generally resist any suggestion that privatized drug testing and prospecting have undergone profound transformation in the last two-plus decades. Instead, they prefer to promote their advantages as simple matters of cost and convenience, rather than any alterations in their conduct of science. The stubborn concentration upon the rationale of cost savings reinforces that tendency and discourages inquiry into the changing character of modern biomedical research. There are at least three reasons to be skeptical of the popularity of narrowly defined, conventional "economic" explanations of the rise of the CRO cited above. First, they tend to divert attention from the actual means through which the promised cost savings came to be realized. In their commercial presentations in the 1990s, CROs frequently compared the cost of their research with that of AHCs, rather than research costs internal to the pharmaceutical industry. Such invidious comparisons with the previous era of drug research in university or other teaching hospitals—implying that CROs were intended primarily to displace academic science—only obliquely concede that pharmaceutical research has been reengineered to a different set of specifications. Second, one of the major selling points of the CROs, again highlighted in their commercial presentations but escaping conventional wisdom, is that data concerning the conduct of drug trials have been rendered more dependably proprietary. In other words, if the sponsor did not want the data to see the light of day, then it would not. It might even provide plausible deniability in the case of legal complications. Publication was only one ancillary goal of drug research; making inconvenient unwanted data disappear was another (Lurie and Zieve 2006, 94). Consequently, information on the conduct of CRO research has become even more inaccessible to concerned outsiders than similar information would have been under the earlier academic regime—a standard feature of the current commercialization regime. A third reason for remaining skeptical about "cost savings" explanations is that when pharmaceutical executives were surveyed about the reasons for their own decisions to outsource clinical trials to CROs, they ranked cost savings as relatively low on the scale of importance (Pichaud 2002). Indeed, difficulties with holding CROs up to consistency and quality standards, together with the risks of noncompliance, have given rise to the pharmaceutical catchphrase, "A CRO is only as good as its last contract" (Azoulay 2003). The clients of CROs are fully aware of the reputation of the sector as the locus of inferior deskilled jobs within the pharmaceutical industry.

Nevertheless, in the face of these qualifications, CROs have managed to convert a set of research protocols originally constructed around the prerogatives of the individual scientist, and, to a lesser extent, the concerns of the medical community, into a second set of protocols more suited to controlling the developmental cycle and marketing of new pharmaceuticals. The CROs have conjured up a set of research practices that are more effectively adjusted to the traffic and rhythms of corporate privatized science. They maintain confidentiality of data; evade prolonged, inconvenient, and costly treatment regimens irrelevant to the testing program; minimize patient interaction; and alter research protocols in a manner that augments the bottom line. They even have rationalized and sped up the very process of composing a journal article that can gain acceptance at the *Journal of the American Medical Asso-*

ciation or *the New England Journal of Medicine.* Another infrequently mentioned consideration is the possibility of arm's-length management of legal liability for adverse outcomes (Petryna et al. 2006, 58fn24). These facts help to explain the undeniable success of the CROs in capturing the bulk of clinical drug trials away from AHCs.

Recent medical literature skeptical of the virtues of CROs glimpses that something is being done differently than in earlier AHCs (Shuchman 2007), but these authors frequently find it difficult to put their finger on just where the real differences lie. Perhaps this can be attributed to the fact that much of the medical literature is more inclined to discuss the failures of CROs as local "pathologies of pharmaceutical science," or perhaps "ethical lapses," than as conscious structural consequences of a wider commercialization imperative in the pharmaceutical sector. These commentators seem unwilling to entertain the troubling notion that we are getting exactly the kind of science we have been willing to pay for. If we can avoid viewing these phenomena as the dubious behaviors of a few morally challenged individuals, or transgressions of the terminally greedy, and instead approach them as structural changes in the organization of science, then it may become possible to regard them as harbingers of the future shape of privatized science.

In the first part of this chapter, we found that the biotech model ends up producing different kinds of candidate molecules, ones better aligned with the business plans of Big Pharma and the general exploitation of intellectual property. In this section, we shall discover that the CRO has had much the same kind of effect on downstream research, producing a different sort of "clinical trial" than had been run previously through academic hospitals, and consequently, a different kind of "data" about health outcomes.

The only time it seems CROs get much attention (outside of the business press) is when some sort of dramatic medical disaster has happened, like the ill-fated London Phase I trial conducted by Parexel for the biotech TeGenaro (Goodyear 2006). But, in fact, such serious adverse events are a symptom of much more profound changes to the process of clinical research, sometimes called "medical neoliberalism" (Frank 2002; Abraham 2007b; J. Fisher 2008; Cooper 2008a, which are often surveyed in the medical literature without making explicit their fundamental connections to the spread of the CRO. The medical and legal literatures have discussed the new regime of industrialized research under five headings: (1) new approaches to research on human subjects; (2) restructured controls over disclosure and confidentiality; (3) management of intellectual property, especially in the case of research tools; (4) transformations of the role and functions of publication, and the systematic appearance of ghost authorship; and (5) reordered goals of scientific research. Here I present a brief overview of each category of transformation, combined with an analysis of the ways in which each phenomenon is linked to vital modern CRO functions. The moral of each tale reinforces that encountered elsewhere in this volume: Alterations in the organization of research eventually change the character of the scientific output.

Research on Human Subjects

Research on human subjects in the pharmaceutical industry has perhaps been the most vexing and contentious source of problems in the drive to speed up the drug developmental cycle and, as such, has drawn substantial attention in the popular press.[42] Rendering research subjects more pliable has always been a touchy proposition for medical "reformers." Examples such as the perverse "experiments" of Nazi doctors on concentration-camp inmates in World War II, the Tuskegee Institute syphilis experiments, or the plutonium injection experiments commissioned by the Atomic Energy Commission (Goliszek 2003) remain paradigmatic cases of twentieth-century "science" gone haywire. Responding to a conviction that not every scientist could be trusted to treat subjects in a humane fashion, the U.S. Congress passed the National Research Act of 1974, requiring that every institution that accepted federal funding must set up an institutional review board (IRB) to monitor the treatment of human subjects. The U.S. Department of Health and Human Services was also mandated to provide oversight and guidance for the IRBs. The need for specialists to staff these academic IRBs gave rise to the job category of the "bioethicist," with one of the few academic growth areas in the discipline of philosophy in recent decades, "medical ethics."

Until 1981, local IRBs, which at that time were housed mainly at universities and nonprofit institutions, oversaw clinical trials. Yet the whole concept of a locally based volunteer IRB had been predicated on an older, obsolete version of lone scientists running small-scale, self-contained clinical trials at geographically isolated AHCs, with the local academic community as their conscience and oversight—a model rapidly becoming supplanted by CROs. There existed a myriad of reasons for pharmaceutical firms to be dissatisfied with local IRBs: They imposed idiosyncratic protocol guidelines; they had no appreciation for cost and speed considerations; their legal status was too uncertain; and so forth. Consequently, in 1981 the FDA permitted the creation of independent IRBs in order to "provide oversight to investigators doing FDA-regulated research in their offices or in institutions too small to support an IRB" (Forster 2002, 517). This led, in turn, to the creation of a novel occupation—dedicated bioethics consultant—which gave rise to contradictions of its own. This new for-profit niche market provided services to CROs from their very inception.[43]

Why would this become so important for the CRO model? The short, harsh answer is that CROs were created partly to alter the nature of medical care in pursuit of lowering costs of the conventional clinical drug trial. What this meant in practice was shifting the center of gravity of the enterprise from figuring out what was specifically wrong with the individual patient to the efficient and timely production of standardized *data* for FDA protocols.[44] Yet this could not be accomplished in any direct or open manner, since that would run the risk of violating everything from the Nuremberg Code to the Helsinki Declarations to local tort law, not to mention the inevitable adverse publicity that would be bad for business. Instead, ways would have to be found to get around standard AHC practice, which meant, among other things, having your protocols blessed by an IRB that understood your business plan.

The advent of independent IRBs thus held several advantages for CROs: Compared with independent IRBs, local academic IRBs sported more lengthy mean approval times: thirty-seven versus eleven days (Parexel 2003, 139). Anything that sped up the process was resonant with CRO objectives. Local IRBs are regulated by the National Institutes of Health, the Office for the Protection of Research Risks, and the FDA, whereas independent IRBs would only have to conform to FDA requirements. Independent IRBs proved capable of financial expansion commensurate with the volume of research reviewed. Some academic IRBs have been known to supervise more than a thousand clinical trials simultaneously, devoting no more than two minutes of discussion per study. Local IRBs were saddled with the same level of funding, regardless of the volume of research supervised, placing severe strains on them as the volume increased (Forster 2002; Ghersi et al. 2004). Further, academic bioethicists in local IRBs would tend to raise questions about conflicts of interest and include severance of ties in their recommendations, a possibility that was anathema to the pharmaceutical industry.[45]

The independent IRB, with its bioethicists for hire, proved a boon to the CRO industry, fortifying it with a substantial competitive edge over rival research institutions. As a prerequisite for the full privatization of clinical research, ethical oversight of a "rationalized" human subject research had itself become transformed into a fungible commodity. Unlike CROs, AHCs and other nonprofit institutions did not fully benefit from the creation of independent IRBs. According to the preamble of the FDA regulation approving independent IRBs, "A sponsor-investigator who is unaffiliated with an institution with an IRB can comply with this requirement by obtaining review at an institution whose IRB conforms with these regulations or by submitting the research proposal to an IRB created under the auspices of a local or State government health agency" (quoted in Forster 2002, 517). Most AHCs already had their own IRBs, complicating or precluding their use of independent IRBs. Universities thus faced a contradiction: While they were happy to encourage bioethicists on their faculty to moonlight to augment their salaries, their own AHC researchers were prevented from using for-profit IRBs to expedite their research.

From Big Pharma's point of view, even this competitive advantage did not completely come to grips with what they perceived as the panoply of drawbacks of human research. A string of high-profile failures of IRBs in the 1990s, which led to federal sanctions at Oklahoma, Johns Hopkins, Duke, Colorado, and other universities, suggested that human subjects' oversight was likely to attract even more costly regulation. Numerous surveys showed that the media coverage of problematic clinical trials discouraged average Americans from taking part in clinical research. Starting in 1980, the FDA had accepted drug applications supported by data from foreign clinical trials as a way to get access to larger populations of patients. By 1987, it was permitting NDAs consisting *solely* of data from overseas trials, when the countries involved met certain preconditions. The requirements were further loosened in 2003, as I discuss in the final section of this chapter. Such overseas trials could conveniently circumvent the more onerous restrictions on treatment of human subjects imposed on U.S. trials, although the CRO industry tends to point to conformity with the bureaucratic letter of "good clinical practice" rules, and stresses the advantage of availability of "treatment-naive" subjects unavailable in developed countries. This unintended consequence of U.S. restrictions on human subject research provided yet another competitive advantage for the nascent CRO industry: Unlike an AHC, a CRO was not tied down to a particular geographic locale or academic setting. Furthermore, it could engage in regulatory arbitrage, using its superior economic and political clout in poorer, less-developed countries to negotiate lower costs.[46]

The muckraking literature on "foreign bodies for sale" has grown in the medical and news media; it periodically has broken out into major policy controversy.[47] One estimate around the beginning of the twenty-first century put the percentage of foreign test results in FDA deliberations at 37 percent (Datta 2003); a different estimate puts it at 32 percent in 2006 (CenterWatch 2008, 92). More recently, the Office of the Inspector General (OIG) found estimates of foreign trials as high as 80 percent for approved applications for drugs and biologics in FY 2008 (OIG 2010). Even more disturbing, the OIG uncovered evidence of early-phase clinical trials prior to filing an IND conducted overseas: these entirely evade any regulatory oversight. A different estimate, based on the clinical trials registered by the twelve largest U.S. pharma firms, places the percentage of their offshore trials at around half

(Cooper 2008a, 82). While the trend has many implications, I shall focus on how it has been implicated in the way CROs have reengineered scientific protocols. First, they have enhanced the speed of Phase II and Phase III trials: Many of the factors counseling caution and deliberation in developed-country trials are obviated or ignored in Eastern European, Chinese, or Third World trials. Putting it delicately, physicians in these countries have no qualms about conducting "rescue studies" for me-too drugs: a prospect that might be regarded with some disdain by U.S. physicians (Petryna 2007, 35). Countries such as China and India encourage what they consider to be Western-level provision of medical treatments by offering direct subsidies to the CROs. Informed consent in such situations often is impossible, so a level of coercion of patients prevails that would be unthinkable in the developed world. Significantly, some of these countries waive or reduce the requirement of preclinical animal trials, further truncating the drug-development profile (Shah 2003; Sharma 2003). While the FDA uses various tools to police domestic clinical trials, with foreign research it can do little more than disallow the results. In more than 90 percent of the cases, the FDA has not been notified that foreign clinical trials have been initiated, and it has essentially no control over their conduct. Consequently, many outside analysts tend to question the quality of the data generated in such clinical trials.[48]

Another unintended consequence of the commercialization of human subject research is that patients in developed countries have begun to realize that CROs sometimes paid recruitment fees to physicians of $12,500 or more per subject (Drennan 2002). Patients in the developed world have begun to rethink their own roles in clinical trials and are beginning to demand direct payment in order to participate (M. Fisher 2003, 260). If scientists are entrepreneurs, why can't patients be entrepreneurs, too? Not surprisingly, such demands are strenuously resisted by the CROs; obviously they provide further incentive to shift Phase II and Phase III trials overseas, where patients are far less obstreperous.

Disclosure and Confidentiality

Much of the medical literature anguishes over the problem of "conflicts of interest" and yet pays insufficient attention to the dynamic interplay between CROs and AHCs. Conflicts of interest may trouble academics and lawyers, but they do not seem to present obstacles for many CROs. One motive for the innovation of CROs was to better accommodate research to the imperatives of strengthened intellectual property, and in particular, to only release data if the sponsor sanctioned it and otherwise keep it under wraps. Once CROs entered the arena in a big way, AHCs found they could no longer maintain older vintages of "open science."

According to some estimates, one-third to one-half of the clinical trial contracts in the later 1990s with AHCs such as the Massachusetts General Hospital (Bodenheimer 2000) or the Geffen School of Medicine at the University of California, Los Angeles (Kupiec-Weglinski 2003), contained restraint clauses, confidentiality provisions, publication embargoes, and a host of other legal controls over proprietary information. Fiduciary officers of AHCs once regarded it as their duty to renegotiate such clauses, but their efforts to set themselves up as a last line of defense for open science had curious consequences. It has been demonstrated that scientists with industry support are more likely than those without it to deny others access to data or research materials (Blumenthal, Campbell, Causino, and Seashore Lewis 1996, 1737). Yet in recent years, many university administrators have succumbed to pressures to accept restrictions on proprietary information in exchange for funding. One consequence has been a growing conflict at AHCs between technology transfer officers and university officials, with the former being more willing to condone restraints on contracts, because of their experience in dealing with patent attorneys (Eisenberg 2001, 239–241). Another consequence is that AHCs have attempted to "reform" their practices to better resemble those of the CROs, in order to recoup lost pharmaceutical contracts (Campbell, Weissman, Moy, and Blumenthal 2001; Pollack 2003). With these practices, universities have managed to invoke the ideal of open science while proving unable to maintain it in practice.

The pharmaceutical companies have not hesitated to exercise their legal powers to restrain disclosure. Although few clinical researchers experience such crude attempts at force majeure to intimidate them to trim their research to conform to company demands, it has been well documented that "using financial, contractual and legal means, drug manufacturers maintain a degree of control over clinical research that is far greater than most members of the public (and, we suspect, many members of the research community) realize" (Morgan, Barer, and Evans 2000, 661). They do this through selective disclosure and restraint on almost every aspect of the clinical trial process within the CRO. The net intentional result is that the science that appears in the journals is the science that has been selectively released by the sponsors. The most replicated finding in the last twenty years of meta-analyses of published clinical studies is that industry funding is highly correlated with results favorable to the drug owned by the study's sponsor (Sismondo 2008). Here are some examples: One survey of research papers on the cost-effectiveness of six oncology drugs showed that "pharmaceutical company sponsorship of economic analyses is associated with reduced likelihood of reporting unfavorable results" (Friedberg, Saffran, Stinson, and Bennett 1999, 1453). Another study (Stelfox, Chua, O'Rourke, and Detsky 1998) found that 96 percent of authors supporting the use of calcium-channel blockers had financial ties to the drugs' manufacturers. A similar result was found for anti-inflammatory arthritis treatments (Rochon et al. 1994). One examination of published surveys and meta-analyses of drug efficacy found that "studies sponsored by pharmaceutical companies were more likely to have outcomes favoring the sponsor than were studies with other sponsors" (Lexchin, Bero, Djulbegovic, and Clark 2003, 1167). Another survey took into account industry affiliations of the academic unit (such as equity ownership) and individual corporate and consultant relationships and found that "approximately one-fourth of investigators have industry affiliations and roughly two-thirds of academic institutions hold equity in start-ups that sponsor research at the same institution . . . these articles showed a statistically significant association between industry sponsorship and pro-industry conclusions" (Bekelman, Li, and Gross 2003, 454).

Some commentators have been offended by the implication that trained clinicians and research physicians can so easily be swayed to produce skewed scientific results on demand, but others insist that the problem is not that investigators are crudely falsifying the data or otherwise abandoning their commitment to truth. They concede that when research is spread over vast numbers of clinicians and disparate geographical sites, then there are simply too many individually small but cumulatively decisive ways for the data to be biased in a "positive" direction. Such sources of bias include the selection of subjects, strategic choices about how to treat drop-outs, protocols for handling and reporting side effects, deciding whether to use placebos instead of competing treatments, deciding the administration of rival doses, decisions about what constitutes a drug's efficacy (sadly, there are rarely clean "cures" for most of the syndromes in question), and decisions about when to end a trial.[49] While such biases have always beset clinical trials in one form or another, the privatization of science tends to insulate them from internal and external critique. As one researcher, Dr. Curt Furberg of Wake Forest, has observed,

> Companies used to do ten trials and then just pick the two they liked the best to submit to the regulatory authorities. Then this was stopped, and the agencies demanded to see all the data. The companies then needed to have more control, and the whole issue of academic freedom hit them. So to avoid this they have taken two routes—the use of clinical research organizations (CROs) and developing countries . . . CROs bypass the issue of academic freedom altogether, as the CRO wants to please the company it is working for and so constitutes no safeguard whatsoever. And the developing countries have no money, so an industry

> dollar goes a long way. Investigators are very anxious to please. (quoted in S. Hughes 2002, 6)

Another source for the bias in results is the rarely acknowledged "sweatshop" character of work in CROs. Compared with their counterparts in large pharmaceutical firms, researchers in CROs are lightly trained, poorly paid, and discouraged from exercising any initiative, which is why they have extremely high rates of turnover (Azoulay 2003). Curiosity is not conducive to the health of the bottom line. As one outsourcing manager admitted,

> There is a line-by-line definition of the CRO's responsibilities. That means that the CRO is less likely to notice stuff that might be going on at the sites. There are no incentives for the individuals at CROs for capturing "soft data," unlike here, where you get rewarded at every level. At a CRO, you might work for two or three sponsors at the same time. So it's all about hard deliverables. Anything beyond the contract you do not get. (Azoulay 2003, 22)

Conflicts of interest have recently become varied and baroque in clinical trials, even as they have become more pervasive. One major perplexity has been to come up with an adequate definition of conflict of interest in a world in which distinctions between academic and corporate institutions have tended to dissolve. Ironically, because the contractual relationships between firms and employees are more formally codified in the case of CROs, formal conflicts of interest may actually technically be less intrusive than in supposedly disinterested academic clinics. However, this may not be cause for optimism.

The problem with a conflict of interest in science is that it turns out to be a Pandora's Box: Once opened, it is nearly impossible to close (Slaughter, Feldman, and Thomas 2009). "Disclosure" offers no panacea, since it is unclear what precisely must be disclosed and to whom and under what circumstances. Should direct payments from the industry sponsor to the researcher be disclosed under all circumstances? Should it be expanded to include stock ownership or, even trickier, stock options? What if the investigator has an executive relationship, or sits on the board, of the sponsoring company or some interlocked firm? Does it cover indirect payments, such as consultant fees, honoraria, trips to resorts, "gifts"? What if the sponsor supports students or others designated by the researcher? These and other questions have been raised on a regular basis in the past two decades. To stem the tide, most universities have clad themselves with some form of conflict of interest policy, but there is no standardization from one institution to the next (Cho, Shohara, Schissel, and Rennie 2000) and no serious enforcement. In the next chapter we shall discover this is true of the NIH as well (Lederman 2010). Indeed, one study demonstrated that fewer than half of the clinical investigators interviewed at UCSF and Stanford could even correctly state the provisions of the conflict-of-interest policy at their own institution (Boyd, Cho, and Bero 2003).

Perhaps what is remiss at the university level can be rectified at the publication level. At least that seems to have been the rationale of the International Committee of Medical Journal Editors (ICMJE) when they promulgated the "Uniform Requirements for Manuscripts Submitted to Biomedical Journals" in 2001.[50] Unfortunately, this noble crusade by the journal editors to expose author ties to sponsors did not sufficiently take into account the larger forces transforming the very structure of publication and authorship in privatized clinical science. One study suggests that this well-intentioned attempt to legislate disclosure has not succeeded because "academic institutions routinely participate in clinical research that does not adhere to ICMJE standards of accountability, access to data, and control of publication" (Schulman et al. 2002, 1339). Worse, the *New England Journal of Medicine* was embarrassed in June 2002 into revoking its prohibition against authors of review papers having any financial ties to the drug companies whose medicines were being assessed. The reason the *New England Journal of Medicine* gave was that they could no longer find such putatively independent experts (Newman 2002). If conflicts of interest have become so ubiquitous, then disclosure can do nothing whatsoever to address the systemic bias that besets pharmaceutical evaluation of clinical trials.

The whole question of the role of conflicts of interest in science is mesmerizing due to its labyrinthine complexity in the new regime; it deserves closer attention than philosophers and science policy analysts have accorded it in the past.[51] The fundamental stumbling block seems to be the tendency to cast the problem as a matter of individual responsibility (Frangioni 2008), rather than as a structural problem in the organization of science. In the conventional treatment, truth is a communal goal that is impeded by biases clouding individual judgment. The weakness of this diagnosis is that no successful scientist believes that he or she is biased, however much he or she might believe it about others. Some interview transcripts evoke this viewpoint:

> It's a delicate thing. You have to decide for yourself. For example, I'm getting money from [a pharmaceutical company] for a study I'm working on. They also have me on speaker's bureau. I feel comfortable with this relationship as long as the slides I use are my own, and I'm speaking about my own research and opinions. I don't think the information I present has anything to do with what [the company] wants me to say.

> The system can be, and is, abused. Some people do give canned talks prepared by the companies that are paying them . . .
>
> $10,000 here or there is not a big deal. Personal financial relationships with sponsors are necessary for growth. People have to look at the big picture and see the benefits that come from academic–industry relationships . . .
>
> There is the risk that I become a complete whore and begin saying things that I don't believe. I'd hope I'd recognize this if it were happening to me, but it is hard to know. The risk to the public is one of fraud—scientists say something is true when it isn't. I don't think conflict-of-interest rules can mitigate either of these risks—it is basically up to the individual investigator to act ethically. (Boyd et al. 2003, 772–773)

The problem with the privatization of research is not that people may have personal biases or special interests, or even that they develop self-serving rationales for reinforcing them. No one should be surprised at this. Disclosure has become an issue in modern biopharmaceutical science because one (academic) set of social structures for navigating the shoals of human cognitive weaknesses was slowly being traded for another entirely different (corporate) one. In the interim, it remained in some actors' interests to blur the distinctions between the two.

Although conflicts of interest would seem to be pervasive in the CRO sector, we have seen there is little evidence of handwringing over disclosure and confidentiality, while in the meantime the CROs have gradually taken over clinical drug testing. At the most superficial level, this follows from the fact that analysts understand that researchers at CROs are first and foremost *employees,* and their motives are expected to be subordinate to the objectives of the firm. The firm, in turn, must be unyielding in its insistence on the immediate objective to supply clinical data to the contractor in a timely and cost-effective way that meets FDA and NDA requirements. The individual employees of the firm may perhaps bear their own personal conflicts of interest with these specific objectives of the firm—ranging from their own idle curiosity to concern over patients' general well-being to bureaucratic infighting to conceptual biases—but no one would ever expect such conflicts to be rectified by codes of ethics, medical journal strictures, or the intervention of academic committees. The CRO's predominant objective is simply to deliver a predefined product on time and under budget. By squeezing out all the scientific curiosity and serendipity of discovery out of clinical trials—that is, less euphemistically, deskilling the workforce—we simply get as output from the process what we expected at the outset: nothing more, and nothing less. In a smoothly operating firm, there are no rogue elements or hidden agendas; if they crop up, then you fire the employee. In CROs, conflicts of interest are not perceived as a problem requiring special remedy or concern, *because the new format has built-in means to discipline them.* This is a direct consequence of the sea change in the organization of science.

But since the CROs have undercut the AHCs in performing clinical trials, the university hospitals subsequently signed onto the neoliberal playbook by drawing the conclusion that it was they who must reform and come to imitate the CROs in structure and conduct (Blumenstyk 1998). Amazingly, a number of academic centers conceived of the notion that they could concoct a quasi-for-profit entity from within their AHC to sink to the lower common denominator of CRO practice, without at all impacting their own academic missions. First, Duke University formed the Duke Clinical Research Institute (DCRI) in 1996;[52] Pitt conjured Pittsburgh Health Research, Inc.; the University of Rochester formed the Clinical Research Institute, and so on. University CROs as wholly owned for-profit subsidiaries based in AHC legacy units are now ubiquitous. Of course, not every public university could follow suit, because many state universities are prevented from competing with private firms by their state charters. Nevertheless, today it is often a mistake to posit a constitutional distinction between AHCs and CROs, as the former has absorbed the latter, often through direct imitation or wholly owned subsidiary. The university has thus indeed literally become more like a corporation, but only as a consequence of now producing the same brand of just-in-time science as the CRO.

Intellectual Property and Research Tools

In Chapter 4, we discovered that the fortification of IP in the globalized privatization regime would never remain sequestered in application to the final consumer goods that tumbled out the hopper of the new knowledge economy in the privatization regime. Once participants came to subscribe to the neoliberal tenet that the Market was invariably the supreme information processor, then it was inevitable that intellectual property considerations would suffuse most of the practice of scientific research. We have observed in this chapter that biotech firms have had recourse to attempts to make money off ownership of upstream research tools, if only because they possessed so few meager sources of revenue over the longer haul. Yet this wouldn't happen, if indeed various devices did not exist to enforce ownership in the downstream research process. The initial recalcitrance of academic researchers to honor IP in the research process thus presented a serious obstacle, which had to be overcome in order for a thoroughgoing pecuniary coordination to take charge. Technology transfer offices within the university structure have sought to impose a certain modicum of discipline, but here we might like also to point to the CRO as another important factor in the spread of novel practices and standards up and down the pharmaceutical supply chain.

Knowledge of the chemical composition and perhaps the manufacture of a drug is not sufficient to turn it into a patentable therapy: One must understand the myriad of considerations from the etiology of the disorder to the correct dosage to the predictable side effects to the interaction with environmental/ecological variables and the nature of delivery systems that will administer the treatment. Knowledge of chemistry and biology shades into knowledge of medicine and clinical practice. The biotech firm and the CRO may ideally occupy opposing poles of this continuum, but in reality, they interpenetrate and commingle in ways that belie their supposed modularity in the outsourcing process. Back in the day when all these aspects were subsumed under the umbrella of the older pharmaceutical firm, these interdependencies could be negotiated and coordinated through bureaucratic hierarchies and stratified divisions. In the modern regime, fragmented firms are supposedly coordinated by contracts, cross-licensing, and the market. In a model so heavily dependent upon the salience of patents for profits, it was necessary that biotechs could exert contractual controls over clinical research, particularly through the instrumentality of MTAs. In that scenario, the biotech firm had no future unless it could exert downstream control of its product line through a parallel set of arrangements. This might function adequately in its relationship to Big Pharma, in instances where the biotech simply opted out of downstream development, but it encountered more friction in its relations with the clinical elaboration in the old-fashioned academic hospital. This was sometimes phrased as the complaint that "academic medical centers have a bad reputation in the industry because many overpromise and underdeliver" (Covance officer, quoted in Bodenheimer 2000, 1540). In order to persist, the biotech model had to conjure a doppelganger like the CRO in order to navigate the riptides of upstream and downstream confluence that constituted drug development.

Hence it is critical to observe how the biotech model and new paradigm of CROs acted as logical complements in the pharmaceutical industry arsenal. Biotechs and CROs had to learn to cooperate with one another in order to render their knowledge bases fungible and valuable. Sometimes the CRO itself made use of the novel IP regime to differentiate its product, while at other times the pharmaceutical firm did so because the CRO made such access significant and profitable. Neither alone proved necessary and sufficient to expand control over the research process; together they were an unbeatable combination.

One indicator of the critical role of the CRO in ensuring proprietary data control has been the spread of so-called electronic data capture (in the jargon of the business). Academic clinical trials tended to be more open-ended, with less standardization of the kinds of records kept, partly due to the wider-ranging curiosity of the investigator. Patient records were involved, to be sure, but they fell under the category of general personal medical records. If intellectual property were involved, one would tend to think of it at the level of the journal article or perhaps the treatment regimen. But the CRO is selling something distinctly different: highly standardized data on therapeutic markers and surrogate "end points" (often rather than real health outcomes), all in the context of proprietary information on the drug and its dosages. Since the clinical process was becoming delocalized and depersonalized, as explained above, data had to be captured and transformed into specialized predefined formats strongly hedged round with property protection. Thus, what at first appeared as a mere technological improvement—the computerization of patient data records—in fact constituted another of the major innovations of the privatization regime in the extension of intellectual property, on a par with other watersheds covered in previous chapters, such as TRIPS, Bayh-Dole, and the transfer of copyright control from authors to journal publishers.

Put bluntly, people had to lose control over information concerning their own bodies if research were privatized (Skloot 2010). Academic hospitals were not all that interested in facilitating this expropriation, which is why CROs loomed so very important in the framework of the new regime. Evidence shows that it has been CROs and Big Pharma who have led the way in introducing electronic data capture in clinical medicine (CenterWatch 2008, 340). The real upsurge in this practice dates only from the last decade (338). Of course, the Internet was a great boon in this process, allowing the collation of comparable data from far-flung foreign clinical sites, another major competitive advantage CROs had over AHCs. In most cases this has been accomplished with proprietary software, locking up access by cladding it with yet another layer of IP. Hence if there might be data that the sponsor does not want the outside world to see, then it will disappear from the scientific record, as surely as if it had never existed in the first place. All the well-meaning legislation to force registry and disclosure of clinical trials will never bring it out into the light of day.

Another key function of the CRO has been to facilitate the writing and elaboration of contracts and MTAs, in order to control the proprietary IP of both biotechs and Big Pharma. This has been illustrated by a number of key developments in the deployal of MTAs in order to engross and police research tools. Chapter 4 related the story of one of the more important contemporary CROs, Charles River Labs, and its role in the extension of patent protection to the Oncomouse. It is significant that in that case it was a CRO that ventured to transgress prior practice in the distribution of murine research tools, and not the Harvard lab of Philip Leder, even though it was he who had filed for the original patent. CROs pioneer the paths to privatization. In another example, it has been a growing trend that MTAs often conflict in their restrictions when two or more encumbered research tools are being used in a particular setting (Rodriguez et al. 2007, 260). As we have shown that universities rarely bother to pay attention to such fine points, increasingly it has been CROs who have taken upon themselves the task of negotiating the contractual thicket.

However, the key feature of the globalized privatization regime is that ownership of knowledge should become stabilized, so that firms are not discomfited by repeated challenges to their intangible assets. Here CROs occupy a market niche not filled by any other actor in the medical system. More significantly, unlike an AHC, a CRO contractually stipulates that it will not seek to patent research tools arising from its research.[53] It simply provides predefined *data;* it does not augment the body of *knowledge.* It is consciously and constitutionally *incurious.* The CRO thus demonstrates a proven advantage over its academic counterparts in what it can provide its clients: end-to-end management of a package of the IP consequences of research tools and data capture. No surprises means no eruption of unwanted research.

The Vicissitudes of Publication and Authorship

It would be a mistake to approach scientific publication as if it were simply a matter of disseminating newly minted information, perhaps after the fashion of the "new information economics." The publishing of research in the modern context performs many interlinked functions, perhaps too many. For instance, the appearance of a paper in a particular journal signals something about its significance; the names appended to the paper lay claim to whatever benefits may accrue to the publication, at least within the parameters of current IP restrictions. Publication also furnishes an option for "scientific credit," a contested and controversial entity in the best of times. Beyond that, the privatization regime has opened up new horizons in the functions of publication.[54] However beset with multiple functions, the commercialization of science has wrought havoc with older notions of the scientific author and her roster of publications. It is symptomatic of the orientation of modern economics that the literature on the privatization of science seems to have passed this phenomenon by, even as it has been the subject of extensive debate and anguish among researchers in biomedical fields.

Once again I take issue with David and Dasgupta (1994), by noting that the commercialization of science has not only had an impact on the level of disclosure of findings, but is also slowly changing the very meaning of the "scientific author." Their neoclassical model treats the scientist/author as an invariant independent entity. The simple integrity of the rational authorial agent is something that modern economists would probably tremble to question, even in their most fetid nightmares. And yet, far from being some passing postmodern fantasy, editors of some of the most prestigious medical journals have found themselves impelled to convene special conferences and retreats to debate the vexed question of "What is a scientific author?"

The first exploratory discussion was held in Nottingham, United Kingdom, in June 1996. Consequently, the fifth revised version of the ICMJE Uniform Requirements for Manuscripts Submitted to Biomedical Journals was promulgated in 1997; a follow-up conference sponsored by leading medical journals was held in February 1998 in Berkeley, California. There, the ICMJE formed a Task Force on Authorship, which met in May 1999.[55] Further attempts to clarify the expected "role of contributors" were promulgated at the *Journal of the American Medical Association* in 2000 because "rules regarding authorship, for example, those of ICMJE . . . were commonly ignored and flouted," although this initiative was not made uniform at other journals (Rennie, Flanagin, and Yank 2000, 89). This initiative was a reaction to a study by Rennie in the latter half of 1997, in which it was found that a full 44 percent of the names on the bylines of papers in the *Lancet* did not qualify for authorship even under a lenient interpretation of the ICMJE criteria (Yank and Rennie 1999). What provoked this flurry of activity? It was first sparked by a number of high-profile cases of scientific fraud in which certain reputable authors sought to repudiate faulty published papers that had their names appended as coauthors, on the grounds that they had not sufficiently monitored or supervised the empirical procedures that had been exposed as bogus. This led to reconsideration of the phenomenon of "gift authorship" or "honorary authorship," where famous or otherwise influential figures were listed as coauthors, even though they had not contributed "significantly" to the project (Bhopal et al. 1997). Subsequently, journal editors such as Frank Davidoff found that some authors submitting papers refused to tone down their interpretations of results, or even take full responsibility for provision of data sets, because of prior but unacknowledged conditions imposed by their industry sponsors. Further inquiry revealed that many listed coauthors, when contacted, would refuse to endorse the full text of published papers due to lack of agreement over methods, statistical analysis, interpretative commentary, directions for future research, and so on (Horton 2002). Embarrassing cases arose in which clinical data provided in the published papers differed substantially from those reported to the FDA (Okie 2001). Things appeared to be coming undone: Authorial voices seemed to have become unhinged from authorial identities. At that point, it became apparent that other IP considerations had also influenced attributions of authorship, particularly as scholarly journals contended with copyright issues in connection with electronic publication, only to realize that stakeholders other than the putative authors also held rights over texts and supporting data. However, the straw that broke the editors' backs was the phenomenon of ghost authorship: the practice through which researchers agree to put their names on texts that had been composed by unnamed third parties, who held final control over the content of the manuscript. Instances of ghost authorship had begun to surface in transcripts of trial proceedings of lawsuits against pharmaceutical companies.[56]

It should be insisted at the outset that some forms of ghost authorship had been tacitly condoned within the scientific community long before the practices we shall describe were a gleam in the eye of a Big Pharma rep. For example, it was pointed out in Chapter 2 that Paul Samuelson ghosted a major part of Vannevar Bush's ur-text in science policy, *Science—The Endless Frontier*. Many don't realize that the preponderance of those authoritative tomes emanating from the National Academy of Sciences are ghostwritten by NRC staff, even though famous scientists are frequently listed as authors (Oreskes, Conway, and Shindell 2008). One might be tempted to deplore such practices, but they rarely seemed engineered specifically to mislead their readers. That is one way contemporary ghostwriting in science has graduated to an entirely different plane, particularly in the biomedical literature.

The incidence of the "guest-ghost syndrome" in the medical literature would now appear to be approaching levels rarely seen outside of sports autobiography. The specter has been raised that the medical journals are teeming with the "non-writing-author/non-author-writer" as major personae (Sismondo 2007). As one can readily appreciate, the extent of this phenomenon would intrinsically be difficult to gauge because of its very nature: Imposture and concealment would not be worth the effort if it could be readily unmasked. Testimony to the concern of the medical community was an elaborate attempt to measure the extent of guest and ghost authorship by surveying all the authors in six different journals in 1996 (Flanagin et al. 1998). In the aggregate, 19 percent of the papers had evidence of honorary authors, 11 percent had evidence of ghost authors, and 2 percent seemed to possess both. Curiously, the prevalence of guest/ghostwriters did not differ significantly between large-circulation and smaller-circulation journals. In a Web-based survey focusing on the 1999 reports of the Cochrane Library (an international organization devoted to maintaining systematic reviews of risks and benefits of particular therapies employing a common methodology), a research team found that 39 percent of the reviews had evidence of honorary authorship, while 9 percent had evidence of ghost authorship (Mowatt et al.

2002). Other evidence of ghost authorship has been exposed through the disciplinary actions of medical journals: For instance, in February 2003, the *New England Journal of Medicine* retracted a paper it had previously published because several listed authors insisted that they had little or nothing to do with the research (E. Johnson 2003). In a different survey design that focused on a specific drug rather than on specific publication outlets, Healy and Cattell (2003) began with information from an unnamed medical communications agency that specialized in the drug sertraline. Using Medline and Embase, they then collated a list of all publications mentioning sertraline from 1998 to 2000. They were able to establish that fifty-five published papers had been coordinated by the agency, whereas forty-one had not. However, only two of the fifty-five papers actually acknowledged that the agency had provided "writing support."

Recent evidence produced as a by-product of litigation demonstrates that ghost authorship is all but ubiquitous in biomedical journals. The documents disclosed in a case against Wyeth over a hormone replacement therapy have been made available on a public Web site, revealing each step along the way as to how favorable articles are constructed.[57] This has prompted one of the experts in this area, Sergio Sismondo, to estimate that 40 percent of all scientific articles on new drugs are ghostwritten. A surplus of ghostwritten papers conjures the need for a plethora of outlets: Separate litigation in Australia has disclosed that Elsevier, the for-profit publisher of journals, was producing *at least six* ghostwritten journals, entirely sponsored and controlled by the industry (Singer 2009a; Goldacre 2009; Grant 2009a,b). For instance, the *Australasian Journal of Bone and Joint Medicine*, was admitted by Elsevier to be a "sponsored article compilation publication, on behalf of pharmaceutical clients." Ghostwriters are apparently so prolific they give birth to ghost journals, especially if all that matters is the thin veneer of legitimacy.

Certainly this proportion might seem outlandish to anyone who considers himself a scientific author (Gotzsche et al. 2007), but a factor omitted in expressions of dismay would be the contemporary dominance of CROs in clinical pharmaceutical research. One major reason for the epidemic of guest/ghost authors in the contemporary medical literature is the rise of the CROs in pharmaceutical research.[58] For CROs, the logic of ghost authorship is quite straightforward. The *raison d'être* of the CRO is to fragment into its component parts and rationalize many of the scientific functions previously performed by the academic clinician or professor of medicine, and one of those functions is authorship. The CRO exists to modularize and subject such functions to the logic of commercial efficiency. The attribution of authorship in Big Science has presented a number of practical problems (Biagioli and Galison 2003), which may partly be attributable to the increased scale of research, but the privatization of research also introduces some special considerations.

The basic fact to remember about CROs is that they are not themselves oriented toward academic authorship: The doctors administering clinical trials for pay are uninterested in authorship, as are the bioinformatics specialists, the patient recruitment team, the in-house statisticians, the site management staff, and the host of other specialists employed by the CRO to organize and conduct drug trials. Because IP is stringently controlled, and personnel turnover is so high in a CRO, it would be quixotic for an employee of a CRO at most stages of a research contract to actually expect to garner credit for a publication. Their careers do not stand or fall by the number of journal papers listed on their curriculum vitae. Furthermore, the modern production of clinical data is governed in the first instance by the requirements of FDA approval; academic publications may be viewed, more often than not, as "infomercials" that aid the marketing of the drug. The confidentiality provisions and publication embargoes covered earlier in this chapter reveal that dissemination of scientific information is subordinate to a larger agenda in the world of corporate science: Commercialized research needs to be subjected to selective disclosure and closely controlled discussion. How better to deploy the required discretion than to hire commercial ghostwriters to produce the desired texts to order? And instead of trying to censor or otherwise muzzle obstreperous academics who participate in the research after the fact, how much more "Pareto optimal" to provide them with preauthored drafts of clinical summaries that are structured to highlight the results deemed useful by the drug company—papers that they can then proceed to publish under their own names, the more readily to further their own academic careers? And if *JAMA* refuses to play along, you can just inter it in some convenient ghost journal from Elsevier.

The way medical ghostwriting works was nicely illuminated by a 2003 Canadian Broadcasting Corporation report. The team interviewed a number of ghostwriters who insisted that their identities be kept confidential, so the broadcast reported an interview with an unidentified writer whose annual salary exceeded $100,000 per year, and who said that a paper in a top medical journal would net him payment in the neighborhood of $20,000.

> [Writer]: I'm given an outline about what to talk about, what studies to cite. They want us to be talking about the stuff that makes the drug look good. [Interviewer]: They don't give you the negative studies? [Writer]: There's no discussion of certain adverse events. That's just not brought up . . . As long as I do my job well, it's not up to me to decide how the drug is positioned. I'm just following the information I'm given. [Inter-

viewer]: Even though you know that the information is often biased? [Writer]: The way I look at it, if doctors have their name on it, that's their responsibility, not mine. (E. Johnson 2003)

Popular and journalistic outlets often approach such phenomena as instances of the breakdown of ethical standards and editorial oversight, but this unduly personalizes what is clearly a structural phenomenon. Such commentators (and even the medical editors of the ICMJE) are still operating within the mind-set of an older conception of science, in which authorship credit in journals is framed as a reward for scientific effort, linked to an identifiable personality: In that world, the buck stops at the author. But the CROs participate in an altogether different kind of economy, in which various claims about drugs are being "sold" to regulators, doctors writing prescriptions, and increasingly, to the patient end-user. Should these claims of efficacy be challenged, they could then potentially be litigated in a court of law and negotiated in terms of monetary liability of a corporate entity. The "responsibility" in question is not that of some free-floating intellectual to an abstract "republic of science," but rather that of a commercial corporation to its shareholders, the regulators, and (to a lesser extent) its customers. When medical editors propose something resembling "film credits" be appended to a paper reporting clinical trials in order to reveal where the buck stops, they have begun to address the complex realities of collaborative science but have unaccountably neglected the realities of commercialized science. One should not prematurely confuse the two. Especially for the CRO, there exists no single person or small number of people whose probity stands planted firmly behind the information disseminated (after all, mostly they are merely employees; many have moved on even before the project was completed; and corporate officers are not personally liable for product negligence); there are only the contractual obligations of the corporation. As the anonymous ghostwriter put it in his interview, it's just not his problem. The scribe who puts her pen to paper is just one more employee, enjoying the same social obligations and dispensations as the laboratory technician (with probably commensurate job security).

This whole system of dispensing with the academic author, except as someone willing to put in a cameo appearance and lend his name to publication after all the work has been done, has evolved much further over the last decade. Although for obvious reasons there are no hard statistics, the proportion of ghostwritten articles in biomedical journals today is far higher than those estimates reported in the 1990s (Gotzsche et al. 2007). The reasons we know this are (a) in each new large-scale tort case over drug damages, there is overwhelming evidence that most academic articles in the most prestigious medical journals regarding the drug in question were ghost- and guest-authored (Ross et al. 2008; Saul 2008; Singer 2009b); (b) when some academic, out of obsolete notions of personal integrity, resists delegating the authorship in any large clinical trial to the CRO in charge, then he is fired by his university employer (Revill 2005); and (c) of greater significance, we now have the first evidence of how the entire ghost author industry works, including the voluntary complicity of medical journal editors (Sismondo 2009).

Scientific ghost authorship is now such a big business that it has its own professional organizations, the International Society of Medical Planning Professionals (ISMPP) and the International Publication Planning Association. Once one acknowledges that scientific publications no longer exist for the "author," then managers can organize the entire process of turning results into words and publications so that it is subordinated to the objectives of the paymaster of the research. For the pharmaceutical industry those objectives are fairly clear: to influence stock market analysts in believing that the firm has a fully stocked drug pipeline (whatever the actual truth); to convince the FDA to allow the novel drug to be marketed; and to convince physicians to prescribe it, perhaps even illegally as an off-label medication. For each one of these objectives timing matters—a favorable publication months after the target date is essentially useless. Hence it is no longer a matter of ghost*writers* so much as it is of ghost *managers* to make the entire process of writing, presentation at meetings, and submission to journals run like clockwork. This constitutes the ultimate triumph of the CRO over the AHC: To rationalize and speed up clinical research, one must manage the entire data flow from well before the clinical trials even start to the last stunning appearance of the triumphal *JAMA* article, something no academic hospital could ever pull off. And do not be fooled: The editors of all the major medical journals are in on the secret, as Sismondo demonstrates, since they regularly address the ISMPP on the best ways to prepare and submit journal articles. That is why when editors profess they are *shocked, just shocked,* to discover that articles they have published are ghostwritten, they are no more believable than Claude Rains in *Casablanca.* All the ICMJE mandates on disclosure of material participation in authorship and financial support are little better than a joke under the current system of ghost management, when the author identity can be adjusted to fit the circumstances.

What Sismondo (2009) and others demonstrate is that ghostwriting and ghost management are so deeply ingrained into the practice of modern clinical research that no one suffers upon being caught out: not the cameo academic authors, even when they are exposed as having taken anywhere from $750–2,500 per article simply to have their name grace its masthead; not the journal editors when they are excoriated by the press; not the publication planners themselves, who are resigned to the fact that their work is torn between marketing and meeting the formal criteria of scientific protocols; and certainly not the biotechs and Big Pharma (Moffat and Elliott 2007). No one thinks they are doing anything wrong here! Indeed, if the publication manager has done her job well, there is really nothing left for the academic author to do (and they know it)! Welcome to the modern world of privatized commercial science.

This really is the apotheosis of the neoliberal view of the world: Commoditized knowledge becomes effectively disengaged from previous archaic craft notions of personal responsibility and individualized expression, to find its most efficient incarnations and uses. The Market is the best information processor known to mankind. Resistance is futile. This is nicely captured by a quote from a publisher of a respected medical journal:

> We spend a lot of time trying to re-educate our journal editors . . . We're saying that you have to change your instructions for authors. You have to reflect the changing mood of the times. And yet we still get journal editors who say, "This journal frowns on ghostwriting" or something similar. "This journal will not accept papers that have writing support." And actually what we're trying to say to them is, "Fine, you may have that view but what you're actually doing is driving it underground. It's far better to be transparent and get this out into the open." (Sismondo 2009)

The Feedback of Ends upon Means

What is the real purpose of a clinical trial at the dawn of the twenty-first century? If you would answer "knowledge of the effects of various treatment regimens," then you would be missing much of the tendentious activity of CROs. The bottom line for CROs, as already suggested, is the facilitation of the approval process for new drugs by the FDA for the pharmaceutical industry, and timely provision of data for the investment and marketing purposes of their clients. Their vaunted advantages over the AHCs reside in the efficient performance of these functions, as they would readily admit. Yet it would be too hasty to simply point to gross numbers of drugs approved and average speeds of development cycles as indices of global success (or failure) of privatized science. As insisted earlier, the major irony that haunts the recent reorganization of clinical research is that all the infusions of corporate funding and all the stress on efficiency promulgated by CROs have at the end of the day produced fewer and fewer truly *new* drugs—that is, drugs that are not merely new molecular entities relative to those previously under patent protection, but are also substantially different from anything that had been established in therapeutic regimens. The imperatives of commercialized science for both the biotech model and the CRO have conspired to bring about this result.

The relevance of this phenomenon to this book is that the commercialization of downstream clinical science has actually promoted the growth of copycat, recycled, and retooled drugs. Clinicians and academic researchers in AHCs have generally given a wide berth to research into copycat and recycled drugs in the past (although, it must be admitted, not eschewed them altogether), since, in their view, there were so many more pressing needs in health care than simply the quest of pharmaceutical companies to keep existing blockbuster drugs under patent. Further, from an academic perspective, their scientific interest frequently verges on nil. CROs generally do not share such scruples. Thus, while the modern regime of privatized science is certainly not solely responsible for the phenomenon of copycat drugs, it has certainly been much better structured and intentionally turned to facilitate their development. For instance, drug companies may themselves want to have access to data on the effectiveness of their copycat molecule relative to existing treatments, but they would not want those data made public (Pear 2003). CROs are quite happy to maintain these patterns of secrecy and disclosure. Consequently, vast sums of money have been poured into research into copycat and recycled molecules, which have been of dubious benefit to the overall public health and, arguably, communal welfare. Therefore, when champions of the bracing virtues of commercialized science point to the munificent increase of private investment in biomedical research, it may be prudent to recall that not everything that comes out of a drug assay or clinical trial is knowledge in the conventional sense of deepened understanding of the biological mechanisms nominally at issue.

One might be tempted to aver that this caveat is too draconian, and that at the very least what the pharmaceutical firms and CROs are providing is a vast archive of clinical knowledge of tested molecules, which may provide important clues to further developments in the future, even if they appear at present to be little more than the validation or elimination of a host of copycat or recycled molecules. But even such an attempt to exonerate the privatized regime ignores the fact that, when clinical trials are run for the more restricted purposes of drug development under the modern IP system, then information that does not further those immediate goals is effectively superfluous and, therefore, a source of inefficiency in research. When clinical trials seem to suggest that a line of drug development is not panning out, then the optimal thing to do for commercial reasons is to hastily terminate the trial. Not only does cutting your losses result in a callous and cynical treatment of the patient population, an apparent violation of the Helsinki Declaration, but it also demonstrates that the purpose of clinical research does not include following lines of inquiry wherever they may lead, or contributing to a common archive of "negative" results, which might in the future be incorporated into larger therapeutic contexts. Perhaps this is why the pharmaceutical industry has resisted repeated calls in the past to register clinical trials, so that outsiders would be at a loss to know that a particular regimen or molecule had ever undergone scrutiny. More recently, they claim to disclose their trials under pain of penalties, while using CROs in order to preserve opacity when needed.[59] The vast volume of clinical trial information that never leaks out from proprietary boundaries, much less actually gets published, can never be considered a "contribution" to medical knowledge in any serious way, but it was never intended to be such. In a very real sense, from the viewpoint of the scientific community, it doesn't exist. Far from jettisoning "idle curiosity" as an extravagant luxury, privatized clinical science treats all curiosity as antithetical to efficient research.[60]

If modern clinical trials do not appear to exist to produce scientific information (with the obvious exception of successful trials submitted for FDA scrutiny) as such, then what exactly are they good for? This is where we discover the most pronounced feedback from means to ends. If the dominant objective of privatized pharmaceutical research in the current regime is to produce a stream of copycat versions of successful drugs, then it follows that a major commercial motivation of clinical trials is for advertising, since the key to a successful copycat drug is marketing. Much of this promotion is buttressed by Phase IV clinical research performed by CROs to lend some veneer of justification for the assertions of the efficacy of these patented medicines (generics often don't merit advertising). Because this research is being tailored to the requirements of an advertising campaign, it is increasingly the case that the contracts for the clinical trials, the ghostwriters, publication managers, and the rest are negotiated not by the R&D arms but by the marketing departments and advertising agencies of Big Pharma (Bogdanich and Petersen 2002; Sismondo 2009). While academics might cringe at the prospect, for a CRO this is just one more ripe market opportunity. The line between science and advertising is consciously being blurred in pharmaceutical research, because the regime of privatized science makes it possible and profitable.[61] The CRO is an instrument that helps make this happen.

It gives one pause to observe the extent to which the outward trappings of science can so easily be subverted into occasions for marketing. It has become standard practice to convene all-expenses-paid "medical conferences" in desirable tourist destinations, which turn out to be elaborate sales presentations for new products, all under the guise of the presentation of research papers (Tilney 2003). Other more modest perks are free dinners and honoraria given in the guise of "continuing medical education," liberally sprinkled with promotional presentations (Wazana 2000; Angell and Relman 2002; Angell 2004). An even more insidious practice is to subordinate the protocol of clinical research more directly to marketing imperatives in the form of seeding and switching trials (R. Smith 2003, 1203). Here the companies simply conduct the trials in order to get the doctors to begin to prescribe their drugs. Nonacademic physicians are recruited to take part in trials for which they possess little information or basis on which to judge the research design; they are paid handsomely to participate but they are unsure of the identity of the ultimate sponsors (since funding is channeled through CROs). Also, they never see the "results" of their endeavors. These trials have no particular research objective, with no well-defined question or set of controls; primarily, the physicians are chosen by the CRO or site management organization simply on the basis of their prescription histories. Since the trials go unregistered, no one can effectively ask pointed questions about the validity of the protocols or the absence of subsequent publications. The objective is simply to get physicians accustomed to prescribing the proprietary and costly drug, which is frequently no better than existing cheaper generics.

Hence the lesson of this brief survey of the rise of the CRO is that when scientific research becomes well and truly privatized, then the conceptual line between research and marketing blurs into irrelevance. "Knowledge" becomes indistinguishable from PR and spin. This is the manifest logic of the marketplace of ideas. CROs are not natural, parochial phenomena, but manufactured, and therefore exportable phenomena. Consequently, there exists the possibility of commercialization of science eventually exporting the CRO model to areas outside of clinical drug trials. Some of the research problems now endemic to the biopharmaceutical sector are not all that peculiar to that sector, and likely they have begun to spread to other sectors. One might identify trouble with human subjects in areas like environmental services or psychological experimentation, problems of confidentiality and disclosure in the commercialized defense industry, the need to control "research tools" in chemical or software industries, the spread of ghost authorship and ghost publication management to information technologies and social policy, and the need to provide marketing innovations for IP in an entire range of wholesale and retail settings. Once the CRO becomes solidly identified with successful institutional innovation in the control of research, there seems to be no special reason why it might not pop up wherever corporate research capacity needs to be outsourced and reengineered, albeit modified to better conform to local concerns.

Thus, it seems a safe prediction that the commercialization of scientific research will continue to occur—stemming from structural changes wrought by CRO-like entities and other devices goading academia to further reengineer its own scientific organization. Consequently, the new phenomena of research we have identified in the pharmaceutical sector may also become more prevalent elsewhere. Unless the structural changes taking place in the corporate sector receive comprehensive attention, the denizens of the university will never come to understand the future implications of the commercialization of scientific research.

Pharmageddon: Rise of the Chinese CRO

The ultimate irony of the new knowledge economy is that after having reengineered the process of scientific research to render it faster, cheaper, and more efficient, and having extended its global reach, then American scientists discover to their chagrin that the way has thus been paved to offshore and outsource their science altogether. This has been a subtext in our narrative since 1980; one observes it happening now in pharmaceuticals and clinical research. As previous chapters suggested, a national research base tends to follow close on the heels of growth of the manufacturing base. "A recent PriceWaterhouseCoopers survey of 185 senior pharma company executives from the Asia Pacific region found that 58% believe the center of gravity of the global pharmaceutical market will be in Asia rather than North America or Europe in the near future" (K. Brooks 2008). As I noted at the outset of this chapter, by many lights China is the new Eldorado of biotech (Engardio and Weintraub 2008), seemingly to a greater degree than India, its nearest rival. But since the biotech model is inherently unprofitable, perhaps it comes as no surprise that the real action can be found in more profitable corridors of downstream pharmaceuticals, and in particular, the CRO industry. "CROs are flocking to the region . . . Demand in the region includes all phases of drug development, although multinational companies are not allowed to conduct Phase I studies in India and China" (K. Brooks 2008). Perhaps unexpectedly, even the conduct of preclinical animal studies has become a Chinese area of relative competitive advantage. "Nowadays, all sorts of CROs are being formed in China. They are not always located in Shanghai. They are not always focused exclusively on chemistry. They don't always start with a small lab. And the founder is not always Chinese" (Tremblay 2008, 13). China's accession to the WTO and its nominal endorsement of TRIPS compliance seems to have opened the floodgates. Perhaps of equal importance is the passage of legislation in China in 2003 that allowed for simultaneous testing of new drugs in China, the United States, and Europe, even prior to

U.S. FDA approval. This has effectively rendered Chinese clinical trials data fully fungible across all major drug markets (Cooper 2008a, 82).

The appearance of CROs has tended to follow the establishment of major beachheads of Big Pharma in building research capacity in China. GlaxoSmithKline has built one of its largest facilities in the world in Shanghai; Novartis has opened a $100 million research center as well (Ainsworth 2007). Eli Lilly and Roche Holding AG have already established early presences in Shanghai with Zhangjiang Hi-Tech Park, where both companies employ hundreds of researchers. In 2005, Pfizer pledged $25 million for an R&D center in China (while closing centers in the United States) that would focus on study design, data management, and statistical analysis for global clinical trials. In 2006, AstraZeneca announced a $100 million initiative to build a new innovation center in China to focus on oncology research, after founding an in-house CRO in China in 2002.

Since the actual Chinese academic base is considered by some to still be weak, this has been mainly driven by cost and market considerations. "By conducting many experiments in low-cost Asia, the drug companies believe they can run more projects while keeping R&D budgets flat" (Engardio and Weintraub 2008). But there is the further consideration that one function of the CRO was to deskill and modularize the labor used in clinical trials; once accomplished, this facilitates shipping the entire operation overseas to regions with lower-skilled workers. An officer of the CRO Parexel has estimated that operating costs in China came in between 30–60 percent lower than comparable research trials in the United States (K. Brooks 2008). China's more than three hundred CROs have thus been offering services in everything from pharmacogenomics to clinical trials, new drug applications, new drug transfers, and exporting. Currently, the majority of Chinese CROs are mostly small, only providing regulatory consultation, drug application, and clinical trial assistance to overseas pharma firms. A few have been founded directly in cooperation with Big Pharma: an example is ChemExplorer, which was created jointly by Lilly and the Shanghai Institute of Organic Chemistry. Major Chinese CROs are located primarily in Shanghai and Beijing, especially in two large biotech parks: Shanghai's Zhangjiang Biopharmaceutical Park and Beijing's Zhongguancun Life Science Park. Beijing alone boasts more than one hundred CROs. China also has upward of one thousand domestic research institutions that claim to be associated with biotech and pharmaceutical science.

The functional division of labor between biotechs and CROs in the United States does not exist in quite the same fashion in China. This may be due to the fact that Big Pharma has so far been able to conduct itself more on its own terms in the Wild East. It doesn't hurt that China, to a greater extent than even the United States, has embraced a neoliberal conception of healthcare provision (Cooper 2008a, 84). Most of the population has been stripped of previously guaranteed health coverage and is therefore prepared to accept clinical trials participation as a stopgap. Besides CROs, numerous research institutes supported by the Chinese government or by drug companies have proven open to contract research opportunities. According to one source, in 2006 approximately 145 CROs and 165 medical institutions had obtained licenses from the SFDA to conduct clinical trials (Cooper 2008a, 85). For example, Tianjin Pharmaceutical Institute specializes in drug metabolism studies; Shanghai Pharmaceutical Industry Institute specializes in toxicology studies; and Shanghai Institute of Materia Medica (affiliated with the Chinese Academy of Science) has been collaborating with GlaxoSmithKline on the development of a chemical compound database since early 2005. The sector has matured to the point that substantial data are only just beginning to become available, particularly because Chinese biotech firms, true to form, are beginning to launch their own IPOs on the NASDAQ and the NYSE.[62] However, venture capital as conventionally constituted in the United States rarely exists in that environment; instead, various arms of the state more frequently provide seed capital (Frew et al. 2008, 47), further blurring the customary public/private divide.

Consciously or not, the Chinese biotechs have learned some hard lessons from their American predecessors. "Several of China's small and innovative biotech companies generate revenue by selling noninnovative products, providing services and/or outsourcing their early products . . . Selling noninnovative products, such as biogenerics and simple diagnostics, is a low-risk strategy for entering the health biotech space" (ibid., 49). SiBiono GeneTech has claimed to be the first company in the world to market a gene therapy product, but Western medical experts have been hesitant to accept that regimen. Most others have concentrated on less innovative me-too drugs or diagnostics. One of the attractions of China for Big Pharma has been its wide-open orientation: Science policy sports some of the character of the Wild East in China. The Chinese SFDA (their indigenous version of the U.S. FDA) was recently beset by scandal, when its former director pled guilty to accepting bribes in exchange for drug approvals, and he was executed in July 2007 (Barboza 2007). Numerous debacles concerning tainted or false pharmaceutical ingredients continue to dog the Chinese industry. Serious quality control seems not to be a priority, compared to the drive for market share. It just goes to show the dominance of the neoliberal mind-set that not one Western drug company has been so perturbed or distressed by these scandals that it has decided to refrain from offshore outsourcing to China.

III

Where We Are Headed

6

Has Science Been "Harmed" by the Modern Commercial Regime?

Viridiana, by this time, feels like reaching for some Valium. Now she can appreciate that the institutions that shape scientific research really have changed over the course of her lifetime, that the strengthening of intellectual property has been a prime culprit (but only an intermediate cause), and that the biotech model of support of science is fundamentally flawed. Her world has been privatized. But something deep down inside makes her hesitate: Might not Science be so powerful an individual calling that, despite all odds, it can still find a way past all obstacles and still produce progress? Maybe Science is just *stronger* than all the fetters that threaten to drag it down. Viridiana, who really does not enjoy being depressed, still wants to entertain the idea that, like Love, Science will find a way. In this, she is joined by a number of recent writers who have devoted entire books to the current problems of commercialized science. Here are just three, out of a profusion of expressions of faith:

> The facts do not support this view of science as handicapped by a lack of sharing and publication . . . Science continues to make rapid and amazing progress despite financial pressures to maintain secrecy . . . While financial issues can have a negative impact on publication and data sharing, they are not likely to completely destroy the ethics of openness or undermine the progress of science. (Resnik 2007, 102)
>
> Academe's role in the convergence of public and proprietary research may result from shifts in the character of science and from an institutional order that increasingly necessitates new strategies for simultaneously managing both academic and commercial worlds. The university's centrality is comforting. Far from being destroyed by commercialization, the academy has become the obligatory passage point for research. (Owen-Smith 2006, 77)
>
> Overall, for protecting the integrity of science and reaping its benefits for society, wholesome developments now outweigh egregious

> failings—though not by a wide margin. Nonetheless, the changes and trends are hopeful. (Greenberg 2007, 258)

Whenever Viridiana encounters such testimonials, she desperately wants to believe—really, she does—but she can't help but notice that the evidence is lacking. The question remains: How would we know if science was being harmed by the full panoply of changes wrought by the commercialization of science? These authors appear as stymied by the conundrum as she is.

For an answer, one might have been tempted to turn to the organizations that exist to lobby and plead the case for the health of science, such as the National Academies of Science (NAS) or the American Association for the Advancement of Science (AAAS). More often than not, in those precincts one encounters repetitive unfounded testimonials to the strength and vitality of science, rather than any serious diagnosis. When the NAS in particular has not been prognosticating more of the same in the near future, it periodically indulges in the most hallucinatory neoliberal fantasies:

> Firms in the United States have mastered wave after wave of new technologies, from aerospace and electronics to pharmaceuticals and nanotechnology. These fields of endeavor have been built on strong foundations of new knowledge and understanding of the physical, mathematical, and biological sciences and of engineering. They have benefited from the establishment over time of a highly supportive national innovation system (NIS). The combination of mastery of the scientific and engineering foundations and the smooth functioning of its NIS has enabled the United States to move effectively in little more than a century from an agricultural, to an industrial, to a post-industrial society. As the 21st century has unfolded, however, radical new challenges and opportunities suggest that the United States is on the threshold of a new era in the development of advanced societies. I call this new era the "post-scientific society."
>
> A post-scientific society will have several key characteristics, the most important of which is that innovation leading to wealth generation and productivity growth will be based principally not on world leadership in fundamental research in the natural sciences and engineering, but on world-leading mastery of the creative powers of, and the basic sciences of, individual human beings, their societies, and their cultures.
>
> Just as the post-industrial society continues to require the products of agriculture and manufacturing for its effective functioning, so too will the post-scientific society continue to require the results of advanced scientific and engineering research. Nevertheless, the leading edge of innovation in the post-scientific society, whether for business, industrial,

> consumer, or public purposes, will move from the workshop, the laboratory, and the office to the studio, the think tank, the atelier, and cyberspace. (C. Hill 2007)

One way to begin to understand why the NAS and the AAAS, whom you might have reasonably expected to be the guard dogs of the integrity of the scientific process in the United States and defensive bloodhounds rooting out scientific corruption, have turned out to be the lapdogs of the globalization regime is to realize that in the case of the former—a quasi- governmental organization that is restricted to only a few luminaries elected by their peers—and the latter—which is a broader-based private NGO run by a self-selected cadre of similarly closed elites, which supports its publication *Science* through advertising[1]—are largely populated in their higher ranks by the kinds of scientists who owe their preeminence to having mastered the modern academic/industrial divide to their own enrichment and benefit. For example, when I wrote this, the president of AAAS at the time was David Baltimore. He was then the Robert A. Millikan Professor of Biology at Caltech, where he served as president from 1997 to 2006, and he is the recipient of the 1975 Nobel Prize for Physiology or Medicine for the discovery of reverse transcriptase, which transcribes RNA into DNA. Dr. Baltimore was a founding member of Ariad Pharmaceuticals and is currently a director of the firms Amgen, BB Biotech, AG (a Swiss investment company), Alnylam Pharma, and MedImmune, Inc. He is also the survivor of a serious crusade to call the integrity of his own laboratory practices into question (Judson 2004; Crotty 2003), and an apologist for the commercialization of science (Baltimore 2003). One readily can appreciate that worthies like Baltimore who head up these organizations do not regard themselves as suffering in any way from recent alterations in the structure and conduct of science. Consequently, projecting their own experience onto the entire institution, they have a hard time taking seriously the notion that science as a whole may have been degraded in some respects under the modern commercialization regime. Their biases are then reproduced in the publications that the NAS and AAAS promote and sanction.

So if the largest professional organizations of scientific researchers in the United States have passed on serious evaluation, how about the government? Since the popularization of the very notion of a "Republican War on Science,"[2] the health of science has become somewhat of a political-party firestorm, even though much of the smoke and most of the heat derives from the supposed clash of science and religion, than has been consciously attributed to the market privatization of science. Nevertheless, in the last few years, certain agencies of the federal government have revealed their awareness of recent possible maladies of the U.S. science base, all the while seeking to evade the ire of their political masters in pointing them out.

One example was provided by an embarrassing incident with the Office of the Inspector General (OIG) of the U.S. Department of Health and Human Services—the agency that oversees the NIH—issuing an audit in 2008 wherein it found that the NIH had done very little to monitor and police the rules on financial conflict of interest of its scientific researchers, even though it had rules governing such conflicts on the books for years (Lederman 2008a). Thus the intra-bureau conflict brought into the open the wink-and-nudge culture of the contemporary commercialization of federally guided science, which simply accepted the neoliberal premise that there was no way that industry funding could bias or otherwise pervert the outputs of scientific research. When the Inspector General asked the twenty-four component institutes of the NIH for the conflict of interest reports supplied by their university grantees from 2004 to 2006, nearly half were unable to do so. Worse, the other half could not provide accurate statistics on the reports tendered.[3] The OIG also suggested that the information that the universities did bother to provide, when they were compliant, was essentially useless. A follow-up in 2009 revealed NIH grant recipients at the universities were not serious about reporting conflicts of interest (OIG 2009). The response of the NIH hierarchy was to resist the suggestions of the OIG that it should better enforce its own stated rules on conflict of interest for scientists. Hence it became apparent that frequent bland assurances issuing from the bowels of the NIH that there had been negligible impact of conflicts of interest upon biomedical science were essentially groundless, because the agency tasked with oversight basically just didn't want to know about the extent of the problem. Here the attempt to measure the contemporary state of science revealed that quality measurement was inaccessible under the current regime.

The pressure on the NIH has not relented, but the "solution" has proven elusive. In May 2010, it proposed some further changes in federal regulations to somehow square the circle. As Francis Collins declared in the press release announcing the changes, "Partnerships between NIH-funded researchers and companies are essential—they have been and they will be . . . we believe it is essential to tighten up the situation to make sure we are obtaining and maintaining the public's confidence" (quoted in Lederman 2010). Note well it was all cast as a matter of appearances, and not about monitoring the possibility that the quality and amount of science may have been compromised, which had motivated the critique in the first place. The proposed rules foresaw expanded reporting requirements to suggest how financial interests might just impact not only the project in question, but also the full panoply of "institutional responsibilities." It almost sounds as if each recipient would be asked to compose something like the book you currently hold in your hands—but that couldn't be serious. The changes would also reduce the payment trigger from the previous $10,000 to $5,000. Again, unless the NIH anticipates becoming the main statistical agency tracking every aspect of the commercialization of the university sector, usurping data now collected by everyone from the AUTM to OECD's *Biotechnology Statistics* (2006b) and Parexel's *Pharmaceutical R&D Sourcebook,* then it simply is not going to happen.

Perhaps an even better recent example of government concern has been the bible of the American science policy set, the National Science Board's *Science and Engineering Indicators.* Past versions of this biannual compendium would predictably plump for ever-increasing sums of government largesse to be channeled into science, nominally in order to maintain the "competitive edge" of American science and industry, even as it hastened to assure us that the American science system was the best in the world. If one examines the text in a recent installment (2008a), everything looks pretty much "situation normal" as usual, with no serious bad news to perturb our complacency. Yet, simultaneous with its publication, the National Science Board felt impelled to issue an unprecedented "companion" pamphlet to the *Indicators* (2008b), which minces no words in identifying a whole range of disturbing developments buried in the mountain of statistical tables:

> The National Science Board observes with concern the indicators of stagnation, and even decline in some discipline areas . . . A decline in publications by industry authors in peer reviewed journals suggests a de-emphasis by US industry on expanding the foundations of basic scientific knowledge . . . In addition, in this century the industry share of support for basic research in universities and colleges, the primary performers of US basic research, has also been declining. Likewise, Federal Government support for academic R&D began falling for the first time in a quarter century . . . industry support for its own basic research has been fairly stagnant in this century, and its support to academic basic research in the US has remained flat, and declined in share of that support for academic R&D to a level not seen in more than two decades. Basic research articles published in peer-review journals by authors from private industry peaked in 1995 and declined by 30% between 1995 and 2005 . . . The drop in physics publications was particularly dramatic: decreasing from nearly 1000 publications in 1988 to 300 in 2005 . . . Other fields where the US declined to near parity with the EU-15 [subset of the European Union] in recent years are biology and chemistry . . . By current measures, the US trade balance across all high technology

> sectors significantly declined over the last decade . . . Of five high technology manufacturing industries identified by the OECD, only that for aerospace had a large positive balance of trade in 2005 . . . critical data on trends in off-shoring R&D by US industry are presently lacking . . . foreign affiliates of US multinational corporations performed $27.5B in R&D abroad in 2004, a rise of 4.7B from 2003, the largest annual increase since a 22% rise in 1999. This amount nearly equals the R&D by affiliates of foreign countries in the US ($29.9B). (NSB 2008b, 1–5)

This looks really quite alarmist coming from the official body of the government tasked with responsibility for the health of science, especially since it had previously been loath to criticize the corporate sector for its management behavior with regard to scientific R&D. Some journalists (e.g., Broad 2004; Lederman 2008b) had been beating the drum of decline for a few years, but till now, government-based commentary had been absent. Perhaps the strident level of distress in that document was permitted airtime precisely because among the narrow coterie of people consulting the *2008 Indicators,* even fewer would bother to read a poorly advertised and separately distributed companion. It was just another bit of news no one wanted to hear. The press conference releasing both documents hardly even registered in the newspapers or even the Internet; the vexed question of whether astronauts had ever gotten drunk on space flights got more attention on the "science" beat around then. The level of distress that could be revealed to the public was itself a matter for controversy and trepidation behind the scenes, however, as evidenced at a National Science Foundation (NSF) conference held on November 7, 2006:

> Dr. [Susan] Cozzens listed the following questions that she thought Dr. Marburger, the President's science advisor, would ask about the leveling off of US publications: Is it an artifact of the data? Why is it happening? Where did all the inputs go? Should I be worried about it? Is it a threat to the US economy? . .
>
> [Participants] expressed concern the study findings might be used out of context or be misinterpreted. They urged caution in drawing implications from the study results. Participants raised concern that the disparity between US research inputs and publication outputs could possibly be used as a rationale for cutting funding for research. A presumption behind such cuts might be that declining research productivity (as measured by publication output) indicates that marginal, relatively unproductive projects are being funded and that these could be eliminated without significant loss. (NSF SRS 2007)

One of the more telling symptoms of the uneasy predicament of the government has been that the science policy community acquiesced in basically allowing dubious bibliometric measures (such as gross number of papers published, publications/GDP, and numbers and rates of citation) to be used as meaningful statistics of science "output," just so long as they presented the U.S. science base in the best possible light; but once the indicators started turning against them, there was an unseemly rush to repudiate them or, at minimum, discover a newfound enthusiasm to explain them away: "It was generally agreed that the share trend in publications was not a cause for concern . . . flattening of absolute numbers of US Thomson ISI database articles does not mean that the US science system is weak" (NSF SRS 2007, 8).

There are questions potentially of great moment lurking here; truth or consequences not clarified by generalist journals like *Science* or *Nature* nor by journalists who have taken it upon themselves to advocate one side or another of the Big Theme that Science (at least in America) might be undergoing decline in our lifetime.[4] The $64 billion question that the public is justified in asking is, Did we really get value for money when we encouraged science to become more like a marketplace? Surely we can all agree that the fruits of science will never be completely recognized until we are all long gone, that the advancement of science moves in convoluted and enigmatic ways, and that the Owl of Minerva only takes wing at dusk. But that does not preclude the possibility that interim bulletins from the front lines could provide Delphic indicators of the eventual long-term consequences of the reorganization of science. Proponents of a new economics of science should not shrink from considering this prospect.

Therefore, I propose to begin to explore those gray areas where science policy mavens have feared to tread. Those technical issues surrounding each potential indicator themselves are extremely complex, with caveats and qualifications hedging interpretation each and every step of the way, and therefore often dull. I must admit up front that no one person or organization has ever measured the "success" or "health" of science in a way that gained the assent of a majority of those curious about the evolution of human knowledge. Economists in particular have had paltry little to offer in this regard, or so I have argued in Chapter 2. But past debacles should not be taken to presage permanent defeat. At the top of the list of devoirs of a new economics of science stands the requirement to systematically survey the evidence that science has suffered modulation and reconstruction over the course of the modern regime of commercialization—something, significantly, no one has yet bothered to attempt.

The pitfalls are legion: What are the correct categories for dividing up the phenomenon? Should one seek to characterize the whole of science, or should one instead be restricted to a preset roster of disciplinary boundaries? (This ignores the complication that science under the current regime increasingly tends to breach long-standing disciplinary boundaries.) Should one treat science at the national level, as in the literature that speaks of "national systems of innovation," or should one assume a more global gaze? (Contempt for the nation-state as analytical entity abounds in the new regime.) Should one evaluate the quality of modern science on its own internally supplied terms, or should one appeal to more external success indicators? (Can one trust a self-identified set of "acknowledged worthies" to render a serious self-critique?) Should temporal comparison be reduced to a few manageable quantitative indicators, or should one concede that a "science of science" remains an impracticable dream for the near future?

Keeping each of these considerations close to the surface of consciousness, I shall attempt in this chapter to gauge the health of modern American science from a diverse set of perspectives and measures. The net result will not resemble anything like a statistical test (reject the null hypothesis of decline at the 5 percent level of error) but will look rather more like a composite portrait of something that lacks a definitive essence but nevertheless is clearly changing in character over time. Sometimes it will make sense to hew to national definitions of scientific activity—particularly when encountering the question of American decline, or whenever I am driven to rely on data collected by state entities. In other instances, I shall instead opt to treat specific disciplines as units of analysis, and the scientists themselves as defining the social space of their endeavors. Mostly in this chapter I restrict consideration to the time frame of the regime of globalized commercialization (1980 to the present), with only a few brief backward glances at the Cold War regime. This patchwork quilt of indicators will be divided into the following arbitrary chapter sections: (1) quantitative bibliometric output measures; (2) examples of just-in-time science; (3) the junk science movement; and (4) patent degradation. Although such a motley collection of indicators would never congeal into a seamless whole (much less a "science of science policy"), it is my conviction that the preponderance of these sorts of evidence can evoke an interim evaluation, a consilience of outlook, that so far has proven elusive in the new economics of science.

The Silence of the Labs: Bibliometrics as Output

Suppose we provisionally accepted the notion that the proximate purpose of scientific research is publication—that is, the dissemination of novel ideas and findings throughout a larger community through some sanctioned or respected vehicle. If that were the case, and further, if we trusted the vehicle to impose some (undefined) modicum of quality control, then it might seem reasonable to simply *count* the number of articles whose authors were identified with a particular geographic entity or institutional affiliation, and then claim that those counts "measured" the immediate outputs of science. Now, from the birth of this construct of "bibliometrics" onward,[5] it was objected that the unit of a "publication" was artificial and arbitrary in the extreme. (One still hears such cavils at tenure and promotion meetings.) Partly in response to this, and consequent upon the military pouring substantial funds into library science in the 1950s, these counts have been augmented by citation statistics: that is, individual "articles" could be rated by the numbers of citations aimed their way in other articles, obviously drawn from what were deemed a relevant subset of publication outlets. Variables such as these constitute the bread and butter of bibliometrics.

Few realize the extent to which the major source of worldwide bibliometric data of this sort is a private for-profit company (or possibly two: see below), and thus the exact composition and content of the data discussed below is proprietary, and more than a little secret, in keeping with the character of the modern regime of science organization. Eugene Garfield and Associates began two pilot projects in the late 1950s with NIH funding: The first was a machine-readable patent citation index, and the second was an index for published literature on genetics. When the NIH declined to continue support for the latter database, Garfield formed the Institute for Scientific Information (ISI) in 1958 to build and maintain national citation databases for the natural sciences, which would then be rented out to universities and business. He first privately published the *Science Citation Index* (*SCI*) in 1963. Since then, the ISI has continued to offer bibliographic database services, which began as a wonderful tool for researchers but which has morphed of late into something else. ISI's specialty is computer-automated citation indexing and analysis, a field pioneered by Garfield. ISI maintains citation databases covering thousands of academic journals, including continuations of *SCI* (its longtime print-based indexing service), as well as the *Social Sciences Citation Index* (*SSCI*), and the *Arts and Humanities Citation Index.* Its product *Web of Knowledge* allows those with online access to follow citation links across something on the order of 16,000 journals, to retrieve articles cited or citing a specific initial article. But broad-based does not mean truly inclusive (Testa 2003); as both Garfield and later ISI representatives have stated unequivocally, the range of journals to incorporate into the database is an *economic* decision it reserves to itself. Sometimes this is justified by an appeal to the so-called Bradford's Law, namely, the observation that a small number of journals publish the bulk of "significant" scientific results. However, the problem of the lack of openness of the decision process remains. ISI was acquired by Thomson Scientific and Healthcare in 1992, was first known as Thomson ISI and is now Thomson Scientific. It is currently a component of the multibillion-dollar Thomson-Reuters Corporation, formed from the 2008 merger with Reuters, a major provider of financial data and business journalism.

This minimal background is required reading because of the ways in which the *Web of Knowledge* and Thomson ISI have been transformed under the pressure of historical events to serve less and less as a helpful tool for researchers (Charbonneau 2006); Larsen and von Ins 2010), and more and more as a contrivance for the discipline and evaluation of institutions and a proxy output measure for the research. In other words, what started out as something harmless, rather like a thesaurus, has turned into a sharp-edged audit device wielded by bureaucracies uninterested in the shape of actual knowledge and its elusive character.[6] In particular, Thomson now retails its products as a one-stop shop for corporate monitoring and control of intellectual property in its many incarnations. Many university ranking exercises and bureaucratic audits are based solely on the Thomson datasets (Labi 2010). Threatened on the one hand by Google Scholar, introduced in 2004, which is (after all) free, and much easier to use for the average researcher (Noruzi 2005), and on the other by the ongoing consolidation of academic publishing by a very few conglomerate firms doing battle with numerous "open source" e-journal start-ups, Thomson has increasingly sought to position the *Web of Science* as another high-cost, limited-access database in a portfolio attuned to the special needs of corporate and administrative clients—thus, the logic of the merger with Reuters. As Ellen Hazelkorn has suggested, "This is the monetization of university data, like Bloomberg made money out of financial data" (in Labi 2010).

Once data provision becomes privatized, it usually follows that a form of oligopolistic competition sets in, and that has happened in the realm of bibliometric statistics. In 2004 the conglomerate Elsevier decided to parlay its vast ownership of journals into its own bibliometric and citation database called Scopus. Its coverage and principles of inclusion are very different than Thomson's, although it trumpets the fact that it covers roughly 16,500 journals to Thomson's 11,500.[7] Thomson and Scopus have very different business models, it seems. Thomson claims to identify and cover the "best science" by its own internal criteria, whereas Scopus is more user-driven, soliciting suggestions for ever-greater coverage, and providing technologically savvy features like RSS feeds and interoperability with ProQuest products. Nevertheless, it still seems that Scopus is not yet the primary source for many bibliographic exercises, even though it has been demonstrated that Thomson's *SCI* covers a decreasing proportion of the universe of scientific publications over time (Larsen and von Ins 2010). More germane to our current concerns in this chapter, Google and Scopus are both useless for the sorts of longer-horizon historical questions I shall pose here, because both are of such recent genesis and change the base from which they sample so frequently.

Ranking individuals, departments, academic institutions, corporations, and the like, according to their "productivity" as well as their possible relevance to targeted intellectual property (IP), has become Thomson's stock in trade. Google Scholar, by contrast, tends to pick up many journal citations in languages other than English missed by the *Web of Science*; but it also makes no distinction between "sanctioned" journals and all manner of other motley publication formats on the Web. Specific source documents appear and drop out of the Google universe almost randomly. *Nature* reported one instance where a single author (in management science) clocked in 815 total citations in the Thomson database; 952 citations in Scopus; and 2,226 citations according to Google Scholar (Van Noorden 2010, 865). Many scientometricians would thus reflexively look askance at Google Scholar, suggesting that it did not "measure" much of anything coherent. In any event, it is impotent to answer historical questions.

The upshot of these considerations is that everyone who in the past sought to use bibliometric measures as indices of scientific output has been effectively forced to make use of Thomson's products. The NSF does so almost exclusively in its *Science and Engineering Indicators;* consequently I shall follow their lead, albeit gingerly, gingerly, since it will be imperative to point to the flaws of these measures at every step along the way. What I am straining toward is to identify some plausible quantitative indices of changes in science publications over the rise of the regime of global privatized science. As every statistician knows, this is only possible if the underlying basket of things measured stays relatively invariant over time, or if that is unattainable, changes in the basket are controlled and documented so that they can be linked backward to earlier measurement data. Yet this is precisely something Thomson refuses to do (although its officers might retort that it is not Thomson's job to provide that service).

The question I shall set out to explore in this section is what is happening to the number and distribution of scientific publications over the range of the globalized privatization regime, the very issue raised in the 2008 companion to *Science and Engineering Indicators.* Thomson acknowledges that this could be accomplished by tracking a fixed set of journals, but Thomson instead constantly revises what it considers to be the "core set" of scientific journals by its own opaque decision criteria, without providing adequate comparative data, which would permit overlaps of old and new core sets. And the rate of change as to the core set is not trivial: reportedly, Thomson's core set encompassed 4,460 journals in 1988 and 5,262 journals in 2001.[8] It is not that the provision of statistics based on an invariant core set is in principle impossible; it is just that because Thomson is not forced to do so by the government, and corporate clients are uninterested in an invariant core set, then they simply neglect to do so.

For some time, the NSF had outsourced the production of bibliometric indicators for *Science and Engineering Indicators,* and hence the reprocessing of Thomson ISI data into fixed core sets, to a separate private firm, namely, CHI, Inc. Because the Thomson set of journals would generally change between each edition, and there were sometimes further complications of journals merging or ceasing publication, it happened that CHI was forced to rebuild the fixed core set anew each cycle for constructing the NSF tables. When Francis Narin retired and sold CHI, the NSF seemed to lose interest in the process, and no one was left minding the store.[9] This illustrates some of the problems of government outsourcing key decisions over knowledge gathering and collation to private firms. It seems the definitions of "science" sought by Thomson's clients bore little relationship to serious definitions of output sought by contemporary scholars of science policy and science practice.

Thus the distinction between a concertedly fixed journal set and expanding journal sets is something no bibliometrician worth his or her salt could ignore, but that does exhaust the fine points of the attempt to quantify publications. Another tricky distinction is what counts as "an article." Thomson says it counts articles, notes, and reviews, but not letters to the editor, news pieces, editorials, or anything else it considers ephemera. If we were casting our glance back to nineteenth-century science, such distinctions would grow ever more tenuous but luckily that won't be an issue here. What does become apparent is that the numbers of articles in a year's individual journal volume grow and contract at different rates in different disciplines, as do numbers of published pages: In itself, this could constitute an interesting indicator of the economic expansion or contraction of research pursued, as well as the level of subsidy involved in the dissemination of research findings. In Tables 6.1–6.3, I give some broad indication of the vicissitudes of American journals in chemistry, mathematics, and sociology over the course of the twentieth century, using a standardized counting procedure.

We often take it for granted that scientific research is a nearly unstoppable juggernaut, inexorably expanding the production of knowledge and pressing against the bounds of human limitations, but one thing to take away from perusal of these tables is that inexorability is neither necessary nor guaranteed. Crises like the Great Depression hurt some sciences but not others; mathe-

Table 6.1. Volume of article production: *Journal of the American Chemical Society*

Year	Number of issues	Number of articles	Number of pages
1900	12	107	414
1910	12	186	872
1920	12	294	1,358
1930	12	804	2,690
1940	12	907	3,574
1950	12	1,415	5,891
1960	24	1,392	6,500
1970	26	1,056	7,733
1980	27	1,014	8,118
1990	26	1,417	9,846
2000	51	1,298	13,040

Note: Numbers of pages in 1900, 1910, 1920, and 1930 are adjusted to reflect changes in page sizes.

Source: Liu 2003.

Table 6.2. Volume of article production: *American Journal of Mathematics*

Year	Number of issues	Number of articles	Number of pages
1900	4	26	388
1910	4	24	401
1920	4	18	286
1930	4	58	922
1940	4	67	912
1950	4	66	867
1960	4	49	943
1970	4	54	1,230
1980	4	45	1,206
1990	6	41	1,082
2000	6	49	1,308

Source: Lui 2003.

matics tends to expand at a relatively modest growth rate, even though it does not depend on large-scale investment in infrastructure, unlike, say, chemistry or physics. Changes in the way that publication is subsidized or otherwise funded also play a role.[10] But speaking of physics, which experienced the most effusive rates of expansion in the middle of the twentieth century, it now appears that, at least with regard to the journals associated with the American

Table 6.3. Volume of article production: *American Journal of Sociology*

Year	Number of issues	Number of articles	Number of pages
1900	6	42	864
1910	6	50	864
1920	6	29	806
1930	6	71	1,080
1940	6	44	938
1950	6	48	622
1960	6	59	662
1970	6	46	1,222
1980	6	70	1,592
1990	6	49	1,640
2000	6	40	1,840

Source: Liu 2003.

Physical Society, that expansion has now ceased, at least from the data presented in Figure 6.1.

One lesson to be drawn from this bird's-eye survey of American journal publication is that the changing nature of scientific journals matters; the state of their health should be a necessary prerequisite for understanding any subsequent bibliometric exercises. The current struggles over the extent to which the scientific journals of the future will be edited and promulgated over the Web (including specialized preprint archives) rather than through older paper channels, and the way in which that technological transformation intersects with the battle over whether scientific journals will be owned and controlled by individual professional scholarly societies or university presses versus a few large corporations such as Springer, Reed Elsevier, Wiley, and Kluwer, are an essential backdrop for any long-term comparisons. Since 1980, privatization has spread to the very heart of the scientific enterprise, namely, the vehicles through which new ideas are conveyed to the appropriate target audience.[11] It should go without saying that the whole setup of Thomson ISI is better suited to track privatized corporate publication of scientific results than the more anarchic Web publication and impending hybrid cyber-media.

Different disciplines also vary astoundingly widely in their publication practices, something that must be kept in mind when dealing at the global levels of aggregation so characteristic of bibliometric practice. As one might expect, the acceptance rates in highly prestigious journals tend to fall in trend over time, even given the counter-trend of expansion of pages per year

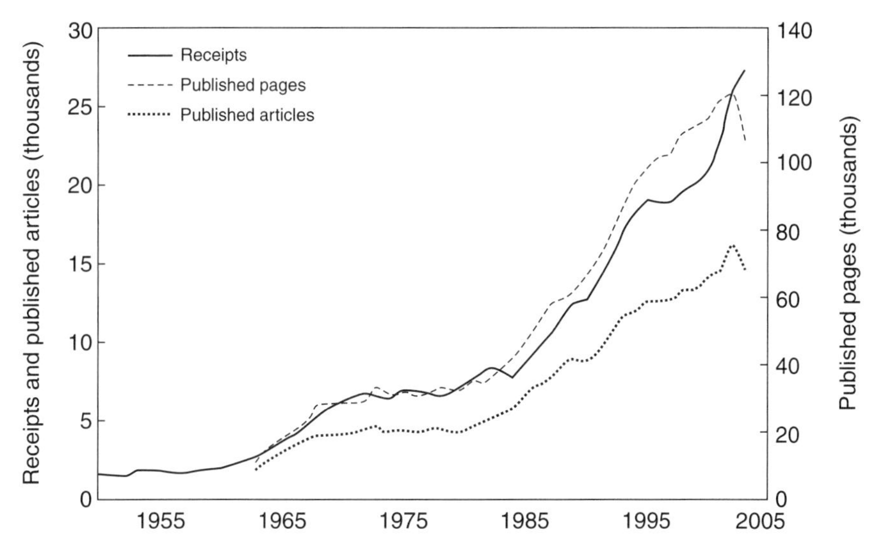

Figure 6.1. Article counts and page counts from 1955 to 2004. Note that *Physical Review Letters* was added to the original *Physical Review* in 1958. In 1970 *Physical Review* split into parts A, B, C, and D; Part E was added in 1993. In this graph, "receipts" means submitted articles. *Source:* Martin Blume, Editor, American Physical Society.

documented in the previous tables. But acceptance rates *between* disciplines vary to an even greater degree, as illustrated in Table 6.4. Here I compare the acceptance and submission rates for the *American Economic Review,* the main publication of the American Economics Association, and *Physical Review,* the flagship publication of the American Institute of Physics.[12] They differ by orders of magnitude, with the *PR eight times* more likely to publish a submitted paper than the *AER*. One should not interpret this divergence as evidence so much of a differential stringency of resources (although it cannot be altogether discounted), as it is an index of the degree of a hermeneutics of suspicion of the respective editorial staffs toward their membership base. The American economics profession has always been on guard against any whiff of heresy or conceptual deviance on the part of the writings of its members, far more so than the physics profession. In the latter, having successfully negotiated the apprenticeship of the PhD and (in many instances) a further lab apprenticeship stood as sufficient guarantee that your ideas were worth serious consideration within the physics profession. This sociological point is directly relevant for bibliometric inquiry, because the physics profession journal structure exhibits a flatter hierarchy, tending to distribute publication among its active members far

Table 6.4. Acceptance rates in economics and physics journals

	Economics papers in *AER*			Physics papers in *Physical Review*		
	Submitted	Published	% Accepted	Submitted	Published	% Accepted
1980	641	127	19.8	4,043	3,009	74.4
1981	784	115	14.6	3,936	3,129	79.4
1982	820	120	14.6	3,770	3,052	80.9
1983	932	129	13.8	4,023	3,142	78.1
1984	921	138	14.9	4,309	3,154	73.2
1985	952	128	13.4	4,697	3,566	75.9
1986	987	123	12.4	5,264	3,932	74.6
1987	843	99	11.7	5,840	4,211	72.1
1988	844	100	11.8	6,217	4,776	76.8
1989	946	116	12.2	6,590	5,210	79.0
1990	911	100	10.9	7,118	5,162	72.5
1991	884	110	12.4	7,652	5,439	71.0
1992	950	108	11.3	8,472	6,043	71.3
1993	900	94	10.4	9,371	7,040	75.1
1994	953	91	9.5	10,443	7,575	72.5
1995	929	88	9.4	10,951	7,579	69.2
1996	976	85	8.7	11,173	7,583	67.8
1997	976	66	6.7	11,280	7,652	67.8
1998	900	71	7.8	11,821	7,845	66.3
1999	927	79	8.5	12,063	8,188	67.8

Source: Scheiding 2006.

more equitably in fewer ranked publications, whereas the economics profession instead forces publication distributions to be spread out more broadly across many less highly ranked (and commercially owned) journals, by severely restricting access to ranked core journals, thus producing higher "peaks," but many more papers fall outside the bibliometric net.[13] The overall bibliometric result would be that a shift from journals following the physics model to those following the economics model (perhaps as a result of the progressive privatization of scientific journals publication through time) would clearly put downward pressure on all the usual bibliometric measurements: article counts, citation counts, and their various derivatives.

The final caveat to be covered here concerns who or what counts as an "author" in bibliometric practice. The contemporary scientific author in the regime of globalized privatization has become more of a corporate than a corporeal entity, in many senses of the term. Derek Price once predicted that the single-authored scientific paper would become extinct by 1980. Although this reveals the track record of prediction for scientometricians has not been much better than the abysmal record of economists, there still persists an element of truth in the notion that science has become less and less the province of the lone individual savant, pecking away at his laptop, alone in his office.[14] Instead, authorship has become more like an honorific state of grace, parceled out sparingly among teams of researchers, on principles that may or may not have had any direct relationship to the effort or ingenuity that was put into the final article. We have already encountered some problems this has wrought when attempting to map "authorship" onto intellectual property and issues of responsibility (in Chapter 5 on the phenomenon of ghost authorship, for example). Here I am more narrowly concerned with the impact the growth of multiauthored articles has had on bibliometric exercises.

Once authorship attribution in individual articles starts rising above fifty or so participants, bibliometric measures start to lose their meaning. In instances where the average ratio of words per author per article sinks down below thirty or so, clearly, scholarly attribution comes unmoored from the idea of words having personal inspiration.[15] Far from being a fluke, articles generated by a battalion of researchers have only become more common in the modern privatization regime, although the citation of Price signals that this trend extends back well before 1980. (It certainly helps explain the gargantuan publication totals inscribed in vitae of scientists in biomedicine and experimental physics.) Some idea of their prevalence in the core set of journals chosen by Thomson ISI can be gleaned by a graph they provide, reproduced as Figure 6.2.

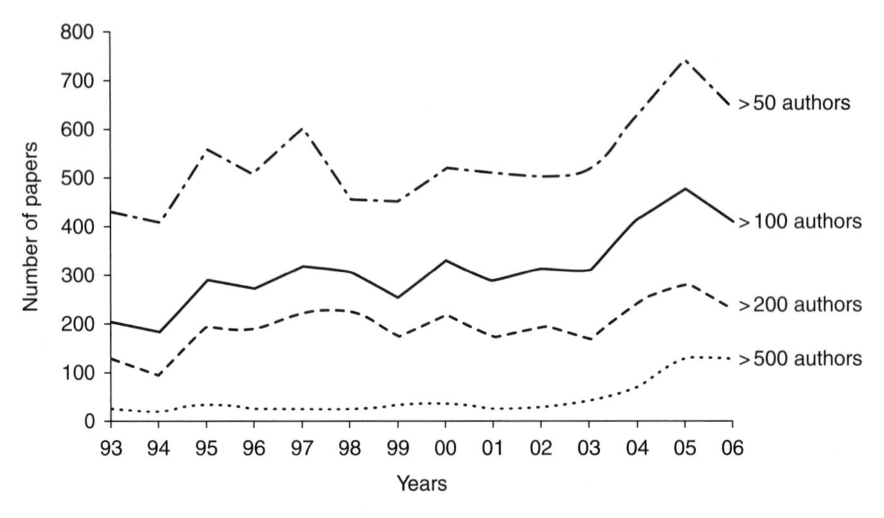

Figure 6.2. Multiauthor papers, 1993–2006. *Source:* http://www.sciencewatch.com/nov-dec2007/sw_nov-dec2007_page1.htm.

We could simply capitulate to defeat at this point, or alternatively, we could impose one more jury-rigged adjustment to the Thomson (and NSF) figures in order to link articles to people, universities, corporations, and nations. The standard options are (1) to ignore the problem by assigning one unit of authorship to every person named as such in the article linked to their reported institutional affiliation (the whole count method) or (2) to assign a fraction of "authorship" equal to the reciprocal of the number of authors to each person identified (the fractional count method) (Gauffriau, Larsen, and Maye 2008). If we are interested in output indicators over time, it should be apparent the whole count method is without merit and utterly misleading, because it vastly expands the countable universe of article attributions as a function of the proliferation of multiauthorship. (Again, I aver here that I decline to say anything cogent about the modern nature of "true" authorship.) The whole count method also "double counts" articles with multinational authors. Thus in this volume I shall only report statistics using the fractional count method, even though that may in some instances lead to somewhat peculiar numerical quantities of "authorship" and its related measures.[16]

Having gotten those pesky preliminaries out of the way, I can now ask, What has happened to scientific publications in the recent past? (All data are from Thomson ISI, unless otherwise noted.) The data comprise the aggregation of Thomson's *Science Citation Index* and its *Social Science Citation Index*—that is, the disciplinary definition of "science" is here construed quite broadly. The first generalization that can be noted is that the U.S. share of total global scientific articles published is falling. At the same time, the European Union countries began to enjoy greater representation in the scientific literature, although their proportion appeared to peak around the turn of the millennium. By contrast, the presence of Asian scientists has been enjoying almost unchecked expansion, as displayed in Figure 6.3.

Should there be any doubt concerning the location of these declines in publication share, it so happens that they can be similarly observed in most individual areas of the physical sciences over the same time horizon, as illustrated in Figure 6.4.

Because the immediate objection might arise that, even though the U.S. share is slipping in total it might nonetheless continue to maintain its lead in quality-adjusted superior work, some bibliometric inquiry has looked at shares of the most highly cited articles in the world. Contrary to expectations, one discovers rather similar continuous declines in U.S. shares, both in total and broken down by individual discipline (NSF SRS 2007, 15–17). It is nevertheless still commonplace to argue that if one controls for shares of active researchers in the total civilian population, and adjusts in various ways for what one considers "quality" publications, then the United States still comes out way ahead of the European Union in bibliographic measures (Dosi, Llerena, and Labini 2006, 1454). But it is unclear why proportions of civilian populations should matter to this issue, and as for quality indicators, they are extremely slippery, as I shall argue later in this chapter.

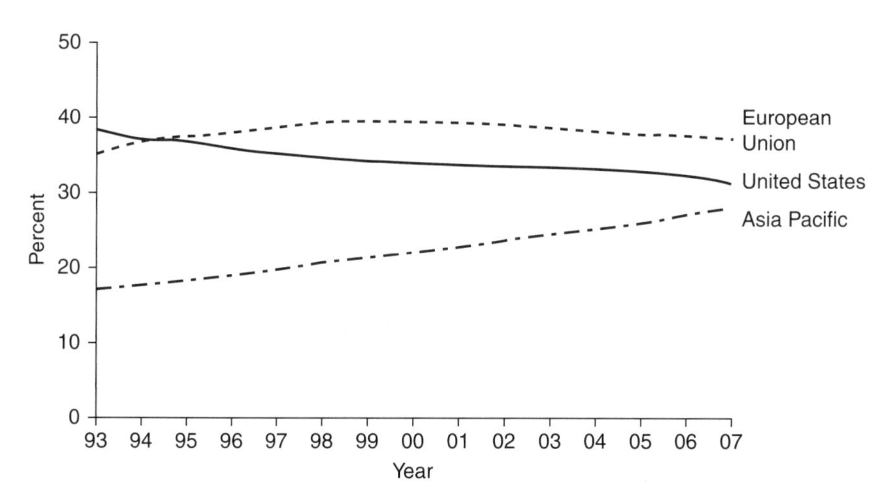

Figure 6.3. Percentage of world science, all fields, 1993–2007. *Source:* Thomson Reuters Science Watch, http://www.sciencewatch.com/nov-dec2007/sw_nov-dec2007_page1.htm.

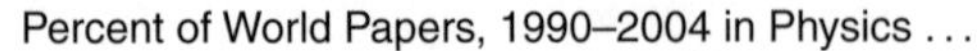
Percent of World Papers, 1990–2004 in Physics . . .

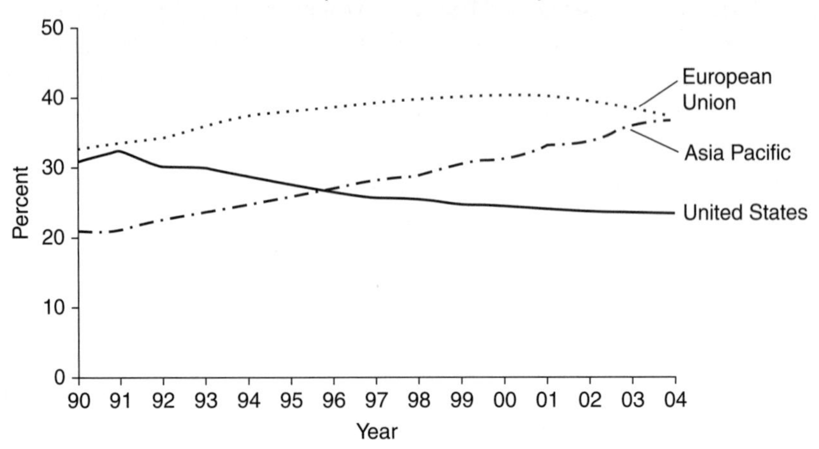
Percent
Year
European Union
Asia Pacific
United States

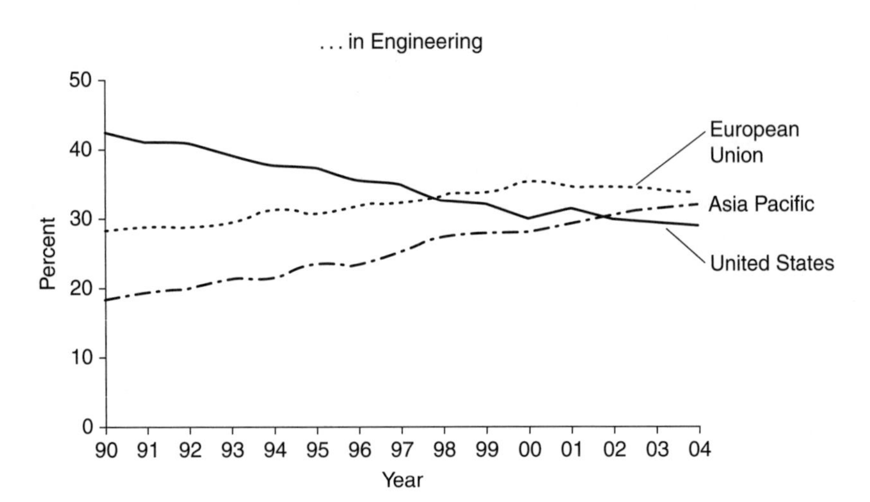
. . . in Engineering
Percent
Year
European Union
Asia Pacific
United States

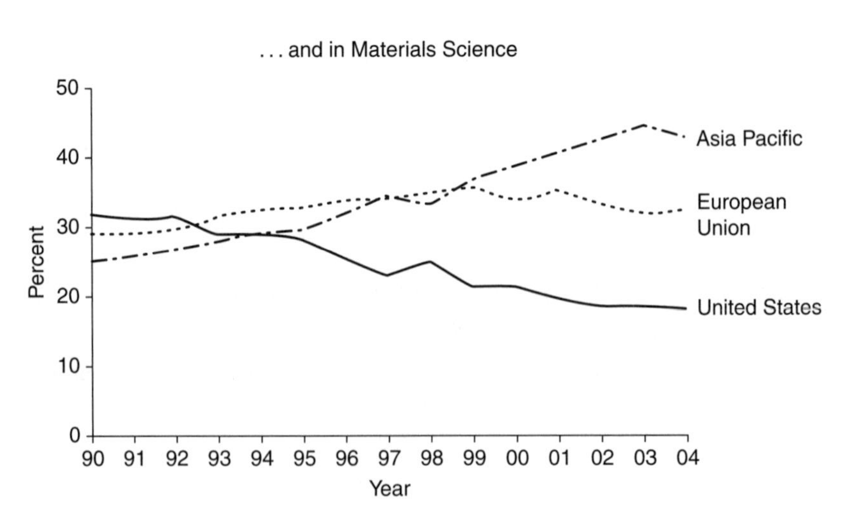
. . . and in Materials Science
Percent
Year
Asia Pacific
European Union
United States

Another knee-jerk response has been the assertion that much of the growth of articles in the Thomson database comes from the addition of new journals and that, because it has been suggested that newer journals are naturally more inclined to publish manuscripts from Asia, the shift in rankings comes from an irrelevant bias within the dataset. An examination of this particular suggestion by specialists in bibliometrics has shown that new journals were not especially biased against American authors, and in any event, the size of the effect was too small to account for the decline in the U.S. presence (Shelton, Foland, and Gorelskyy 2009). However, there remains the very real problem of the faithfulness of representation of the expanding universe of scientific publication of the Thomson data; fixed-basket index numbers are a constant source of dispute in economics when it comes to price indices, and such qualms would be equally pertinent here. Yet jiggering the basket for political purposes is of commensurate concern and is relevant to the questions posed here.

There are, of course, many other possible reasons for the United States to slip from world preeminence in output of scientific articles: One that immediately occurs to most people is that as the world becomes more developed and technologically advanced, it was inevitable that the United States could no longer continue to dominate science in the way it had done since World War II. In this view, U.S. supremacy was a historical aberration, bound to erode as other countries caught up in development status. This has been more or less the benign interpretation of globalization and was the predominant view said to emerge from a series of interviews of scientists conducted by the NSF:

> If US researchers figure less prominently in the journal literature, the reason does not appear to be because they are reporting their findings in ways that bypass the journals. Those interviewed consistently reported that research done in other developed countries and in several emerging Asian economies is getting better and becoming more abundant. In their view, improved capacity overseas is more likely to account for the increased share of S&E papers from foreign institutions than changes in what Americans have been doing. In an expanding literature, they see a continuing, even growing, American presence, but more marked growth occurring in other countries. (Bell, Hill, and Lehming 2007, 2)

But perhaps the casual impressions of individual high-profile American scientists are not exactly the best sources of intelligence concerning the global evolution of the modern privatization regime. (Recall the AAAS problem

---

Figure 6.4. National proportions by scientific field. *Source:* Thomson Reuters Science Watch, http://www.sciencewatch.com/nov-dec2007/sw_nov-dec2007_page1.htm.

broached above.) We can further explore the Thomson ISI database by looking at what has been happening to the absolute number of scientific articles originated by U.S. authors, using the fractional count method. Here the interpretation grows more controversial. The NSF and many science policy scholars have taken the position that absolute levels of U.S. scientific publication have "flattened" for a decade or more.[17] But an equally persuasive case can be made that we are living through a period of absolute decline of published scientific output in the United States (Hicks 2007; Tijssen 2004; Adams 2007). I will briefly examine some of the evidence.

As indicated above, Thomson ISI does not maintain a fixed journal set over time for constructing a consistent time series. Using that as an excuse, the recent *Science and Engineering Indicators* (NSB 2008a) presents an estimate of total U.S. and European Union scientific papers (reproduced in Table 6.5 as columns 4 and 5, respectively) from 1995 to 2005, showing a minor U.S. decline, with a sharp recovery in 2003–5. But these aggregate numbers cannot betoken what the NSB commentary claims for them, since everyone openly admits that Thomson has sharply augmented the journals in its base set toward the end of the period: The fact of more added journals would by itself guarantee more total papers attributed to the United States. Yet there is some cause to question why *S&EI* chose to treat that expanded set version as definitive, because in another NSB publication (Hill 2007) they did manage to provide some data from a fixed journal set measure extracted from the Thomson database, although only extending from 1988 to 2003 (reproduced as columns 2 and 3 in Table 6.5). Comparison of the two versions is instructive. In the fixed set, the U.S. contraction is more severe and long-lasting, and it never recovers to the level attained in 1992. The fixed set for European Union counts suffers some attenuation after 1999, but seems to recover much more readily than the United States. For purposes of further comparison, an independent attempt by Chemical Abstracts to count papers attributable to the United States versus the European Union is reproduced in Table 6.5 as columns 6 and 7. Although the peak and timing of U.S. decline diverges from that in columns 2 and 3 (as one would expect from a different journal sample), once again the U.S. decline is not reversed in the data time frame. When one sticks to the legitimate statistical desideratum of a fixed journal set, one invariably detects absolute declines in U.S. scientific publications from the mid-1990s onward. The absolute decline of American publications is also evident from some data from key individual journals like *Physical Review*.

This raises the uncomfortable prospect that the NSF has selected data in *S&EI 2008* explicitly chosen to disguise or otherwise sugarcoat the absolute decline in bibliometric measures of U.S. scientific output.[18] There is the very

Table 6.5. Absolute numbers of published scientific articles, by different sampling schemes

Year	U.S. fixed journal set	EU fixed journal set	U.S. variable journal set	EU variable journal set	U.S. papers in ChemAbs	EU papers in ChemAbs
1988	172.2	131.3			106.6	83.3
1989	180.4	140.5			110.4	84.7
1990	181.3	142.4			110	83
1991	182.2	144.6			132.2	100.7
1992	183.1	154			117.8	95.9
1993	180.1	152.4			126.2	107.1
1994	178.6	162.1			158.1	121.1
1995	179.3	165	193.3	195.8	152.1	134.4
1996	174	169.3	193.1	195.8	160.2	139.9
1997	167.1	169	189.7	208.9	151.5	135.8
1998	165.6	174.3	190.3	214.7	136.6	128.8
1999	163.7	174.6	188	217.1	141.2	133.8
2000	159	170.9	192.7	222.6	136.3	127.8
2001	160.9	170.1	190.5	220.4	139.2	134.7
2002	154.2	164.3	190.4	221.7		
2003	165.6	173.9	196.4	224.8		
2004			202	230.4		
2005			205.3	234.8		

Sources: All data are in thousands, using the fractional count method.
Cols. 2 and 3: (Hill et al., 2007, App. Table 2)
Cols. 4 and 5 (NSB 2008a, App. 5–34)
Cols. 6 and 7: (Hill et al., 2007, App. 1)

real danger of the politicians in charge poised and ready to shoot the messenger, as broached in the introduction to this chapter. The problem seems to be that the period since 1995 experienced substantial increases in U.S. academic and industrial R&D expenditures (inflation-corrected), expansion of the American academic R&D workforce (although not tenured positions, as I have repeatedly pointed out), and yet output as measured by bibliometric statistics (modulo all the qualifications we have tried to make explicit) has been falling in trend. This unmistakable bibliometric contraction, by the way, will further cast doubt upon the message thought to be inherent in the rise in patent statistics that are often promoted by cheerleaders for the modern privatization regime as proof of concept of commercialization of research (an issue explored later in this chapter).

One of the more insidious effects of the attempts by official bodies to mask the decline is the suppression of some insistent questions that would naturally arise once the reversal in trend had been acknowledged. Perhaps the most obvious question that imposes itself is, Why is there a reversal in the mid-1990s? One possible answer was provided in Chapter 3. The bloodbath of corporate in-house R&D, which happened right around that same time, and the subsequent move to outsource applied corporate research, goes quite a long way in explaining the sharp break in the bibliometric trend in a decade that was otherwise relatively devoid of events foreshadowing dire alterations in the science base. This was pointed out a few years ago in an important but overlooked article by Tijssen (2004). After registering substantial growth in corporate scientific publications in the 1980s and early 1990s, a renaissance that had actually increased the corporate share of global published scientific output, the situation was sharply reversed in the mid-1990s. Moreover, there was also a national component to this great transformation: "American and European companies appear to be downsizing the basic research efforts or at least restricting the dissemination of results in international journals, whereas their Asian competitors seem to focus more strongly on basic research activities and/or publish more research articles in peer-reviewed international journals" (ibid 709.). Further light has been shed on this phenomenon by James Adams (2007). In a sample taken of 823 firms from 1990 to 2004, he shows that the scientific publication records of the top 200 firms in R&D expenditures in the United States in 1998 fell *continuously* from 1992 by almost half, whereas the publication record of the remaining 600-plus firms rose. The top 200 clearly consist of the oldline diversified firms with legacy in-house labs, whereas much of the rest were situated closer to the biotech model. The difference in scale between the two subsets was so great, however, that it resulted in a substantial fall in publications for the entire aggregate sector throughout the period. This reiterates the point that the shedding of in-house R&D capacity on the part of the nation's largest firms was an epoch-making event in the history of the U.S. research base, a transformation that has not been offset by the rise of boutique for-profit small businesses. I shall shortly return to this point: The sectors most responsible for the global privatization regime are also the ones most responsible for the U.S. publication decline.

Because we are dealing in rather large unwieldy aggregates, perhaps the decline has been localized in just a few areas, while the changing composition of the sciences may otherwise offset the blip, with new, more vital areas picking up the slack after a transition period. Again resorting to the fixed journal set data neglected by the NSF, reproduced in Table 6.6, we recognize that this scenario does not appear to capture the current dynamic. Biology and clinical medicine, the favored sons of the privatization regime, have experienced contracting publications since the early 1990s in the United States; rather more unexpectedly, so have mathematics and the social sciences. Indeed, one of the more striking facts about U.S. modern science is that the contraction has occurred across the board, with only a few minor exceptions.

The person who has thought longest and hardest on this issue is Diana Hicks (2007, 2008). She has doggedly insisted that the phenomenon should be considered as one of decline, not retrenchment; because of its broad scope

Table 6.6. U.S. publications, various sciences, fixed journal sets

Date	Clinical medicine	Biology	Physics	Mathematics	Social sciences
1988	53,512	12,730	17,883	3,856	8,303
1989	55,592	13,421	18,923	3,972	9,041
1990	55,516	13,922	18,907	3,239	8,917
1991	55,428	13,664	20,098	3,329	8,682
1992	55,516	12,885	19,632	3,429	8,535
1993	55,794	12,061	18,946	3,394	8,389
1994	55,362	12,134	19,652	3,185	7,719
1995	55,535	11,857	18,815	3,013	7,704
1996	54,220	11,266	17,992	3,076	7,383
1997	52,948	11,050	17,146	2,788	7,047
1998	52,421	11,057	16,945	3,072	6,984
1999	52,631	10,059	17,017	3,017	6,917
2000	50,186	10,970	15,965	2,984	6,612
2001	50,569	10,864	16,376	2,834	6,826
2002	47,782	11,416	15,809	2,771	6,369
2003	51,278	11,824	17,115	2,895	6,778

Source: Hill et al. 2007, App Table 2, from data supplied by Thomson ISI, fractional counts; units are single articles.

and relative obscurity, it requires something more than armchair speculations. She considers some recent candidates, such as the facts that the United States has an aging research cohort (Mervis 2007), or a general authorial trend toward more collaborative work (Bell et al. 2007). There are many excuses that one hears, such as the possibility that the newly formed journals, perhaps codifying cross-disciplinary research, often contain the real cutting-edge science (and escape the net of Thomson ISI); or that science is becoming Web-acclimatized with preprints, e-journals, and other such phenomena (ditto); but she dismisses these possibilities. Interestingly, she also discounts the effect of patenting on academic publication, based on a single article (Agrawal and Henderson 2002); here she may have been a little too hasty, because the fact that most professors who patent at MIT also prove to be star publishers may not tell us all that much about the cumulative effects of the patent system upon aggregate publication in both the corporate and academic spheres. Instead, Hicks boldly adopts the position that the U.S. science base is now falling behind both the European Union and the Asian Pacific powers, and because funding and research personnel are expanding more rapidly in those other areas, U.S. scientists are slowly but surely being nudged out of the core set of academic journals.[19] In essence, she has been arguing that the U.S. quality is not degenerating as such; it is just that certain key rivals are getting better at science at an accelerating rate, and in the process they are thus displacing American authors from core journals over time.

Without necessarily disagreeing with much of her stress on global forces, I would venture to suggest that she does not take the analysis far enough. The downturn in U.S. publications clearly started in the business sector, and it preceded much of the Chinese buildup of a science base, and in particular, it occurred in areas of science closely related to high-quality in-house R&D programs. The decimation of this sector was not a simple case of the non-U.S. industrial sector getting better; it was a clear instance of the U.S. resource base becoming demonstrably hobbled and starved. Examples are legion, but one can simply point to the history of Bell Labs, perhaps the preeminent instance of world-class in-house corporate research in the United States. AT&T spun off Bell Labs as Lucent, after letting it deteriorate (Endlich 2004). Lucent was not able to survive in the new regime merely by doing world-class science, and it was subsequently bought up by Alcatel. Alcatel, in turn, has gutted whatever research capacity remained, in favor of turning the formidable patent portfolio of the desiccated husk into a bag of patent troll curses, to rain down fear on the economic landscape (Brumfiel 2008). In just one case, Alcatel won $1.52 billion in damages from Microsoft, for Bell work on the MP3 format. Inevitably, article output has suffered, just as if one had decided to dismantle Stanford and sell off its capital assets to make a quick profit, then turn what was left into a patent troll.

Yet the notion that the decline in publications can be completely accounted for by the shrinking of the corporate sector would be too stark and simplistic a proposition. I argued in previous chapters that in the new globalized privatization regime publication is neither a desideratum nor a right of a scientist: rather, it is something that may be permitted or blocked, encouraged or reined in, depending on the strategic plans of powerful actors in the science bureaucracy. In Chapter 4 I described some of the extraordinary fetters placed upon scientists who stumble into a research area studded with intellectual property, which include prior publication restraint, confidentiality clauses, disclosure agreements, and the like. The notion that the proliferation of these fetters has had absolutely no impact on the rate of scientific publication in America beggars belief. Indeed, there is already some evidence that, in the sample set of the top 110 U.S. universities, total scientific publications have leveled off since 1995, although publication rates at private universities have nonetheless continued to increase (Adams 2007, 77). Some such phenomenon is more or less what one would have expected of the privatized regime; the research output of state universities continues to be hobbled by commercialization demands and the relinquishment of their public service orientation, whereas a few rich, private universities thrive in the hypercommercial environment.

And here we discover the second major cause of the American publication decline, this time located in the university sector. A further overlooked variable in the bibliometric equation is that the proportion of faculty in most universities situated below the top ten who are actually tenured and therefore expected to publish has fallen drastically since the 1970s, from an estimate then of about 57 percent to roughly 35 percent in 2003 (Gravois 2006). The displacement of tenured faculty by temp gypsy faculty has only accelerated since then, with the gypsy category expanding at a rate of 7.2 percent, while tenured faculty expanded at a paltry rate of 0.4 percent (Lederman 2007). Although definitions of categories of temporary workers differ dramatically across colleges, making comparisons difficult, one report estimates that full-time tenured or tenure-track instructors at all public research institutions fell to 41 percent of all instructional staff in 2005 (JBL Associates 2008, 3).[20] In 2006, American universities attained the dubious milestone that, for the first time in history, the majority of full-time employees in higher education fell into the category of administrative/managerial functions, rather than teaching and research faculty (Jaschik 2008). Administrators do tend to prattle on about excellence, but they rarely contribute to the cutting edge of research. Research faculty have progressively been replaced by part-timers and other nontenured categories, which have proliferated in the modern university. Thus the cadre from within the modern university from which one would expect a stream of publishable research has become a substantially smaller proportion of total university employment in the modern privatization regime. None of this story depends upon the (correct) observation that some other countries have vigorously augmented their science base and are consequently enjoying an upward trend of publications. In other words, much of the slowdown in bibliometric output measures in the United States is directly attributable to the consequences of the modern privatization of science for the U.S. university sector.

The catchphrase that the "privatization of science is a global phenomenon," and therefore it must be incongruous to insist upon its differential impact on a particular nation, should be discounted in the current context because of the peculiar status of the United States within the world system of science in the last half of the twentieth century. First, the United States was in most dimensions the first to pioneer the neoliberal reorganization of the science regime, just as it was the first to fashion the military shape of the Cold War regime. Most developed countries tended to follow its lead, although all indications are this may change in the near future. Indeed, insofar as other developed countries opt for the neoliberal script in all its particulars, including the commodification and privatization of education,[21] the enforcement of TRIPs-mandated fortification of IP, the imitation of Bayh-Dole style giveaways to corporate interests, and the offshoring and outsourcing of research functions in pursuit of the most "efficient" production of science, then in all likelihood they will also experience a decline in article output in the foreseeable future.

Second, the U.S. industrial research base has been heavily outsourced to other lower-cost countries, which has given them a onetime boost while at the same time sharply diminishing U.S. domestic article output. The official data on corporate outsourcing found in *S&EI 2008* are deeply flawed, as explained in Chapter 1, but some indication of the scope and extent of the cross-border outsourcing movement in the first decade of the new millennium can be gleaned from data presented in Thursby and Thursby (2006a) and reproduced in Table 6.7, which was based upon interviews with 250 heads of corporate research departments in 15 separate industry classes. The numbers of firms locating new R&D facilities (within a three-year time frame of the interviews) in the United States, Western Europe, China, India, and the rest of the world are individual table entries. The most striking aspect of these tabulations is that they suggest, were it not for European interest in tapping into the American base, according to these executives there would be *no new industrial research capacity built in the United States.* After the financial crisis of the fall of 2008, that prospect does not seem so very

Table 6.7. New corporate R&D facilities, by home office and facility location

Home office	United States	Western Europe	China	India	Other	Row Total
United States	0	19	30	9	13	71
Western Europe	14	10	23	9	12	68
Other	0	0	2	0	2	4
Column Total	14	29	55	18	27	143

Source: Thursby and Thursby 2006a.

far-fetched. Hence the contraction of U.S. scientific output is closely linked to the collapse of corporate investment in it. The second arresting aspect of the table is the strong movement of corporate R&D to China and, to a lesser extent, to India. This net export of American research capacity by itself could account for not only a modicum of the American decline in output, but perhaps for a fair proportion of the rising publications of China and East Asia as well. It is not merely that the scientific capacities of East Asia have been growing, but also that the investment in China by U.S. companies has had as one of its consequences an increased Chinese propensity to publish in English-language outlets. Hence the changing rates of scientific output are much more closely related to various novel features of the neoliberal regime than the benign version of globalization would allow.

Quality Degradation of Science under the Modern Regime

Resorting to counts of bibliometric data to get at gross movements in scientific output is one thing, flawed and controversial to be sure, but ultimately, it is something an economist can wrap his mind around, and perhaps even endorse. Yet venturing that one step further, poking around in order to inquire whether the modern privatization regime has influenced the "quality" of the knowledge being produced, is a quest fated to lose friends and influence. The tension, of course, derives from the whole productivist orientation of economists: Everything is cast upon a procrustean bed of more or less, in a headlong drive to quantify, with no serious consideration of the seemingly discredited Hegelian transformation of quantity into quality.[22] Yet it is precisely there, in the suspicion of a corruption lurking at the heart of science that is directly related to its transformation into a marketplace of ideas, and *not* in the concerns in the previous section that there are fewer and fewer academic papers being published, that seems to emotionally resonate with Viridiana Jones, not to mention the Cassandras bewailing the modern commercialization of science. Perhaps the world could somehow manage to get by with fewer published papers in *Science* or *Nature,* but what will happen when the scientific knowledge we get for our money is itself degraded on a regular basis? If we all were forced to vigilantly monitor and regularly audit the quality of science generated under the new regime because of the intervention of market considerations, then commercialization would indeed have come at an exorbitant price. That would pose economic "efficiency" considerations sorely neglected under the current neoliberal theory of knowledge.

The reason such an inquiry will undoubtedly call down the wrath of the righteous upon me is that it triggers the temptation to bloviate about "the" scientific method, the enduring need for unstinting "trust" in science, the inbred love of truth among scientists, the lamentable sad cases of "a few bad apples," and the like. These kneejerk reactions immediately fill the atmosphere with such a toxic fog that we tend to lose all sight of the real intent and purpose of raising the issue. Hence our intention is to (at least temporarily) take what is designated as research output at face value, without digging deeper. Thus, for instance, the problem of scientific fraud will not be central to my current inquiry (Goodstein 2010). Rather, I want to set out some relatively tractable notions of "good science," starting from some aggregate measures, and then ask what has happened to them.

Philosophers have turned out not very helpful in this regard, primarily acting as though there abided generic abstract benchmark characterizations of reliable knowledge, meanwhile drawing the bulk of attention away from the social conditions and structures that might improve or degrade the process of validating and augmenting the quality of knowledge.[23] Perhaps even more disturbing, the profession of science studies has become absorbed with the prospect of "democratizing" science and diminishing the power of expertise, to the neglect of consideration of the degradation of the knowledge base that might possibly ensue as a consequence of "opening up" science to various constituencies. And then there is the consideration that no academic would feel comfortable entertaining the notion that history would conclude that the very quality of knowledge had deteriorated under their watch or that of their peers. It would be one of the most damning things you could say to someone dedicated to the life of the mind.

So heaven and earth stand arrayed against the project pursued in this particular section. But I shall heedlessly press onward against the disdain of the madding crowd, because there persist phenomena here that need to be analyzed under a single framework. In each of the subsequent sections, I explore impressions that the knowledge produced under the auspices of the current regime has been diminished in quality, and I briefly point to evidence frequently cited to fortify such impressions. The distress over the debasement of science will fall under three quintessentially American headings: just-in-time science, junk science, and the degradation of patent quality.

Just-in-Time Science

At various junctures in this volume, participants in the modern regime of science organization have consistently pointed to one aspect that has proven an irritant: the speed-up of the tempo of modern production of results in commercialized science, and the tendency to promote the quick and dirty result over the calm and measured finding. Nearly every proponent of the new regime cannot help but deplore the lackadaisical attitudes of the ivory tower academic and disparage their slack concern for the imperative to get those new discoveries out the door and into the world as soon as possible. We have observed that the CRO has captured the clinical trials market share away from the academic hospital because it can produce the data for pharmaceutical companies in record haste; that large multinationals shed their in-house corporate labs because they didn’t stock the product pipeline in a timely fashion; and that one reason the linear model was repudiated in the science policy community by the 1990s was that the implied postulate of long and variable lags between basic science and product development was simply rejected as unnecessary. There is no denying that the logic of commercial activity has been the edict to shorten the temporal gap between conceptual innovation and the consumer, just as the logic of modern lean production has been to accelerate throughput while minimizing the time inputs need to be kept in inventory. Indeed, that is Alfred Chandler’s basic story of the rise of the mega-corporation in his *Visible Hand*. But the nature of the speedup is palpably different when knowledge is treated as a fungible output, as in the neoliberal vision of the marketplace of ideas.

The logic of the modern speedup in science is to minimize or dispense with everything that may slacken the pace of the production of novelty. But speed can only be of the essence if the terminus is fixed and perdures independent of the quest for knowledge. That is why neoliberals love to portray the history of science as one damn race after another—the race to element 113 (ununtrium), the race to the double helix, the race to the Higgs boson.[24] However, the more we learn about the actual processes, the more we come to appreciate the real bonanza of knowledge does not come from the preconceived target entity (which in some sense is already “known”), but rather from all the unexpected things that turn up along the way. Scientific research is above all a process that has to remain open to the serendipitous, the unexpected, the incongruous, and the unanticipated. Hence it is a perverse misrepresentation of the process of research to insist that accelerated acquisition of certainty is the secret to scientific success. Indeed, the stress on competition in the portrait of science as a dash to the golden finish line tends to obscure the ways in which labs working on similar problems have to negotiate a strange amalgam of rivalry and cooperation in order that their findings may be accepted and validated within the wider disciplinary community. Historians of science and SSK scholars have understood this much longer and much better than the neoliberals: "The history of science teaches that every great step in the direction towards demonstrating a final reality shows that this reality turns out to lead in a quite unexpected direction" (Bachelard [1928] 1987, 284). A better metaphor for scientific research might be the coordinated gathering of information for a map during an expedition: The prepared mind will notice relevant and unexpected features along the way, which eventually constitute the net addition to knowledge, even though everyone realizes that the terminus is the production of the cartographic project. Surveys take time; any fool can sketch a featureless diagrammatic map on a napkin.

Thus one way that the modern commercialization of science might have a deleterious impact upon the quality of science is through the forced inducement of quick and dirty techniques to produce attenuated results on schedule, under budget, and within the parameters of the contractual relations. It is precisely this phenomenon I shall hypothesize underlay the further complications described in the next two sections, namely, junk science and patent degradation. Yet observational evidence bearing upon just-in-time science is extremely hard to get at directly.

Perhaps one of the least well-understood examples of just-in time science was the supposed race to complete the sequencing of the human genome between Celera Corporation of Craig Venter and the government-funded Human Genome Project (HGP).[25] As is well known, the arbitrary finish line was engineered in 2000 by Bill Clinton to make it appear a tie, so as not to unduly embarrass any of the contending parties. However, instead of privatized science having the salutary effect of merely speeding up the project, as was baldly asserted by Gary Becker (2000), a neoliberal economist and MPS member, it actually served to degrade the quality of the reported genome.

Perhaps the most common statement one encounters concerning the race between the government HGP and Celera was that "competition from the private sector was good for the outcome." This bald travesty concerning events as they transpired just goes to show the extent to which the neoliberal conception of the world has come to trump the actual history of science. In fact, with the passage of time we can now see clearly that the attempt by a private firm to upstage the government project was bad for Celera *and* bad for the quality of the science performed. The former argument is illustrated by the fact that after 2001 Celera careened from one failed business plan to another, finding that it could not consistently make money as a database service, then as a start-up biotech/drug research company, and lately, as a diagnostics firm offering cardiovascular- and cirrhosis-oriented testing services.

In July 2008 its holding company, Applera, divested itself of the firm, and its offices in Maryland were shut down.[26] Celera is still losing money, but not by doing basic biological research anymore. The weaknesses of the biotech model in general were covered in Chapter 5, so I won't belabor that point again. Here I want to show how Celera reveals the pitfalls of just-in-time science.

In spite of the massive press attention devoted to the "race to sequence the human genome," it is distressing just how little factual information concerning the technical side of the contest between Celera and the HGP seems to have seeped into general consciousness. First off, Craig Venter was not personally responsible for innovating or inventing much of anything. The basic set of operations used to read or "sequence" the genome was invented back in 1980, and it persists down to the present. Frederick Sanger innovated the principles of the "shotgun," which include amplification of the DNA (by splicing and growing it in bacteria); extraction and shearing of the DNA; reading the short sections by electrophoresis and gel separation; searching for overlaps in sequences to stitch together the truncated bits; and then assembly of the linked sequences into consensus "contigs." The reason one needs to shred or shotgun the DNA in order to sequence it is that the gel separation can only handle lengths up to roughly a maximum of a thousand base pairs (although ~five hundred was commonplace), whereas something like the human genome contains more than three billion base pairs. The innovations relevant to our story are technological developments that permit the automation and acceleration of one or more components of the process in order to render the reading of something like the human genome more feasible.

The two alternative variations of the shotgun in the late 1990s were the "hierarchical shotgun" (or BAC walking) favored by the HGP and the "whole genome shotgun" (WGS) favored by Celera. Suffice it to say that the HGP adopted the former protocol because it was slower and more methodical, repeatedly alternating coarse-grained chromosome mapping with fine-grained sequencing; it was more labor-intensive, but it did allow portions of work on the genome to be farmed out to geographically separate institutions, an important political consideration at that time. Furthermore, BAC walking lent itself to phased release of individual subsets of genome portions as they became available, something to which the HGP had committed itself. Venter, on the other hand, was a big believer in using computer algorithms and raw computational power to sequence the entire genome in one fell swoop in the whole genome shotgun. Neither technique was radically conceptually new, and the latter had been considered and rejected by the HGP in 1997. It was true that by the later 1990s the WGS was certainly quicker and somewhat cheaper, but that was largely because it was *inferior* by the standards prevalent at that time. In particular, it was understood that the larger the genome, and greater the proportion of long repeat sequences, the *worse* would be the quality of reads from the WGS vis-à-vis BAC walking. When Venter praised the method, it had been in the context of his sequencing of smaller bacterial genomes, which exhibit many fewer repetitive stretches, but it was precisely the case of the mammalian genome that many felt was unsuitable for the WGS. Indeed, the current consensus is that the WGS by itself is still an inferior technology to apply to mammalian-sized genomes.[27]

So why did Venter press ahead with the WGS, and subsequently garner the reputation of being the shunned maverick rebel genius behind a radical improvement over the fuddy-duddy HGP and all previous approaches? This is where the new regime of just-in-time science came into play. In order to mount the WGS method, massive computer automation of the sequencing process was required; the NIH had already decided it wasn't worth it, given it had opted for BAC walking. In order to pay for it outside the HGP, Venter had to entice his venture-capitalist backers with a treasure chest of valuable intellectual property locked away in the human genome. But the gold would only stay valuable if the HGP didn't get there first—no one at Perkin Elmer would plausibly make big money off development of WGS per se—so here we observe how quick and dirty science became wed to a neoliberal crusade to undermine government science policy. Crudely, Venter and his start-up, Celera, had to foster the trappings of a race between his own speedy, glitzy technology and the galumphing HGP simply in order to stay in the game; once they managed to accomplish this (partly by winning over journalists in the popular and business press), the threat of expropriation of the human genome effectively forced the HGP to join in the "race," to consequently degrade the quality of its own reports to block Celera's blipkrieg and partially adopt WGS techniques, so as not to be upstaged by Celera. Perhaps the reader can begin to grasp why a fake photo finish needed to be manufactured out of whole cloth by Bill Clinton in 2000, merely to put an end to an untenable situation.

In a standard neoliberal strategy, Celera had access to the open archive of HGP sequence data in real time, while keeping its own findings proprietary and secret.[28] The appropriation of public scientific findings while asserting a commercial right to engross and control the sum total of public and private results, along with a simultaneous insistence that commercialized science was just naturally more "efficient," was a mockery and a scandal. However, contemporary science journalists seem to have been reticent in pointing that out. Venter muddied the waters by claiming that Celera sequence data (vintage 2000) would be freely released to the public, although that never really happened until 2005, and then only when Celera had decided its database business model had failed and after it had unceremoniously fired Venter. Furthermore, far from getting the data out into the world with greater alacrity and free of charge, in the interim Celera actually placed so many encumbrances on access to its Web site that it generally could *not* be used for further research anyway (Eisenberg 2006, 1024). "Celera's game plan, from the beginning, was a classic bait and switch scam. In this scenario, the company's strategy was to use the promise of free, unrestricted access to the data to undercut support for the public project and thereby set the stage for a lucrative monopoly in selling the sequence on a fee-for-service basis" (Olson 2002, 935; Bell 2003). The only glitch was, once the clientele twigged to the fact that the quality of the Celera draft was so poor, even this scam didn't pan out as planned. But they did succeed in undermining public science in the interim.

Even though the "working draft" of the human genome was announced in the "all must have prizes" photo finish in June 2000, actual "publication" in February 2001 revealed that the genome was nowhere near finished. Something on the order of 83 percent of the genome had been "finished" by some loose criteria and around 119,000 "serious gaps" remained (Olson 2002, 938). Because most researchers couldn't get comprehensive access to the Celera version, a serious comparative evaluation between the two drafts was then out of reach.[29] But more significantly, at that juncture, Celera lost interest in the cleanup activity that would be required to bring the draft up to snuff for a number of reasons. First, it had been losing money on its database business for a couple of years; second, the cleanup would inevitably involve BAC walking, the technique it had so vocally and publicly shunned during the "race"; and third, Celera researchers had already moved on to grab at more IP, using the WGS to sequence the genomes of the mouse and Drosophila fly in 2001. In fact, during the transient era when Celera was still selling itself as a fast sequencing company, it earned a reputation as a rip-and-run outfit: "They haven't completed the experiment . . . In Drosophila, Celera did a whole genome shotgun and then Gerry Rubin's lab spent two or three years on the process of trying to finish it, and they're still trying to do it . . . They didn't finish directly from the whole genome shotgun data, so what was more cost effective?"[30] A few years' experience taught most participants that the WGS as practiced by Celera was quick and dirty just-in-time science, good enough for the undiscerning patent office, but not nearly of the quality that would underpin further genetic research. The cleanup was inevitably left to the publicly funded remnants of the HGP. Indeed, the problems presented by centromeres and telomeres are so vexed, that even down to the present, only something on the order of 92 percent of recent drafts are considered "finished."

Congressman Vernon Ehlers in 1998 accused Venter of doing quick and dirty science (Olson 2002, 937); now we know he was right. Celera did not substantively help in the process of decoding the human genome; it just threw sand in the gears and then absconded when the real hard work had to be done that had been foreseen all along. It tried to pass off a degraded draft of the human genome as an epoch-making accomplishment, then sought to conquer new organisms. It fell afoul of the general fate of biotech start-ups: With few exceptions, you just can't make money by doing basic science in the commercial model. Yet confounding all reason, Venter persists to this day in misrepresenting the history of Celera versus the HGP as David versus Leviathan. Was it the scientist or the buccaneer who asserted, "I think [Celera] is one of the few successes, however briefly, in biotech, and certainly it has to be the best story in bio-information technology"?[31]

Another example of the just-in-time effect can be found in the relatively high-profile debacle of the "discovery" of element 118—something perhaps more people can readily grasp (G. Johnson 2002) than the ins and outs of sequencing the genome. The periodic table starts losing some of its pleasing symmetry around element 90, in part because the heavy elements (called "actinides") are unstable and break down through radioactive decay. After the unlovely footnote appended to the periodic table of elements 90–103, the row resumes with the so-called trans-actinide or superheavy elements, all man-made because they have such short half-lives. Starting in the 1940s, the Lawrence Labs at Berkeley (LBL) used its lead in cyclotrons to discover/produce neptunium, plutonium, and eventually fourteen other trans-actinide elements. But by the 1980s, LBL had lost its position as the world leader in the synthesis of transuranium elements; in any event, the elements were getting harder to produce and detect. Leadership had shifted to the Institute for Heavy Ion Research in Darmstadt, Germany, and to the Joint Institute for Nuclear Research in Dubna, Russia. A scientist named Victor Ninov had helped to produce elements 110–112 at the institute in Germany in the mid-1990s, and LBL hired him in 1996 to reinvigorate an effort to look for very heavy elements beyond 114, partly inspired by a theory that suggested there might be an "island of relative stability" out near elements 116–118.

The point to be stressed here is that this situation seems to conform quite closely to the neoliberal notion of science as a race to a fixed discovery. Normally, slamming atoms into lead to produce new elements was a slow process yielding minute quantities of doubtful identity. Along came a theory (Smolanczuk 1999) that tells us all we need to do is pump up the cyclotron energies, smash some krypton into a lead target, get the decay product detectors working just right, and voilà, you have discovered element 118. Ninov and other team members, seeking to restore LBL's past glories, decided to cut corners and bypass debugging and testing of new equipment and protocols and try and synthesize element 118 in April 1999—that is, they opted for just-in-time science. Partly, this was justified in reference to its privately funded rivals. Because of his past successes, and also because of the speedup and that no one else really understood the software, Ninov was the only team member to deal with the original raw data emitted from the cyclotron, stored on magnetic tape in binary format. Ninov and team thus decided they discovered elements 116 and 118 in one fell swoop, and they rushed to publication (Ninov et al. 1999). But some rival groups had trouble finding any trace of 118, and worse, once colleagues at LBL got the hang of the new technologies, they went back to the original raw data tapes and were perplexed not to find any of the claimed decay events. LBL convened an internal review team, which also came up dry, so they recommended that LBL issue a retraction. However, Ninov as lead author refused to sign the retraction letter, so publication of the retraction was delayed until July 2002, after LBL fired Ninov. The subsequent sequence of outside auditors expressed astonishment that only one person would be the sole examiner for data at such a crucial juncture, but, of course, that simply misses out on the proliferation of vulnerable choke points within the elaborate division of labor in any Big Science experiment. One of the main findings of science studies is that replication of every possible step of a procedure is uncommon in most science because one cannot banish the local contingency within any complex experiment. Nonetheless, since the case was so high-profile, the American Physical Society took the unprecedented step of issuing new ethics guidelines in 2002;[32] but this just looks like the standard defense of any beleaguered bureaucracy, to try and turn a structural problem of just-in-time science into a *personal* problem of individual ethics. The indispensable role of suitable pacing in science is illustrated by the subsequent discovery of element 118 (it was always there all along, right?) by the Russian institute and Lawrence Livermore in 2006, but not by the Smolanczuk method.

It is too bad that most laypeople get their impressions of how science works from TV shows like *CSI* and *Bones*, because the area of forensic science has suffered dreadfully from a bad case of just-in-time disease. A sequence of highly publicized crime lab failures and wrongful prosecutions has led even the National Academy to be spurred to action, denouncing the practices of both forensic technicians and forensic scientists (S. Moore 2009). In reaction, private companies set up to sequence DNA for hire sought to blunt and contain criticism of their work, by appealing to—what else?—proprietary information.

There has been one relatively overlooked attempt to delve into the just-in-time effect beyond the realm of the individual incident or anecdote. In a curiously overlooked article, the legal scholar Jeremy Grushcow (2004) sought to find a measure of the effects of increased secrecy on scientific research. He noted that most countries don't allow open promulgation or publication of discoveries prior to patent applications, so that it would be likely that disclosure in professional settings would be delayed or otherwise embargoed prior to formal submission of application papers. Further, in the federal circuit court ruling *MIT vs. AB Fortia* (774 F.2d 1104), it was held that oral presentation of a paper before more than fifty people at a conference or the distribution of a full paper to at least six individuals constitutes "publication" for the purposes of patent law. In a very labor-intensive inquiry, Grushcow began with the abstracts of papers presented to the 1980–1990 meetings of the American Association for Cancer Research, and he linked each to a subsequently published paper on the same topic by the same authors. He categorized each author as coming either from a university, the NIH, or a particular industry. He also searched the U.S. Patent Office database to find if there was a patent associated with the particular paper. He then measured the "gap" between formal presentation and publication over that time period, which was chosen purposely to explore recent changes in the science regime, such as Bayh-Dole and other legislation. He argued that the longer the gap, the more the possibility of communal scientific conjecture and refutation, and the less secrecy was involved with the particular project. Conversely, the prospect of patentable knowledge would lead to a truncated gap. In a fascinating selection of statistical cross-tabulations, he found that the size of the gap fell by almost half from 1980 to 1990 for university researchers, fell much less substantially for NIH researchers, and actually rose slightly for industry-based researchers. Furthermore, the gap fell to essentially zero for university scientists who sought patents, while remaining at a commensurate level to that of the NIH researchers for the remainder of those university scientists not seeking patents. Indeed, more than half the university scientists filed patent applications *before* their public presentations by 1990, which in effect results in a negative gap. None of this would be particularly startling to anyone who has lived through the advent of the commercialization regime upon universities, but it does provide solid quantitative evidence of the just-in-time effect. The rush to commoditize knowledge tends to shelter research from the previous level of scrutiny by the larger scientific community, at least until it becomes propagated within the commercial sphere (and hence much more entrenched). This effect, I shall argue in subsequent sections, results in qualitative degradation in the character of the knowledge produced. Again, I have no wish to idealize or otherwise gild the Cold War regime—secrecy was rife there also—but with the presentation-publication gap there is a precise before-and-after comparison that reveals that the situation of university-based research is *worse* under the current regime, at least with regard to commercialized knowledge.[33] The effect has been to squeeze out the interactive effects of the interplay of scientific community so beloved of everyone from Karl Popper to Stephen Jay Gould. One might speculate that a future serious economics of science would produce many more empirical instances of this shrinking gap in other fields of research.

Junk Science, Sound Science, Rancid Science

All too frequently, scientists tend to disregard controversies over the validity and quality of science staged in the courtroom, relying on their impression that there is a steep firewall between how science is vetted in the laboratory and how it gets mangled in legal disputes. However, one persistent theme in this volume is that the project to uncork the academic bottle and get research results rapidly out into the world of the modern privatization regime has had all manner of unintended consequences. One of the most significant developments of the last three decades has been the rise of the "junk science" movement, something that has done more than any other phenomenon covered in this chapter to actually degrade the quality of science in the new regime. It has accomplished this not only directly, by elevating dubious knowledge to comparable status to that produced by long-standing academic disciplines, but also indirectly, by weakening those same disciplines through promoting the idea that there is a surfeit of dodgy academic science corrupting the journals and airwaves, fostering the impression that every concerned citizen must gird themselves to be wary of corruption in the vast archive of scientific thought.

The ways in which the junk science movement has morphed into the "sound science" movement would make even poor George Orwell spin in his grave. A comprehensive chronicle of how this happened still awaits its historian, but I shall attempt a provisional sketch here. Basically, the libretto can be decomposed into three movements: The first is from the 1960s to 1980s, the second covers the 1990s, and the third extends down to the present. In the first act, which mostly was confined to the courts, a few industries were dissatisfied with the ways that academic science was being used against them in product liability suits, and therefore they decided to fund some more friendly advocacy research, which could be used as a part of their defense strategies. The major protagonist here was the tobacco industry, but others were also involved early on. In the second act, the scenery opens out to encompass some important neoliberal think tanks that counseled the firms to go on the offensive by publicly attacking their opponents as dealing in "junk science." The unspoken objective of this blitz was to begin to remove responsibility for validation of science from the academics, to render it more amenable to a marketplace of ideas as the ultimate arbiter of truth. The third act moved back into the courts and the legislatures, where new rules instituted in the name of "sound science" in fact ratified the victory of the neoliberal regime.

The first act of the drama is probably the best understood part of the scenario in science studies. Due to later court settlements against the tobacco industry for misrepresenting the health hazards of smoking, we have been granted unprecedented access to normally proprietary records that document the ways in which the industry, in alliance with some PR and product litigation support firms like the Weinberg Group and Hill and Knowlton, developed a novel strategy to block exposure to liability from the harmful effects of smoking.[34] Mimicking a standard refrain among academics that "more research is needed," the consultants had picked up some tips from the neoliberals that one could build an entire "counter-science," even if it was little more than a Potemkin village, and that it might turn out to be even more effective in frustrating litigation and regulation than merely throwing lawyers at the problem.

In the mid-1950s, when evidence began to link cancer to tobacco smoke, the industry started out by founding the U.S. Tobacco Institute, nominally to carry out and fund research on smoking and health. While the institute was recognizably an industry creature, it became the staging point from which to mount an entire institutional campaign that is now widely recognized as setting the pattern for many subsequent citadels of commercial science. As David Michaels puts it, they learned that debating the science turned out to be easier, cheaper, and more politically effective than directly debating the policies themselves. We might rephrase it that they came round to accept that scientific debate was engagement in politics by other means. The key tenets were to promote otherwise isolated scientific spokespersons (recruited from gold-plated universities, if possible) who would take the industry side in the debate, manufacture uncertainty about the existing scientific literature, launder information through seemingly neutral third-party fronts, and wherever possible recast the debate by moving it away from aspects of the science that it would seem otherwise impossible to challenge. As one famous tobacco company memo put it:

> *Doubt is our product* since it *is* the best means of competing with the "body of fact" that exists in the mind of the general public. It *is* also the means of establishing a controversy. Within the business we recognize that a controversy exists. However, with the general public the

> consensus *is* that cigarettes are in some way harmful to the health. If we are successful in establishing a controversy at the public level, then there *is* an opportunity to put across the real facts about smoking and health. *Doubt is* also the limit of *our "product"* . . . Truth *is our* message because of its power to withstand a conflict and sustain a controversy. If in *our* pro-cigarette efforts we stick to well documented fact, we can dominate a controversy and operate with the confidence of justifiable self-interest.[35]

The inspiration was to take one aspect of what many philosophers (from Peirce to Popper to Putnam) had argued was central to scientific epistemology and to expand it into a principle of research funding and management, guided, of course, by explicit self-interest in steering the threatening controversies of the day. At first the practice started small, but again under the model of the neoliberal thought collective, whole rafts of think tanks, "institutes," and labs were founded to carry out various components of the program. Among the most significant were the George Marshall Institute, the Annapolis Center, the Competitive Enterprise Institute, the Center for Science and Public Policy, and the Manhattan Institute, not to mention a range of lesser entities (Jacques, Dunlap, and Freeman 2008). By the 1970s these structures, in conjunction with a few smaller centers founded within universities, began to form a parallel scientific universe, a whole mirror world of white papers and dubious fact sheets and counterfeit journal publications explicitly constructed to mimic academic scientific output while keeping the original funding and motivations obscure. While most observers have some awareness of think tanks devoted to social policy, almost no one has noticed that neoliberals have also developed a phalanx of think tanks to produce the kinds of *natural science* that they felt were warranted by their pecuniary and epistemic interests. This is the apotheosis of the belief in the marketplace of ideas.

One should not think that this vast fabrication of science-to-order was only or primarily limited to one or two cases, or to issues surrounding tobacco, although it does appear that tobacco was the first test case. For instance, ancillary documents from the tobacco settlement reveal Hill and Knowlton providing histories of its early organized intervention in a number of scientific issues, including the link between vinyl chloride and cancer, dioxin and human health, many issues in groundwater contamination, asbestos and its effects on humans, and even an early program of "denial" in the case of ozone depletion by fluorocarbons. One memo included in this release explicitly admits, "Hill and Knowlton was asked by DuPont to calm fears, get better reporting of the issues, and gain two or three years before the government took action to ban fluorocarbons." One can observe a delicate neoliberal cost-benefit analysis of a few more years of profit on one side and scientific truth on the other. The collateral damage began to show up in the orthodox scientific literature: "The contours and content of the scientific literature are directly and intentionally shaped by parties seeking to succeed in litigation . . . if not for the litigation, or fear of future litigation, the body of scientific literature about a particular topic would be quite different" (Michaels and Monfortin 2007, 1142–43).

By the 1980s these purveyors of alternative science for hire had become so successful in the American context that these circumstances forced the innovation of a second phase of the reorganization of science. The more powerful the floating mirror world of neoliberal science became, the more frequent grew high-profile clashes with academic science, particularly in the courtroom. This could not help but foster the impression that many "scientists" in the dock were little better than intellectual prostitutes willing to sell themselves to support any random litigant with deep pockets (forgetting, of course, this had been the lodestar of the tobacco lobby in phase 1). Too much unfocused cynicism about science among the general populace provoked by such spectacles would tend to undermine the entire procedure of using mirror-world science to postpone or block unwanted regulation and thwart liability settlements. Would "science bought and sold" undermine its own effectiveness? The time had come for the construction of an especially poignant version of the neoliberal doctrine of the "double truth" (Mirowski and Plehwe 2009): one for the populace, and another for the elites.

The denizens of the neoliberal think tank the Manhattan Institute came up with the brilliant idea of blaming the academic scientists themselves for purveying what the Institute dubbed "junk science."[36] A big splash was made by the Manhattan Institute scholar Peter Huber's book *Galileo's Revenge* (1991), which put the concept of junk science and the so-called tort reform movement on the intellectual map. Huber's work was directly related to the previous phase of science policy through the funding and encouragement of the tobacco industry (L. Friedman et al. 2005, S17). A lawyer and author of books like *Hard Green: Saving the Environment from Environmentalists* and *The Bottomless Well: Why We Will Never Run Out of Energy,* Huber provided work for his think tank that often promoted mirror-world corporate science, thus his appreciation for its paradoxical character. In a subtle move in *Galileo,* Huber asserted that all sorts of unscrupulous academics were following the lead of some even more unscrupulous attorneys in bringing all manner of frivolous suits to court based on fake or tendentiously distorted science; corporations throughout the land were being bled to death by a thousand unkind cuts. Something, most assuredly, had to be done about this. What Huber was really saying was that it was okay for corporations to unleash their think-tank surrogates to create doubt as their product, but that it was unfair when the plaintiff of the litigation would engage in similar tactics. The solution, as Huber proposed it, was for the courts to take it upon themselves to screen or otherwise banish what Huber and his allies had tarred as "junk science." In other words, Huber sought to reduce judicial competition and reinstitute asymmetry in the doubt production business. This crusade was subsequently conducted under the banner of "sound science" and has been amazingly effective in the United States.

The Advancement of Sound Science Coalition (TASSC) was founded in 1993 as a "national coalition intended to educate the media, public officials and the public about the dangers of 'junk science.'"[37] Phillip Morris covertly created this organization initially to generate scientific controversy regarding the link between secondhand smoke and cancer, although it quickly expanded its remit to include food additives and auto emissions. Steven Milloy, an adjunct scholar at the Cato Institute, became executive director of TASSC in 1997 after a stint of lobbying for the tobacco industry. The close ties of the sound science movement to the tobacco industry reveal how lessons learned in one arena were transmitted smoothly to other sciences and other controversies. Milloy now apparently devotes his time to www.junkscience.com, which combines Fox News videos with position papers promoting corporate mirror science and features ongoing analysis of the global warming "hoax," including a running calculator of how much the Kyoto Protocol has cost the United States since going into effect worldwide.[38]

The sound science movement rapidly spread across the gamut of industries seeking to counter regulation with junk science–style defenses, most notably of late in the area of global warming. The Union of Concerned Scientists has documented that one (very big) corporation, ExxonMobil, had by itself funneled $16 million into the creation of a whole elaborate network of mirror-science units. Journalists, lumbered by their naive notions of "balance," got suckered into lending these front groups credibility, not realizing that they had been largely conjured and promoted by a very few hand-picked Exxon evangelists. One major precept of mirror-world science is that a single spokesperson or just one shell front group is never sufficient for a high-class campaign: "By generously funding a web of organizations, with redundant personnel, ExxonMobil can quietly and effectively provide the appearance of a broad platform for a tight-knit group of vocal climate science contrarians. This seeming diversity of organizations creates an echo chamber that amplifies and sustains scientific disinformation" (Union of Concerned Scientists,

2007, 11). Through their affiliated neoliberal networks, these groups subsequently went so far as to found organizations that would perform the same functions for religious attacks upon other areas of science, like attacks on Darwinian evolution orchestrated through the Discovery Institute and the Thomas More Law Center.

The three major legal landmarks of the sound science movement have been the Supreme Court ruling in *Daubert v. Merrell Dow* (1993), the Shelby Amendment to the Data Access Act (1998), and the so-called Data Quality Act (2000).[39] *Daubert* changed the Federal Rules of Evidence to include specific criteria for evaluating the admissibility of scientific experts and evidence and imposing prior conditions of "testability," peer review publication, acceptable error rates, and whether the science involved has been "generally accepted." It also fortified the role of the judge to act as gatekeeper to decide which scientific evidence could be heard in trial, usually through pretrial motions. In effect, *Daubert* and subsequent rulings— including *General Electric v. Joiner* (1997) and *Kumho Tire v. Carmichael* (1999)—clarify what has been a prodigious change in the law and have directed judges to identify "sound science" and block "junk science." Hence *Daubert* admitted defeat for the previous model of letting juries decide after hearing conflicting arguments on scientific issues: *The world was deemed too rife with bad science.* What has been hard for nonlawyers to understand is how a set of criteria that would appear on the surface to help filter out a lot of the dodgy science generated by corporate think-tank mirror-world units would instead end up reinforcing and fructifying mirror-world science. Indeed, many of the above appellate rulings favorably cite Huber (1991), which begins to signal the twisted way this happened. The short answer is that *Daubert* permits the litigant with the deepest pockets to issue seemingly endless pretrial challenges to the quality of the science of the opponent. Furthermore, given its implicit "corpuscular theory of scientific fact," this places the burden on the plaintiff to validate each and every study referenced in the case. Because so far this method has mostly been deployed in lawsuits seeking damages for product or other liability, it ends up as a major means of delay and sabotage through pretrial motions, and hence it has been clearly biased toward corporate defenders in these lawsuits. In the name of "sound science," *Daubert* restores the asymmetric bias toward corporate science that had been almost lost in the courtroom when science for sale had become something accessible to every ambitious product liability lawyer. "Junk science" has now become the epithet of choice unself-consciously bruited about whenever corporate officers encounter portraits of the natural world that do not conform to their own vision of a smoothly operating marketplace of ideas.[40] But if junk is all so very ubiquitous, even inhabiting supposedly reputable scientific outlets, then has the entire tenor of the scientific research process been compromised? Some neoliberals sound like they might concede this.[41] Or maybe the sound science movement just ended up lowering the mean quality of the random scientific article selected from its proliferation of possible "sources"?

The Shelby Amendment (1998) is yet another one of those superficially well-meaning interventions with neoliberal consequences. It was a minor amendment to the Freedom of Information Act, which extended the applicability of the act to any federally funded grantee—formerly one could only use the FOIA to pry out documents or data possessed by a federal government agency. The supposed motivation behind the Shelby Amendment was to make it possible for "citizens" to check whether federally funded research was really "sound science," by demanding all raw data and notes from the principal investigator. This specious justification would be rapidly refuted by simply asking what kind of person would ever want to do such a thing. The real agenda behind the amendment was to make it possible once again for the corporate mirror world to harass and otherwise challenge any academic science or scientist it didn't like. These legal challenges have become more frequent in the last decade, helping turn any science with policy implications into a treacherous snake pit of threats and legal maneuvers. The sheer asymmetry of the perversion of the FOIA became apparent once one was brought to realize that there was no equivalent "Shelby" for the corporate think-tank mirror world: All the research they funded was completely proprietary and therefore immune from subpoena. Shelby was therefore a brilliant two-pronged attack, weakening both federally funded science and academic science vis-à-vis the corporate mirror world, all in the name of "open science." There is no better example of how appeals to sound science are cynically used to degrade the quality of the entire scientific research base than the Shelby Amendment.

The Data Quality Act (2000) is the legislation most frequently mentioned when accusations are hurled concerning the partisan character of the modern regime of neoliberal science (e.g., Mooney 2005, Chap. 8). Stealthily inserted as a last-minute paragraph in a Congressional Omnibus Appropriation Act, it stipulated that any and all information disseminated by federal agencies must meet certain standards for objectivity, utility, and integrity; peer-reviewed scientific journals were implicitly tarred as insufficiently equipped to impose these standards. Of course, one paragraph was not going to contain the Secrets of the Ages as to how to attain and enforce those epistemic virtues, so the act punted to the Office of Management and Budget as the promulgator of strict guidelines sometime in the future. It did, however, stipulate that once the guidelines were in place, each agency had to have a mechanism in place for "correcting" challenged information; furthermore, it had to report to an auditor the number and nature of complaints received. The bureaucratic nightmare takes a number of turns here we simply pass by,[42] but the net result has been that once the system was put into place, it became possible for corporations who were able to closely monitor the federal regulatory process to intervene at a very early stage through challenges to the "quality" of information being accessed, often using the mirror-world science that those very same industries had generated to sow doubt in the first place. Challenges have been filed in areas such as climate change, WHO dietary guidelines, NIH research on sodium and blood pressure, and our old friend, tobacco. An attempt by the Government Accountability Office to figure out just who had been entering challenges under the Data Quality Act for fiscal year 2003–2004 found that forty-four out of eighty were launched by for-profit entities, and that in thirty-two cases the information was changed or withdrawn. Again, under the seemingly virtuous pursuit of "sound science," we observe the prosecution of political disputes as though they were scientific controversies, and one consequence is that the quality of the actual science has been compromised.

The war over junk science and sound science is nowhere near drawing to a close. The tobacco industry, to many people's amazement, is *still* funding surreptitious research that absolves them of responsibility in a whole range of smoking-related ailments, essentially buying a whole university to act as a front for its "sound science" (Finder 2008). The final act in the junk science saga is the strange spectacle of people who would not normally be seen consorting with the neoliberal think tanks and their sound science crusade now coming to support the notion of what they call "litigation science."[43] Taking their cue from some developments in science studies, they have begun to ask whether staking out boundaries between purpose-built science for litigation purposes and academic science in standard peer-reviewed journals actually makes sense. As they argue, most science grows out of financial interests of one sort or another, so why think that public identification of the specific funders is dependable evidence for the quality of the science produced? Biases can be hidden, and if the patrons are sophisticated, they most probably will be. They also suggest that there is no regular empirical evidence that litigation science is inevitably of poorer quality or that cross-examination in a courtroom is less effective in picking apart the flaws of a piece of research than running the gauntlet of peer review in a journal. Finally, they insist that litigation science is needed because more conventional peer-review science is just too slow to respond to the requirements of social conflicts. In other words, here is where the phenomenon of just-in-time science converges with junk science, all to fulfill an unmet market niche.

This is not the first time we observe science studies beginning to make a pact with neoliberal conceptions of knowledge, but it is certainly one of the more dispiriting.

Patent Quality Degradation

There has been one sphere of activity where one can find economists actively debating about the possible degeneration of the quality of knowledge under the modern regime. (Of course, they exempt their own quality of knowledge from the dispute.) Not surprisingly, their attention has been galvanized by the fact that the profound fortification of IP rights in the modern regime has provoked a backlash. Mostly ignorant of a long history of economic skepticism over the benefits of patents in particular, some orthodox economists have lately taken it upon themselves to buck the neoliberal trend of making excuses for the contemporary fortification of intellectual property.[44] I have already taken a first pass at the problems of patents in Chapter 5; there has been a proliferation of outrageously ridiculous patent awards in the United States, like the ones for the peanut butter and jelly sandwich, the user-operated amusement apparatus for kicking buttocks, the Amazon "one-click" check-out, a penile volumetric measuring device (patent#: US 7062320), a hyper light-speed antenna (#US 6025810), or the method of swinging on a child's swingset. Indeed, the display and disparagement of ridiculous patents has become something of a minor art form on the Internet.[45] But all the buffoonery tends to obscure a deeper point: Numerous changes in the modern patent system have conspired to degrade the quality of the knowledge patented, and since the modern regime has become more oriented toward the commodification of IP, this may have had profound feedback effects upon the quality of science that is being "incentivized" by the brave new world. After all, if you can successfully patent a penile scale, how about a wonky protein receptor, or a faulty algorithm, or a fake molecule? A commonplace error exacerbated by the widespread acceptance of neoliberal economics has been to blithely presume that the "value of knowledge" is essentially identical to its market value, then to equate market value with the observed valuation of the intellectual property. One encounters this misconception repeatedly in the science policy literature, for instance. Yet if patents have indeed become debased to any considerable extent, then we may begin to suspect that something is rotten in the marketplace of ideas, and all equations are off.

It so happens that Adam Jaffe and colleagues kicked off this interesting line of argument with an important paper back in 1998. There they sought to explore the possibility that the "observed increase in university patenting may reflect an increase in their propensity to patent . . . rather than an increase in the output of 'important' inventions" (Henderson, Jaffe, and Trajtenberg 1998, 119). In the period examined, they noted that university spending on R&D had not expanded anywhere near the rate of growth of patenting, so ratios of patenting to spending were rising. Either universities had uncovered great economies in their operation, or else more junk was getting patented. These authors cast about for some measure of the "technological importance" of patents and settled on the number of citations a patent received in other patent applications as a plausible proxy. They also invented a measure of the "generality" of a patent, which boiled down to the number of patent classes that provided citations to a particular target patent. After the collation and manipulation of a rather large data set, and selecting a 1 percent random sample of patents to provide a benchmark for comparisons, they discovered that over the period 1965–1988 university patents generally received 25 percent more citations than their random sample (and also scored higher in "generality"), but that toward the end of this period the university advantage had declined very substantially. Indeed, rather dramatically: By 1987 nearly half of all university patents were attracting no citations whatsoever, a sea change from the earlier years. If one imagined the university sector as the source of basic discoveries from which other derived technologies would emerge, than this might be regarded as a dire development. Jaffe and colleagues concluded that the commercialization of university science, in conjunction with structural changes to law and the USPTO, had made it "economic to patent ideas of lower expected quality" (1998, 125).

This finding was rapidly taken up by the left wing of the European economists of science (Dosi, Llerena, and Labini 2006, 1452), but in retrospect, this was unfortunately premature. While the Jaffe exercise was clearly prompted by observation of the voluminous spate of ludicrous patents being awarded, one can readily observe in the aftermath how quantifrenia came to overwhelm common sense in this instance. The crux of the issue is how one should correctly evaluate the value of a patent, and the quality of the knowledge nominally "protected" by that patent. They, of course, are nowise identical or even necessarily correlated and should never (even provisionally) be treated as such.[46] Patents bear economic value for all sorts of reasons unrelated to the progress of science: They may be used to block rivals from entering your market or to frustrate anyone the holder wishes to prevent from further researching in the area; they may be used as defensive devices against other big holders of patent portfolios; they can be used as defenses in antitrust actions; they can be pretexts for parasitical litigation by patent trolls. Thus it would almost always be a travesty to try and "impute backward" the scientific significance of a discovery from some market indicator like license revenue, or patent renewal fees, or the extent of litigation. Value need not reliably map into significance or importance from a scientific perspective. But the recourse to something like patent citation statistics is nowise superior, *pace* Jaffe.

One must first realize that patent examiners are responsible for the citation data entered into patents—they are nothing like the citations in scientific journals. Examiners append citations to demonstrate that they (or the applicants) have correctly identified "prior art" for legal purposes. Indeed the requirement that patents be "nonobvious" was only thrust on the U.S. patent system by a 1952 statute, and it has since proven very difficult to enforce. Examiners have never been seriously tasked with providing a comprehensive map of the knowledge involved in a patent application. But one entire strain of criticism of the American system is precisely that the burden of proof for rejection rests solely on the patent examiner.[47] Given that the examiners are woefully overworked and understaffed—intentionally so, say many critics of neoliberal stories of patentees as "customers"—and spend an estimate of on average less than twenty hours on a single patent examination (U.S. Federal Trade Commission 2003, 5), there is no way that the resulting patent citations indicate much of anything more than the strategic orientation of the petitioner. Further, U.S. examiners are especially weak on evaluating prior art in foreign contexts or in many out-of-the-way areas of science (Sampat 2004). If an examiner should happen to reject an application, U.S. law allows an applicant to try again an unlimited number of times in a "continuation" application until the PTO relents. The expansion of patents into novel areas like software and business methods has relegated whatever exists of prior art to realms well beyond the competence of existing examiners, if not altogether off the scientific map. So being a patent examiner is hardly an enviable proposition: All forces conspire to have U.S. examiners approve ever-growing numbers of dodgy patents. Indeed, the primary cause of elevated patent citations is simply the pedestrian fact of greater numbers of patents being awarded per examiner; this is one of a myriad of reasons attempts to find significant correlations between patent citations and license revenue generally have failed (Sampat and Ziedonis 2004).

This all might not matter too much in the larger scheme of things, except for the fact that the original recourse to patent citations has now swamped the entire discussion of the decline of patent quality, such that both economists (Verspagen 2006) and lawyers (Mills and Tereskerz 2007) now regularly assert with confidence there has been no decline in U.S. patent quality, which is clearly a non sequitur, and mainly false. The economists most responsible for this reversal of fortune are David Mowery and Arvids Ziedonis (2007). In a sequence of papers they have come up with one excuse after another why either the decline of patent quality didn't happen or else was a spurious artifact that should not give anyone pause. For example, Mowery and Ziedonis (2002) claimed that Stanford and UC Berkeley did not themselves emit patents of diminished quality after 1980, which rang more than a little tinny, since it seemed to be special pleading by faculty from those very same idyllic prosperous campuses when they said that their home institutions shouldn't be blamed for whatever has ensued.[48] They derived these results from patent citation statistics but uneasily admitted that if one instead looked at how many of each school's licenses generated positive income, the proportions fell fairly dramatically after 1980 (2002, 403). They then opted to blame any perceived decline in quality on inferior (read: public) universities that had rushed to cash in on the patent race after 1980, only to reassure us that once those tyro universities learned the ropes, the whole phenomenon would simply melt away (401). Sampat, Mowery, and Ziedonis (2003) essentially reprocessed data similar to the original (Henderson et al. 1998) and claimed that the original finding of diminished patent quality was an artifact of truncated samples and that allowing more time for citations to accumulate resulted in no differential rate of patent citations before and after 1980.[49] This just reiterated Mowery's long-standing refrain that nothing substantive has changed in the university before and after the 1980 watershed of the modern commercialization regime.[50] But patent citation data are not sufficiently robust to support this relentlessly upbeat prognosis, because they cannot reliably "measure" quality.

Eventually, the notion that there had been no discernible change in patent quality was a doctrine that quite simply could not withstand the light of day in the public sphere. Reams of stories of ludicrous patents, combined with the rhetorical force of Jaffe and Lerner (2004) and the attentions of a phalanx of patent lawyers (Kahin 2001; Kesan and Gallo 2006) had led to a political movement to reform the American patent system, a quest that continues to the present day. The problem has been that "reform" became a rag doll tossed around and fought over between different powerful interests,[51] yet it seemed to be in no one's interest to further empower the thought police (e.g., the examiners) to reject patent applications; consequently, the notion that faulty patents harmed contemporary science got lost in the scrum. No one has seriously entertained the notion of weakening patents back to their pre-1980 status.

In the meantime, orthodox economists shifted their position from denying (like Mowery) that the quality of patents had been degraded, to a proposition more resonant with the basic neoliberal worldview: that even if some low-quality patents had been awarded, *it didn't matter from an economic point of view*. The main themes of this narrative were that (a) most low-grade patents don't matter anyway, because they are rarely licensed and even more rarely litigated (in other words, invalid patents afford little "bargaining power" to their owners); (b) it would be better to open up the process of challenge to patents to anyone who wanted to chime in—this was presented as the *avant-garde* option of a Wikipedia-style chat room for patent challengers—and thus tap into the free labor of the wisdom of crowds through some format of Web-based, post-grant review (Graf 2007; Biagioli 2007) and (c) most licenses on weak patents are granted to downstream firms who are not market rivals to the patent holder in any event, so there are few economic incentives to subject those patents to close scrutiny.[52] This is yet another variant of the neoliberal precept that "the market can solve any problems supposedly generated by the market in the first place." For an economist, the market could not possibly under any circumstances adulterate the essence of traded knowledge.

What had gone missing from this formidable phalanx of academic discourse was any acknowledgment that there could be blowback from a badly degraded patent system onto a science base rapidly being reengineered to respond to precisely those incentives dictated by that regime of fortified IP. The fly in the ointment wasn't just that patent examiners were too harried or ill-equipped to detect the quality of their submissions; it was that the median bit of "responsive" research itself had been degraded in the interests of getting those applications rapidly submitted to the USPTO in the first place. "Responsive" research in many instances is research that evades responding to scientific critique, in favor of succeeding at a different set of criteria dictated by the sponsor. But to explore such questions, it would become necessary at minimum to (for instance) integrate patent data with scientific bibliometric data, taking into account the considerations I discussed above: Sticking to what can be directly found inscribed on a patent will not begin to clarify matters, as I have argued. This would be a Herculean task, one well beyond the means of even a well-funded academic unit. Once again, the modern privatization regime insinuates itself, because there currently exist for-profit businesses that purport to provide these services for an exorbitant fee, mostly to lawyers.[53] While these services are attuned to the needs of patent lawyers and technology transfer officers, I am not aware of any academic researchers that have managed to adapt them to the types of questions I pose here.

Consequently there have been precious few forays into examination of changing quality of knowledge using bibliographic data combined with patent data. Perhaps the question resists large-scale generalization, since it so happens that the prevalence of links between academic literature and formal IP vary rather dramatically across the sciences and industrial sectors, as well as over time (Leydesdorff 2004), which in itself constitutes an argument against the one-size-fits-all approach to the evaluation of commercialization of university science. However, the one exercise that stands out in a relatively barren landscape is the paper by Fiona Murray and Scott Stern (2007). Bypassing the commercial data services, they chose to examine two year's issues (1997–1999) of the journal *Nature Biotechnology,* which they reasonably assumed would yield a high proportion of articles implicated in claims for IP. By doing a patent search on publicly available databases, they discovered that just under 50 percent of the 340 articles were linked to a patent grant, on average three years after publication. As has been found by other researchers, the articles associated with patents subsequently received on average more academic citations than those without, but as they point out, this could be due to spurious correlation with external variables, such as university affiliations or higher numbers of authors per article. The main drawback here as elsewhere is the lack of an independent measure of the relative scientific significance of any two given papers, even in the same journal. Instead, they develop an intriguing before-and-after comparison of academic citations for the *same* paper pre- and post-patent grant. Under U.S. law, scientists can only publicly disclose their discoveries one year before filing for a patent; however, publication lags tend to be much shorter than the lag between filing and patent grant, so in effect a lot of commercial science gets to exist in two states: published but prepatent grant (that is, no formal IP protection), and published with explicit patent grant. Given all that has been said above about the low rate of rejection at the PTO, plus the overwhelming role of MTAs in commercial science, one should not overstate the difference between the limbo of pregrant versus certainty of postgrant status; at best there may be a differential awareness among other researchers of the extent to which IP in that area of science had been staked out by some lab or university. Nevertheless, Murray and Stern's regressions (perhaps themselves a little too besotted with econometric wizardry) were said to reveal a 10–20 percent drop in academic citation of the *same* paper after it was granted a patent, in comparison to the preceding period. Those authors interpret their results as supporting Rebecca Eisenberg's notion of an "anti-commons effect" of patents on research, but possibly this constitutes too much of a stretch. What it may indicate is not so much an index of the quality of the research, but rather an attempt to gauge the extent to which the relevant scientific community has been obstructed from freely evaluating the quality of the research by the intercession of the grant of intellectual property rights. A major lesson to be gleaned from this brave attempt at empiricism is that aggregate econometric quantitative inquiry into the debasement of research by commercialization will always be a tenuous and fragile proposition.

It may very well be this is the phenomenon Eisenberg has been trying to highlight all along, but she has been frustrated by her own adoption of the obsolete Arrow/David language of "the tradeoff between the two effects of private appropriation [of science], greater private incentives for research performance and underutilization of research results" (2006, 1016). What is happening is not a trade-off of a little bit of commercialization in order to offset previous "underutilization" of science. It is rather the production of a different sort of science under a top-to-bottom reorganization of the way in which research is funded, judged, and promulgated. Hence it is a mistake to focus so intently on patents when gauging the effects of commercialization on research: Most of the frustration of inquiry these days tends to happen at an earlier stage with MTAs, as mooted in Chapter 4. It is moreover a mistake to rely upon the "author" as a stable linchpin when tracking citations to academic publications, when an unknown proportion of the biomedical literature is infected with ghost and guest authorship. It is a mistake to simply take the Thomson ISI database as an adequate representation of the state of "good science," or the universe of bibliometrics in the twenty-first century. It is a mistake to treat academic citations as pristine "commerce-free" indicators of scientific significance once an entire sector of scientific research has been reoriented in a market-informed direction. And it most definitely is a mistake to ignore the fact that participants in the brave new order voluntarily and readily discuss the commercial manifestation of privatized regime science in its guise as IP degraded in quality, even from the vantage point of its sponsors.

Why Science Is Not a Gamble

A common reaction to the evidence presented in this chapter is that you cannot gauge the harm to science by means of statistics, as I have attempted in this chapter, without taking into account that results are in some sense stochastic and that significance only comes in the format of a distribution of output. Some science outputs are good, and some are inferior. Science, they say, is a gamble, like other risky endeavors: but it can be done economically. Luckily, many social phenomena come in such distributions, and we know how to use probability and statistics to tame the risk and optimize the beneficial outcomes, just like in speculative markets.

It has been known for some time now that many of the key quantitative variables in science—bibliographic measures like citations and published papers per person; social measures like extent of participation in conferences, prizes, and other formal occasions; and economic indicators like license returns per patent—have highly skewed long-tail distributions. They don't fit "normal" (Gaussian) distributions. In practice, only a small number of scientists and a small number of ideas seem to garner the lion's share of attention, discussion, elaboration, and reward in almost every area of science.[54] A myriad of possible explanations of why this has turned out to be the case have been proposed,[55] but there is nothing remotely approaching consensus wisdom, and thankfully, we can pass them by here. What matters in the current context is the fact that these distributions cannot be regarded as Gaussian (or even log-normal) because they exhibit the trademark "fat tails." The consensus evaluation is that they resemble most closely the general class of central limit approximations of random variable distributions known as the Pareto-Levy stable distribution. Physicists are familiar with this class of distributions because they show up in all sorts of self-similar and chaotic processes. Pareto-Levy distributions sport some distinctive attributes that render garden-variety statistical inference more than a bit tricky: For instance, they generally have infinite variances (and infinite moments above the first), and therefore (except in a few special cases, like the Gaussian and the Cauchy) do not enjoy analytical density functions or well-behaved estimators. This is all explained in a previous publication of mine (Mirowski 2004a, Chaps. 11 and 12), so there is no need to go into detail here. However, there are one or two facets of Levy stable distributions that bear directly upon the vision of a new venture capital science model.

Lee Smolin (2006), among others, thinks that a concertedly commercial approach to science will encourage more diversity in research because businesses are more likely to diversify their risk by spreading their investments. This reprises the neoliberal approach to the marketplace of ideas, where one "economizes" with respect to information by subjecting each choice to a cost-benefit calculation. Maybe in the world of simple Gaussian processes one might be able to do so, but this is badly misleading if we can accept that the underlying distribution of scientific outcomes is Levy stable. Take, for instance, a sample from the Cauchy distribution. The distribution of the sample mean is identical to the distribution of each individual item in the sample (Feller 1971, 51). Collecting more information beyond the first sample reading is thus redundant in this sort of world, and the whole notion of "economizing" on information goes right out the window. In other words, here ignorance is "rational," because every notion of the packaging of risk through diversification loses all meaning, and one cannot reduce variance through pooling strategies. Smolin and those sharing his approach are thus caught in a rather nasty cul-de-sac: his neoliberal argument for funding a broad range of research strategies becomes an argument for the *futility* of any such policy of investment diversification, once one takes the statistical data seriously. Indeed, we might elevate this precept to one of the principles of a serious future economics of science: The venture capital practice of trying to anticipate scientific breakthroughs will *always* lose money with a limit approaching probability one, modulo some details concerning the convergence properties of the distributions involved.[56] If Pareto-Levy distributions characterize scientific research, then the marketplace of ideas will predominantly be an elephant's graveyard of failed theories and abandoned researchers.

Now, one might object that Levy stable distributions seem to pervade many financial variables as well, but that does not mean that the Chicago Board Options Exchange has had to shutter its doors.[57] Here the field of science studies offers interesting insight into the actual practices in the world versus economists' orthodox theories of markets in the face of pervasive ignorance and uncertainty. As Donald MacKenzie (2006) has shown, financial market traders at CBOE know that their Gaussian models of risk and finite variance bequeathed them from economists don't really describe the behavior of prices, particularly in times of trouble and turmoil. Rather than reject them outright, say, like some imperious follower of Karl Popper, they instead introduced "fudges" into their systems of calculation to try and take into account the pervasiveness of fat tails and highly skewed returns. Most of the time these fudges are sufficient to allow traders to coordinate their behaviors, but when one finds one's portfolio caught in an infinite variance downdraft, like the case of the failure of Long Term Capital Management (run by Nobel-winning economists) or the October 2008 credit crisis, bankruptcy is the only stochastic absorbing state (barring nationalization). Maybe market traders live on to speculate another day, and maybe they don't; the system has covered both eventualities.

Herein lies a major difference between finance markets and any notional marketplace of ideas. In the financial world, everyone can be drawn to the siren song of that mega-payoff out in the nether reaches of the fat tail, believe they can hedge against disaster, and sell off or otherwise discard the mass of shabby low-return or losing investments that compose the bulk of the distribution. But when one transfers this logic to the university (say, with technology transfer officers lusting after that "home-run" patent that will put them into the black), the lack of congruence becomes apparent. For in order to have some hope of bagging one of those rare high variance events in science, you can't just sell off or otherwise divest the rest of the low-productivity scientists: *You need the entire distribution just to make the research process function.* Science is not beset with reprehensible slackers (Klamer and van Dahlen 2005). Cherry-picking research projects and eminent scientists in the first bloom of fame may only serve to accumulate retrospective reputation; it very rarely eventuates in truly novel further science, as many universities and foundations have learned to their chagrin. The median scientist (paper, dataset, idea) is not just an inconvenience; he/she/it is an indispensable input into the research process. Deans and scientists think they can recognize talent when they see it, but in that respect they are little different from the phalanx of avid gamblers found in any downmarket casino, convinced that they possess a foolproof system that can beat the house. You can't "efficiently" gamble on an optimal portfolio in a Pareto-Levy world, and you cannot economize on information in science.

7

The New Production of Ignorance

The Dirty Secret of the New Knowledge Economy

> [T]here has fallen to the universities a unique, indispensable and capital function in the intellectual and spiritual life of a modern society.
>
> —WALTER LIPPMANN, 1966

Dave, my mind is going. I can feel it.

Let us rendezvous once more with Viridiana Jones in our journey, before it draws to a close. While she has repeatedly experienced the shock of recognition while reading this book, there is no denying she feels that it has gone overboard in some respects. Can the university have really grown so irrational and self-destructive over her lifetime? Why has it enthusiastically bought in to the biotech start-up model of research when that sector as a whole loses money, is inured to more than 80 percent failure rates, and has yet to produce a serious track record of new and innovative cures? Has it really elevated patents a prime indicator of the worth of knowledge (and stipulated it a criterion for tenure in some fields) when the quality of patents granted has simultaneously been degraded? Has it really erected all sorts of barriers to productive research, like material transfer agreements (MTAs) and nondisclosure agreements and preemptive ownership of copyrights, to such an extent that gross aggregate measures of science conducted, such as (suitably adjusted) numbers of papers published in core journals by researchers in the United States, have fallen over the last decade? Have scientists really turned a blind eye to clear portents of intellectual corruption? Have the biosciences really permitted scholarly journals to become quagmires of ghost authorship, ghost journals, and hidden conflicts of interest? Have universities invested all their hopes in their technology transfer offices, when they almost never break even, much less make serious money?[1] Why are universities encouraging a few faculty to become personally rich in corporate start-ups at the expense of aggregate tenured employment in universities at large? Viridiana can't quite bring herself to concede that the push to "commercialize the university" has been one of the biggest Ponzi schemes this side of Bernie Madoff and Allen Stanford.

Yet, it is not simply hand jive. She now sees that many of Walter Lippmann's progeny[2] no longer believe in the "special role" for the university that was a prime motive for her becoming an academic in the first place. Viridiana feels she has a somewhat better grasp now on some of the things that were nagging away at the margins of her consciousness, but somehow, in the end, it hasn't really helped her mood. Like most of her American compatriots, she wants every trenchant critique to end with an identikit list of ten things we can do to fix it, or an ambitious program of "reforms" that assuage our feelings of political helplessness. She has little patience with Big Picture political theory. And anyway, doesn't a sad air of nostalgia hang like a pall over this book? Time and again have I not written as though the Cold War was, if not a Golden Age of Science, at least a time when the median scientist knew that his or her small contribution was part and parcel of a bigger noble quest?[3]

Just when Viridiana was feeling most irritated with this book, she happens to stumble across a paragraph in a book review in the *New Republic:*

> Last year, I published a book describing how right-wing economics had come to dominate American politics. Whenever you write a book about something bad that's happening, you get asked for the solution. I'd shrug and admit that I didn't have one. The questioner would usually look slightly disappointed, so I'd add that nothing lasts forever, and eventually something will come along to change things. The financial crisis might be that something. (Chait 2008)

In a flash of insight, Viridiana gets the parallels with the present volume: It will probably take a system crash before the leaders of today's universities will admit the current wave of commercialized knowledge production has proven unsustainable on its own terms. Until then, bureaucrats will strategically make use of the economic crisis in order to redouble their efforts to privatize the university (Kelderman 2009; Gray 2009). What is a scientist to do in the current predicament? In the meantime, wouldn't there be some merit in drawing up a systematic bill of indictments of the present regime of science management, just as part of an attempt to get clearer on the warning signs, in preparation for what promises to be big changes coming down the pike? Just wait till Chinese scientists start being portrayed in the world press as the intellectual *avant-garde* in research areas deemed "hot": will neoliberals change their tune then? Or perhaps, when a Chinese firm snaps up Thomson Reuters, initially to better control financial reporting, but then as an afterthought to skew bibliometric measures of science more in their own direction? The future, as always, is as yet unclear.

Upon further reflection, Viridiana begins to comprehend why natural scientists do need to get a better grip on neoliberalism as a prevailing theory of the nature of knowledge. When naive but well-meaning crusaders come clutching their "reforms" to "fix" an isolated problem with science, be it the creative commons license or proposals to fund scientific research programs through elaborate prize competitions, be they disciplinary codes of ethics or legislation to append some compromise version of a "research exemption" to contemporary patents, they play right into the hands of their neoliberal opponents. For the neoliberals don't make piecemeal proposals around the edges to "fix" what they regard as deficient aspects of science: They come equipped with an ambitious and comprehensive vision of how to reengineer markets economywide to better conform to their ideal vision of the marketplace of ideas. They reflexively romance the market and revile the regulatory agencies. This has taken place within a phalanx of think tanks engineered to translate that vision into manageable chunks: the Shelby Amendment, the Uruguay Round of the GATT negotiations, the new U.S. cabinet-level post to further strengthen IP control, Supreme Court rulings like *Daubert* and *LabCorp*, the statutory extension of federal student loans to those enrolled in distance-education schools like the University of Phoenix, the bringing of trade in educational services under the aegis of the WTO, and so on. Neoliberals love to dazzle with ditsy science-fiction scenarios, like recent claims that the discovery of physical laws can be completely computer automated (Keim 2009). They are the ones who are upbeat, promising a shiny new world of whiz-bang science right around the corner. The changes they extol are big and bold. They know how to market ideas. The neoliberals are not out to "save" the university: It has been in their opinion a major citadel of resistance to their economic crusades in the past and therefore must be disciplined, not saved. If some fail, it will be no tragedy, wrote Milton Friedman:

> Businessmen, who may be bankrupted if they refuse to face facts, are one of the few groups that develop the habit of doing so. That is why, I have discovered repeatedly, the successful businessman is more open-minded to new ideas . . . than the academic intellectual who prides himself on his alleged independence of thought . . . Self interest has been reinforced by the herdlike instinct of so many intellectuals, by their sheltered environments, in which they talk only to one another. (1978, xi, xiii)

The neoliberals have therefore remained three or more steps ahead of sententious defenders of open science, with the latter always playing catch-up because they didn't understand the nature of the game that they were caught up in.

There is in fact a concise theoretical point that runs like a red thread throughout the chapters of this volume, and the time has come to make it explicit. Not only has each chapter been constructed to introduce detailed evidence that the neoliberal approach to the marketplace of ideas and its embodiment in the regime of globalized privatization is flawed, but indeed each has been carefully selected to illustrate precisely where the flaw resides. The Achilles heel of neoliberalism is that it gets the functions of markets in society all wrong: Markets are not only limited and intermittently unreliable information processors; they can equally well be deployed to produce *ignorance.* As George Stigler admitted, "The marketplace rewards the tastes of consumers . . . whether the tastes are elevated or depraved. It is unfair to criticize the marketplace for fulfilling those desires" (1963, 90). Markets *do* respond to the demands and wishes of those with resources, but the upshot may just as well be the willful intentional production of ignorance for many target groups.

The current modern regime of science organization in many respects is not a new knowledge economy as much as it is an engine of agnogenesis. What are the real consequences of the materials transfer agreement, the modern patent, the contract research organization, for-profit ghost management of publications, the sound science movement, the economics of information, or for that matter, even most of the science policy literature? Do they exist to augment and enhance human understanding, or are they instead bent to rather murkier ends, leaving us all less wise and more confused than at the outset? I can just picture Viridiana (and maybe you, patient reader) starting to back away in horror, wondering just how far this paranoid rant might go. But I would implore Viridiana to provisionally entertain the notion that here resides the blind spot of all previous incarnations of the economics of science, the maggot deep within the fruit. The primal presumption of the neoliberal concept of the marketplace of ideas is that if information is a commodity, then it must necessarily be a "good"; some of it is lower quality than the rest, to be sure, but even if only of infinitesimal worth, more is always better, intones the economist. Just as there are no "negative" prices, there is putatively no such thing as negative information. No one would ever voluntarily pay to become dumb and dumber, would they? Ignorance is therefore like a vast vacuum, the infinite empty space surrounding our bustling little planet, at least for the modern neoclassical economist; any infobit introduced into it can only diminish it, if only to some tiny extent. There is a necessary arrow of time, because people only strive in one direction. Capital accumulation can only be a story of human augmentation. Retrogression, at least for most economists, is not an option. This is such a dowdy simple point, so foundational to the economics profession that it goes unspoken, that it takes a lot of effort to perceive it at the heart of a pervasive worldview and even greater effort to disabuse oneself of the notion.

Initially, I had planned to compose a chapter on the various ways in which formal economic theory treated information,[4] but now I have come to appreciate there is no need for such a technical exercise in this book. Not only would it bore Viridiana silly, but it would also distract her from comprehension of the reasons why the plethora of minor high-tech "reforms" so prevalent in scientists' commentaries on the current regime—increasing mandatory "disclosure" to offset conflicts of interest (Brainard 2008; Slaughter et al. 2009), a Creative Commons amendment of IP to include "copyleft" (Boyle 2008), Web-based open source approaches to biological experiment, prize competitions run through for-profit Web entities like www.innocentive.com (Dean 2008), a Sarbanes-Oxley Bill for science (Michaels 2008), mandates for public registries of clinical drug trials, wiki-based collaborations—none even begin to address the real problem. Techno-enthusiasts like Yochai Benkler (2006), however well intentioned, merely serve to distract attention from the fundamental predicament of modern science. Contrary to what some of my young friends have said, it is not some misbegotten exercise in nostalgia to trace the current degradation of the American science base back to its historical roots in specific political and economic initiatives, and then carefully gauge the extent of the rot over time relative to the previous regime.

America is crowded with a surfeit of overconfident reformers who have little comprehension of what they seek to reform. Whatever their avowed politics, they are inspired by the widespread conviction in the regime of globalized privatization that if you bend knowledge generation to commercial discipline and market organization it will inevitably result in a surplus of knowledge, and, furthermore, "competition" will promptly direct it to its most efficient users. In this neoliberal version of the world, there is simply no way that knowledge could become clotted or otherwise diminished or corrupted by having a price attached.

One discovers this thought pattern in the most curious and unexpected of places, such as in science studies and the history of science, where the median individual would be personally offended if you suggested that they were mere conduits for neoliberal folk wisdom:

> We also are chipping away at the public image, indeed the self-conception, of "intellectuals" and scientists, who typically have argued in universals. These individuals speak and write of what should be valid for everyone, of what is true or just or good. Intellectual history and history of science aspire to undermine these universals by recovering their specificity and locality. (Porter 2009, 642)

This could just have easily been written by Hayek himself; all one needs to add is an explicit appeal to the market test to supersede the discredited universals.

Strangely enough, there currently abides a pop literature that thinks it detects a similar sort of problem in the American body politic, but it misses out on the significance of the means of production of ignorance. Of late, there have been a spate of books with titles like *The Age of American Unreason* (Jacoby 2008), *The Dumbest Generation* (Bauerlein 2008), *The Assault on Reason* (Gore 2007), *The Cult of the Amateur* (Keen 2007), *Does the Truth Matter?* (Geenens and Tinnevelt 2009), *Empire of Illusion* (Hedges 2009), and *Just How Stupid Are We?* (Schenckman 2008). Most of them aren't very insightful. It is easy to dismiss them out of hand as the hyperventilation of the flabby remnants of the 1960s generation; blogs are full of these types of accusations. What better topic upon which to bloviate armed with little or no knowledge than the proposition that people around you are getting dumber, no? And yet there lurks a different and darker possibility, that these philippics are in fact symptomatic of something bubbling up that's a little unusual, if not quite new, in the contemporary knowledge economy. The celebration of Wikipedia is one of its symptoms (Mirowski and Plehwe 2009, 418–428; Carr 2010). Maybe the spread of home schooling, the dissolution of libraries, the strangulation of the sophisticated newspaper, and the Googlization of research have had insensible yet cumulative effects upon the processes of science. Just suppose we are onto something when we suspect that the neoliberal "reforms" of American science and American society over the last two decades have actually fostered the corruption of knowledge output. Set that alongside the parallel phenomenon of the neoliberal "reforms" of both the primary (Apple 2005, 2006) and university educational systems and then a dumbing-down of the average American over the same time frame stops being a scenario pitched beyond the prudent realm of possibility.[5] Here again, Viridiana's creeping uneasiness may indicate something more beyond her own personal misfortunes.

The possibility of the very corruption of our attention spans has occurred to more than a few people of late. Nicholas Carr, in an entertaining article in *The Atlantic* (2008), and in his book *The Shallows* (2010), has suggested that the Internet has a downside, as well as its being a great boon to all of us who used to have to devote a trip to the library every time we wanted to look up a citation. As he writes, certain manifestations of ease in supply of the stuff of thought may also come to shape the process of thought. "My mind now expects to take in information the way the Net distributes it: in a swiftly moving stream of particles . . . I can't read *War and Peace* anymore. I've lost the ability to do that. Even a blog post of more than three or four paragraphs is too much to absorb. I skim it" (2008, 56). The temptation might be to retort that perhaps those weak-willed humanists may fall prey to louche practices, but that would never happen to serious scientists. Yet, in *Science* of all places, there is evidence to the contrary:

> Using a database of 34 million articles, their citations (1945 to 2005), and online availability (1998 to 2005) . . . as more journal issues came online, the articles referenced tended to be more recent, fewer journals and articles were cited, and more of those citations were to fewer journals and articles . . . the number of years of commercial availability [of a journal online] appears to significantly increase concentration of citations to fewer articles within a journal . . . These changes likely mean that the shift from browsing in print to searching online facilitates avoidance of older and less relevant literature . . . If online researchers can more easily find prevailing opinion, they are more likely to follow it, leading to more citations referencing fewer articles. (Evans 2008, 395, 398)

While not exactly the precise same effect evoked by Carr, there is again the phenomenon that, contrary to the usual construction of the marketplace of ideas, augmented choice through increased access actually leads to a *narrowing* of the range and quality of the knowledge shared; explicit commercialization of provision of scientific publication only exacerbates the effect. Both Carr and Evans are tempted to attribute this to a form of technological determinism, indicting something specifically about computers as addling our curiosity, but, of course, it may instead actually be symptomatic of the larger set of neoliberal changes to society I have enumerated in this book. Distressing as it may seem, people may just be getting measurably dumber or, at minimum, less willing to engage with novel complex arguments.[6]

Luckily, I need not overdo playing the part of Cassandra to Viridiana's Laodice, as it seems there are the beginnings of a literature that takes seriously the proposition that various social structures can be dedicated to the *production and promotion of ignorance,* and that if there happens to emerge a flourishing market for ignorance, then the production of knowledge will take a beating. I shall wind things up by spelling out the implications of this proposition in this chapter, but before that, I want to highlight the work of the historian of science Robert Proctor as proposing the study of the production of ignorance under the rubric of "agnotology" (in Proctor and Scheibinger 2008). It is good to have a term close to hand to refer to the active production, maintenance, and manipulation of ignorance, because it will prove necessary to distinguish that phenomenon from two other ancillary connotations of the term: (1) the naive state of nonknowledge, namely, the "vacuum theory" of the economists and (2) the state of selective inattention due to fundamental limitations in our cognitive makeup. Both obviously exist, but they are a part of our individual epistemic predicament; yet precisely because they are constitutive of each of us as humans, they are *not* the subject of the historical case to be made in this chapter.

The capacity to differentiate the three versions of "ignorance" will be a propaedeutic for everything that follows in this chapter. It will turn out that many commentators muddy the modern production of ignorance by shifting indiscriminately between three different connotations. For instance, it is undeniable that the more research we do, the more we discover that we do not know. This is just *Definition 1 (vacuum theory)* and is a perfectly healthy phenomenon, a prophylactic for the besetting sin of hubris so common among intellectuals. Likewise, the exponential growth of scientific publication renders it impossible to aspire to the status of true polymath and stay on top of everything. This is *Definition 2 (bounded rationality)* and is just one important aspect of our cognitive predicament, that attention and memory are limited. Of course, we are impelled to develop rules of thumb (and computer prosthetics) in order to navigate our way through a world far richer than our paltry abilities to grasp it. Who would argue otherwise? Yet, beyond those two phenomena, when whole sets of institutions are deliberately bent to sow doubt, to spew out a fog of contrarian results, to reassure the uneducated that the truth is whatever they want it to be,[7] to treat the unequal distribution of knowledge as the natural dictate of freedom of choice and simultaneously to praise the innate "wisdom of crowds," then a surfeit of ignorance is the inevitable intentional consequence. This is *Definition 3, the manufacture of ignorance.* As I shall argue, it is precisely this third kind of ignorance that neoliberals have theorized and promoted. While neoliberals certainly didn't invent it, they are its most ardent contemporary boosters. Yet in an ironic way, seriously coming to terms with Definition 3 turns out to be also the best argument against subscribing to the neoliberal worldview. From here on out, Definition 3 will be our touchstone referent of any subsequent use of the term "ignorance" with regard to the commercialization of science.

The Neoliberal Will to Ignorance

The roles and functions of ignorance in theories of society have gotten neglected in the rush to praise the new knowledge economy.[8] An earlier generation of sociologists and anthropologists were much more willing to entertain the notion that ignorance (Definition 3) could actually perform certain social functions and that, from the vantage point of certain actors, it was worth the effort to foster and sustain. Émile Durkheim, for instance, argued that solidarity and cohesiveness of groups were better brought about in an indirect fashion and that ignorance therefore was a lubricant that might facilitate integration of individuals into communal activities. More recently, McGoey (2007) has argued that certain bureaucratic organizations, such as regulatory agencies, might find it logical to feign ignorance and conduct one faulty inquiry after another in order to serve implicit unspoken interests. Without either endorsing or rejecting such notions, it is far more important for our current argument to understand how key neoliberals have approached the issue of ignorance. And the premier representative of a theoretically sophisticated neoliberal theorist is Friedrich Hayek.

Hayek is noteworthy in that he placed ignorance at the very center of his political theory: "The case for individual freedom rests chiefly on the recognition of the inevitable ignorance of us all" (1960, 29). Most commentators tend to interpret this as an appeal to ignorance (Definition 1, and possibly 2), but I think they need to expand their horizons. The distinction begins to bite when we take note that Hayek harbored a relatively low opinion of the role of education and discussion in the process of learning, and notoriously, an even lower opinion of the powers of ratiocination of those he disparaged as "the intellectuals." In this, he diverged dramatically from the opinions of one of the early heroes of the neoliberal movement in the 1930s, Walter Lippmann (recall the 1966 quote that prefaces this chapter). This, of course, was the mirror image of Hayek's belief in the Market as a superior information processor:

> Nor is the process of forming majority opinion entirely, or even chiefly, a matter of discussion, as the overintellectualized conception would have it . . . Though discussion is essential, it is not the main process by which people learn. Their views and desires are formed by individuals acting according to their own designs . . . It is because we normally do not know who knows best that we leave the decision to a process we do not control. (Hayek 1960, 110)

Because I have already surveyed the general character of neoliberal doctrine and have scrutinized it in detail elsewhere (Mirowski and Plehwe 2009), for the nonce I more intently focus on the supposed nature of the learning process that "we do not control" and its relationship to ignorance. For Hayek and other advocates of "emergent" social cognition, true rational thought is impersonal but can only occur between and beyond the individual agents who putatively do the thinking. As Christian Arnsperger so aptly put it, for Hayek, "rational judgment can only be uttered by a Great Nobody" (2008, 90). That may seem odd in someone superficially tagged as a methodological individualist supporter of freedom, but it just goes to show just how far ignorance (Definition 3) has become ingrained in American political discourse. The trick lies in comprehending how Hayek could harbor such a jaundiced view of the average individual, while simultaneously elevating "knowledge" to pride of place in the economic pantheon:

> Probably it is true enough that the great majority are rarely capable of thinking independently, that on most questions they accept views which they find ready-made, and that they will be equally content if born or coaxed into one set of beliefs or another. In any society freedom of thought will probably be of direct significance only for a small minority. (Hayek 1944, 164)

For Hayek, "Knowledge is perhaps the chief good that can be had at a price" (1960, 376), but it is difficult to engross and accumulate, because it "never exists in concentrated or integrated form but solely as the dispersed bits of incomplete and frequently contradictory knowledge which all the separate individuals possess" (Hayek 1948, 77). You might think this would easily be handled by delegating its collection and winnowing to some middlemen, say to academic experts, but you would be mistaken, according to Hayek. He takes the position that all human personal abilities to evaluate the knowledge commodity are weak, at best. And this is not a matter of differential capacities or distributions of innate intelligence: "The difference between the knowledge that the wisest and that which the most ignorant individual can deliberately employ is comparatively insignificant" (ibid., 30). Experts are roundly disparaged by Hayek, and he accuses them of essentially serving as little more than apologists for whomever employs them.[9] On the face of it, it thus seems somewhat ironic that Hayek would be touted as the premier theorist of the new knowledge economy. But the irony dissolves once we realize that central to neoliberalism is a core conviction that the Market really does know better than any one of us what is good for ourselves and for society, and that includes the optimal allocation of ignorance within the populace: "There is not much reason to believe that, if at any one time the best knowledge which some possess were made available to all, the result would be a much better society. *Knowledge and ignorance are relative concepts.*"[10]

What purportedly rescues Hayek's system from descending into a relativist quagmire is the precept that the Market does the thinking for us that we cannot. The real danger to humanity resides in the character who mistakenly believes he can think for himself:

> It was men's submission to the impersonal forces of the market that in the past has made possible the growth of civilization . . . It does not matter whether men in the past did submit from beliefs which some now regard as superstition . . . The refusal to yield to forces which we neither understand nor can recognize as the conscious decisions of an intelligent being is the product of an incomplete and therefore erroneous rationalism. It is incomplete because it fails to comprehend that

> co-ordination of the multifarious individual efforts in a complex society must take account of facts no individual can completely survey. And it also fails to see that . . . the only alternative to submission to the impersonal and seemingly irrational forces of the market is submission to an equally uncontrollable and therefore arbitrary power of other men. (Hayek 1944, 204–205)

There you have Hobson's choice: either the abject embrace of ignorance or abject capitulation to slavery. The Third Way of the nurturing and promotion of individual wisdom is a sorry illusion.[11] The Market works because it fosters cooperation without dialogue; it works because the values it promotes are noncognitive (O'Neill 2003).The job of education for neoliberals like Hayek is not so much to convey knowledge per se as it is to foster passive acceptance in hoi polloi toward the infinite wisdom of the Market: "General education is not solely, and perhaps not even mainly, a matter of the communication of knowledge. There is a need for certain common standards of values" (1960, 377). Interestingly, science is explicitly treated in the same fashion: If you were to become an apprentice scientist, you would learn deference and the correct attitudes toward the enterprise, rather than facts and theories (ibid., 112). Of course, Hayek rarely refers to "the Market" as I do here, preferring to refer instead to euphemistic concepts like "higher, supra-individual wisdom" of "the products of spontaneous social growth" (110). Formal political processes where citizens hash out their differences and try to convince one another are uniformly deemed inferior to these "spontaneous processes," wherein, it must be noted, insight seems to descend out of the ether to inhabit individual brains like the tongues of the Holy Ghost: This constitutes one major source of the neoliberal hostility to democratic governments. But the quasi-economistic language testifies that the nature of the epiphany is not otherworldly but more distinctly mundane and pecuniary: "Civilization begins when the individual in pursuit of his ends can make use of more knowledge than he himself has acquired and when he can transcend the boundaries of his ignorance by profiting from knowledge that he does not himself possess" (22).

This language of "use and profit from knowledge" you don't possess might seem a bit mysterious until we unpack its implications for ignorance (Definition 3). I second the analysis of Louis Schneider (1962, 498) that Hayek should be read as one of a long line of social theorists who praise the unanticipated and unintended consequences of social action as promoting the public interest, but who takes it one crucial step further by insisting upon the indispensable role of ignorance in guaranteeing that the greater good is served. For Hayek, the conscious attempt to conceive of the nature of public interest is the ultimate hubris, and to concoct stratagems to achieve it is to fall into Original Sin. True organic solidarity can only be obtained when people believe (correctly or not) they are only following their own selfish idiosyncratic ends, or perhaps don't have any clear idea at all of what they are doing, when in fact they are busily (re-)producing beneficent evolutionary regularities beyond their ken and imagination. Thus *ignorance helps promote social order*, or as Hayek said, "knowledge and ignorance are relative concepts."

I have heard the objection that this characterization is tendentious because it hangs the peculiarities of all of neoliberalism on the idiosyncratic writings of one man. But it so happens that many of the early Mont Pelerin Society members took very similar positions on the marketplace of ideas, some even within the same rough time frame. I have already quoted Milton Friedman disparaging academics. One of my own personal favorites is George Stigler's *The Intellectual and the Marketplace* (1963). Therein he argued that businessmen were better than academics or government agencies at promoting academic freedom and diversity of thought. Intellectual talent works best when it is gathered together in like-minded hothouses, says Stigler, but public education tends to disperse talent, subjecting it to entropic decay. Truth and progress, by their very nature, can only ever be possessed by a small elite, and the marketplace of ideas can serve that function just fine. The masses are optimally stupid because of a rational cost-benefit calculation on their part: "The large mass of the public does not find it economically worthwhile to become well acquainted with the effects of policies which have small harmful effects on each beneficiary"[12]; "I cannot believe any amount of economic training could wholly eliminate the instinctive dislike for a system of organizing economic life through the search for profits" (1963, 95). Instead, intellectual entrepreneurs in privately funded think tanks/universities will churn out the knowledge that elites want and need, perhaps even before they fully realize it; when elites see evidence of the right stuff, they will gladly pony up the funds to support it. "Inquiry has been most free in the college whose trustees are a group of top quality leaders of the marketplace" (87). The marketplace of ideas turns out to be an uncompromisingly closed elite phenomenon, with economic elites funding elite scientists, and the rest of the world safely ensconced in bone ignorance. This was not a doctrine intended just for economics but for all of science. As Stigler said in an address to the Mont Pelerin Society,

> Affairs of Science, and intellectual life generally, are not to be conducted on democratic procedures. One cannot establish a mathematical theorem by a vote, even a vote of mathematicians. [Therefore] an elite

> must emerge and instill higher standards than the public or the profession instinctively desire.[13]

Stigler was awarded the National Medal of Science in 1987. Maybe the National Science Foundation itself has something to answer for in the rise of neoliberal agnotology.

There are at least two salient implications here. The first is that neoliberals are not at all troubled by the contemporary transformations of science and the university described in this volume. *The production of ignorance (Definition 3) is a sound business strategy, not a retrograde intervention.* Take, for example, the junk/sound science movement. So what if the Marshall Institute fills the room with fog about the true effects of secondhand smoke, or the Competitive Enterprise Institute obscures the real impact of global warming, or the Discovery Institute reframes the extent of biological evolution (Oreskes and Conway 2010)? So what if government regulation is blocked by drawn-out legal battles over the kinds of paid science that can be introduced in regulatory interventions? Or, if you like, take the spread of MTAs throughout university science. So what if certain classes of research are frustrated and stymied by reach-through clauses, or that knowledge transfer is slowed to a crawl between researchers in the same fields? Or consider, if you will, the creeping phenomenon of ghost management of scientific papers. So what if the reader has no idea who or what really stands behind the work reported on the printed page? Or take the (implicit) cover-up of the fact that numbers of scientific publications are actually falling in recent decades in the United States, while Americans continue to congratulate themselves on their superior innovative abilities. So what if the general populace is lulled into complacency by the lack of solid statistics concerning research in the United States on the part of the government and the National Academy of Sciences? None of these constitute real symptoms of debility for the true neoliberal. As long as scientists can be cajoled to defer to the Market to decide how knowledge will be subsidized, sorted, winnowed, and allocated, their resulting personal ignorance can only be eventually conducive to the public good, because it makes the system work more smoothly. Indeed, the trademark neoliberal doctrine in science studies since 1980 has been the mantra "Science has always been commercial" (Latour and Woolgar 1979; Shapin 2008b): an utter travesty of the actual history, as I noted in Chapter 3, but the first tenet of the neoliberal credo.

The major point to be savored here is that individual ignorance fostered and manufactured by corporations, think tanks, and other market actors is suitably subservient to market rationality, in the sense that it "profits from the knowledge that the agent does not possess." Paid experts *should* behave as apologists for the interests that hire them: This is the very quiddity of the theory of self-interest. As Schneider explains, "Organic theorists hold that while actors may cojointly achieve important 'beneficent' results, they do so in considerable ignorance and in ignorance of the socially transmitted behavior they are reproducing contains accumulations of 'knowledge' now forgotten or no longer perceived as knowledge" (1962, 500). Burkean conservatism revels in the preservation of tradition, the great unconscious disembodied wisdom of the ages. This is why cries of "teach the controversy" in the schoolroom, "sound science" in the courtroom and stipulations of "balance" in the news media are sweet music to neoliberal ears. Neoliberals strive to preserve and promote doubt and ignorance, in science as well as in daily life; evolution and the market will take the hindmost.

The second salient implication is that, from the neoliberal vantage point, science does not need special protection from the ignorant, be they the partisan government bureaucrat, the craven intellectual for hire, the lumpen MBA, the Bible-thumping fundamentalist, the global cooling enthusiast, or the feckless student. In an ideal state, special institutions dedicated to the protection and pursuit of knowledge can more or less be dispensed with as superfluous; universities in particular must be weaned away from the state and put on a commercial footing, dissolving their distinctive identities as ivory towers. Science should essentially dissolve into other market activities, with even its "public" face held accountable to considerations of efficiency, profitability, and subservience to personal ratification. "Competition" is said to ensure the proliferation of multiple concepts and theories with the blessing of the private sector. The only thing that keeps us from enjoying this ideal state is the mistaken impression that science serves higher causes, or that it is even possible to speak truth to power, or that one can rationally plan social goals and their attainment.[14] Despairing of extirpation of these doctrines from within the university in his lifetime, Hayek and his confederates formed the Mont Pelerin Society and then forged a linked concentric shell of think tanks to proselytize for the neoliberal idea that knowledge must be rendered subordinate to the Market.[15] Little did he suspect just how successful his crusade would be after his death.

Hayek is sometimes portrayed as a postmodern figure who did not believe in capital-T Truth, but, again, I don't think that really gets to the heart of the matter. Equally misguided would be the interpretation that Hayek would only promote the production of instrumentally useful knowledge: "Science for science's sake, art for art's sake, are equally abhorrent to the Nazis, our socialist intellectuals, and the communists. Every activity must derive its justification from a conscious social purpose" (Hayek 1944, 162). Instead, I believe he initiated an important neoliberal practice as advocating a double truth doctrine: one for the masses, where nominally everything goes and spontaneous innovation reigns, and a different one for his small tight-knit cadre of believers. First and foremost, neoliberalism masquerades as a radically populist philosophy, one that begins with a set of philosophical theses about *knowledge* and its relationship to society. It seems at first to be a radical leveling philosophy, denigrating expertise and elite pretentions to hard-won knowledge, instead praising the "wisdom of crowds." Writers such as Malcolm Gladwell, Jimmy Wales,[16] James Surowiecki (2004), and Cass Sunstein (2006) and many science studies scholars contemptuous of experts are its pied pipers. This movement appeals to the vanity of every self-absorbed narcissist, who would be glad to ridicule intellectuals as "professional second-hand dealers in ideas."[17] But, of course, it sports a predisposition to disparage intellectuals, because "knowledge and ignorance are relative concepts." In Hayekian language, it elevates a "cosmos"—a supposedly spontaneous order that no one has intentionally designed or structured—over a "taxis"—a rationally constructed order designed to achieve intentional ends. But the second and linked lesson is that neoliberals are simultaneously elitists: They do not in fact practice what they preach. When it comes to actually organizing something, almost anything, from a wiki to a corporation to the Mont Pelerin Society, suddenly the cosmos collapses to a taxis. In Wikipedia, what looks like a libertarian paradise to outsiders is in fact a thinly disguised totalitarian hierarchy.[18] In the spaces where spontaneous public participation is permitted, knowledge in fact degrades rather than improves. But no matter, because the absolute validity of that knowledge was never the true motive or objective of the exercise; rather, subordination of the overall process to corporate strategic imperatives provides the real justification of the format, as well as its economic foundation. It is all about "optics" and controlling the agenda. It adds up to a double truth doctrine: one truth for the masses/participants, and another for those at the top.[19]

Christian Arnsperger (2008) captured the double truth doctrine nicely by insisting that Hayek had denied to others the very thing that gave his own life meaning: the imprimatur to theorize about "society" as a whole, to personally claim to understand the meaning and purpose of human evolution, and the capacity to impose his vision on the masses through a political project verging upon totalitarianism. It was, as Arnsperger puts it, a theory to end all theories; it was not so different from the "end of history" scenarios so beloved of Hayek's epigones. The doctrine of special dispensation for the elect is one very powerful source of ongoing attraction of neoliberalism for the disaffected, the feeling of surrender to the wisdom of the market by coming to know something most of the nattering crowd can't possibly stomach: Freedom itself must be as unequally distributed as the riches of the marketplace. The ignorance of hoi polloi serves to reconcile them to that brute, bittersweet fact.

Do Economists Really Love Science?

I doubt if there is any branch of social science that ardently pledges its undying love for science more than economics. If imitation really is the sincerest form of flattery, then economics from its origins (Mirowski 1989) has been desperately seeking the approval of the natural philosophers. Most of this affection has historically been unrequited, but that is neither here nor there for our present purposes. The question before us now is whether economists can love science so very much that they might stifle it, or whether they can at least leave it less flourishing than when they found it. By this I don't just mean that "money can induce individuals and organizations to make judgments and decisions that violate research norms such as objectivity, openness, honesty, and carefulness" (Resnik 2007, 77). No, I seek to explore something far more insidious, and thus far more dangerous: the kind of thing immortalized in Oscar Wilde's *Ballad of Reading Gaol* or Ingmar Bergman's *Scenes from a Marriage*. If recent events even remotely mirror these archetypical plot lines, then this has dire consequences for the future of a viable economics of science.

By contrast with the Mont Pelerin neoliberals, I believe that most mainstream neoclassical economists are sincere when they pledge their fealty to science. Most wouldn't freely endorse the tenets that ignorance is bliss, or necessary grease for the wheels of social order, or the inevitable terminus for the vast mass of humanity. If anything, their mathematical models frequently induce them to treat ignorance as a simple deficiency, along the lines of the vacuum theory of ignorance (Definition 1).[20] Economists regard themselves as intellectuals and lovers of knowledge, by and large. Many in the profession think economists are the true heirs of the Enlightenment (that is, in the eventuality they have some notion of what the Enlightenment was). Most of them really do love physics, even if they flunked out or otherwise bailed out of physics programs elsewhere earlier in their careers. Nonetheless, I want to explore the proposition that recent trends have enrolled them into practical complicity with the new production of ignorance, and therefore, by implication, economists may be among the last people on earth to whom you should voluntarily entrust your science base. (I shall deal with the paradox of self-reflexivity soon thereafter.) This, in turn, will lead us to contemplate what sorts of things can be done about the modern globalized privatization regime of science.

Without making a major issue of it, the dodgy track record of economists in their dealings with science has run like Ariadne's thread throughout the current volume. In Chapter 2, we observed how various attempts to "defend" science in the university as a necessary complement to economic growth have been less than stellar in the logic of their empirical and theoretical elaborations. Chapter 3 noted in passing that neoclassical economists have benefited tremendously in postwar America, first by becoming allied with the military during the Cold War regime (Mirowski 2002) and later by occupying a central location within the current regime of globalized privatization. Even though science organization and funding had been transformed from top to bottom twice in twentieth-century America, economists persisted in treating it as though it were one generic phenomenon spanning the entire century. (To be fair, philosophers of science and some sociologists were equally culpable in this.) Chapter 4 marked the turning point, where mainstream economists became enrolled into the defense of key aspects of the contemporary regime, serving as apologists for the strengthening of intellectual property and arguing that science was basically unchanged by the reconstruction of the university and the proliferation of encumbrances upon research such as the MTA. Chapter 5 looked at the track records of biotech and pharmaceutical companies, the poster children for the new knowledge economy, and found things were neither as rosy nor as straightforward as many economists had painted them. Chapter 6 demonstrated that the quality of the science produced in the new regime had been degraded in very specific ways; only here was there a deafening silence from the economists. In the meantime, they argued that it was simply a matter of fiscal prudence that most universities would need to scale back on research and to cut costs (J. Johnson 2009). After this litany of failure, it would appear only prudent to inquire into how the mainstream economics profession could have gotten things so very wrong.

The diagnosis would far exceed in length the etiology of the disease described in this volume, but we have now arrived at a major proposition that accounts for the fact identified in Chapter 2 that modern orthodox economists fervently believe that science is the ultimate motor of the economy, even in the face of decades of controverting evidence (Macilwain 2010). Simple expenditures on R&D, however defined and denominated, do not readily correlate with economic growth. Expenditures on "human capital" do not directly translate into augmented skills and knowledge. Why not? The reason could be simple: If the neoliberal reengineering of science has resulted in a vast ramping up of the production of ignorance, as I have repeatedly suggested in this volume, then it immediately follows that more expenditure on science does not necessarily result in more scientific output. Agnotology destroys the correlation. Promotion of the marketplace of ideas can easily destroy knowledge just as readily as it can augment it; some more sophisticated neoliberals understood this perfectly well, as I have documented above. The reason economists come across as credulous and naive when they prescribe more education and more expenditure on R&D as the panacea for every economic problem under the sun is that they only have room in their models for ignorance (Definition 1), but there is much more in heaven and earth than is dreamt of in their philosophy. As long as they are willing to preach the marketplace of ideas throughout the land, the neoliberals are perfectly happy to let the economics profession maunder in their own ignorance. And this ineffectual morass includes the National Academy of Sciences and the AAAS, both of whom seem to think that politics boils down to lobbying for ever-increasing largesse at the public expense, absent any serious justification, beyond outmoded appeals to the linear model of innovation (or, worse, national chauvinism).

A detailed account of the missteps mainstream economists have made when it comes to modern science policy would undoubtedly exceed the space limits of any book, not to mention the patience of the reader.[21] To avoid either of these, I will simply gesture toward the Big Picture. The trends and tendencies that turned out to be important have all conspired to bring economists to the threshold of what I believe is a fundamental fallacy: that "science" can and should be subject to the very same analytical practices (what they engagingly call tools) that they have applied to any other commodity or situation—in other words, there is nothing particularly special about science, except in the sense that it may require an insignificant tweaking of economic theory (Zamora Bonilla 2008). As the philosopher Uskali Mäki has commented on this move, "Viewing science as an economy means transferring the familiar ideological and political issues from economics to science theory along the dimension of hands-off free markets . . . The capacity of science to reach whatever epistemic or other goals depends on its industrial organization, market structure, regime of regulation, or governance structure . . . But this requires a troublesome translation from economic theory to the philosophical vocabulary of knowledge and growth" (2008). Examples of the tweaks have been covered in Chapter 2; here I confront the deeper proposition that science unproblematically falls into the class of things economists can and should attempt to anatomize and minister unto.

The main trend derives from a Big Picture narrative of the history of economic thought that I have proposed in other work. Restricting my generalizations to Anglophone dominant schools of economics, most would concede that British Classical Political Economy (Smith, Ricardo, Malthus, Mill, Marx) set as its main task the exploration of the physical principles of production and their implications for social organization. By construction, science was something that intruded from "outside" this problem situation, although it might inform the broad outlines of what it was possible in principle to "produce." The rise of neoclassical economics in the late nineteenth century profoundly changed the terms of this setup. Not only did this school attempt to arrogate the status of science by close imitation of the mathematical formalisms of physics (Mirowski 1989), but it also redefined the core of economic theory to be concerned with the efficient static allocation of things between individual agents. The move was underdeveloped along two dimensions: (1) the actual psychological processes of the agents were finessed (if not actually repressed) under the rubric of 'utility', and (2) physical production was downgraded to the tenuous status of a static virtual phenomenon, situating it further removed from any grounding in the natural sciences. In a strange way, neoclassical economics sought to become more "scientific" in outward form (mathematics, imitation of physical field theory) while simultaneously becoming less tethered to the physical sciences in substance and content. The trend continued with the third great transformation of economic theory after World War II, from withdrawal from concern with allocation per se to greater efforts devoted to treating the agent as an information processor, patterned upon various theories of the computer (Mirowski 2002). The story of postwar mainstream economics has been a parade of various attempts (rational expectations, game theory, behavioral/experimental economics, neuroeconomics, behavioral economics) to take economic analysis in a more cognitive direction, but, again, without any serious engagement with the natural sciences (except for continued appropriation of characteristic mathematical formalisms).

One can draw all sorts of implications out from this frame tale (as I have done elsewhere), but the one inference germane to our present concerns is that each subsequent transformation of mainstream economics has only exacerbated a deep confusion about how knowledge relates to the economy, and its attendant adoption of the appropriate stance of this economic theory toward the natural sciences. While it is a trite observation that mainstream economists act like they can superannuate and subsume all other social sciences within their own "paradigm," only a few foolhardy souls have argued that modern economics should also subsume the natural sciences under its explanatory purview. The upshot has been that confusion reigns over the appropriate way to incorporate knowledge into mainstream economics models: the scandal of the "economics of knowledge" is that there is no agreement or standard approach to the putative topic.

Of course there are numerous individual options, each with their own proponents: Chapter 2 covered the Arrovian "public good approach" and various odd attempts to encompass technological change in growth theory, if only because they get mentioned so much when economists turn their attentions to science. But each of those theoretical traditions (and most other more obscure options) encounters in modern economics an equal and offsetting model set that neutralizes most of the conventional wisdom that they advocate. Take, for instance, the work of George Stigler and Gary Becker (both Mont Pelerin members, as it just so happens). In a famous paper (Stigler and Becker 1977) they argue that all differences in knowledge, "whether real or fancied," should be reduced to a production technology within the utility function, so that all economic questions can be reduced to static given preference functions that do not change over the life experience of the person. Of course, such a model has absolutely no relationship to any known school of cognitive science, but that was really never the point. Instead, what this modeling strategy accomplishes is to reduce so-called rationality to an utterly tautologous nonrefutable phenomenon: There can be no such thing as real difference between agents, and certainly no such thing as false knowledge (Vanberg 2004). Anyone who entertains the resulting "human capital" as a serious theory of human knowledge (a) must be blind to the fact that it contradicts other mainstream approaches to knowledge as a "thing" such as the public goods approach cited above; (b) must hold cognitive science in the same contempt as do Becker and his epigones; and (c) must suppress the possibility of the conscious production of ignorance. This is just one example of the unconscious muddles in economics concerning knowledge; there are literally untold bounties of them.

If you can't get your story straight about how to treat knowledge in your core price theory, then you most certainly will have a hard time stabilizing your approach to an economics of science. But then it seems the situation has become more addled and incoherent of late, if such a thing were possible. Over time, because of the successes of the Mont Pelerin/think-tank phalanx in America (and then elsewhere), the median member of the mainstream economics profession grew more neoliberal in her orientation and outlook.[22] One consequence was to ratify almost every economist in the belief there existed a standard approach to the economics of knowledge within the mainstream, however sparse the empirical evidence for such a view. Usually they just elevated their locally favored circumscribed model to pride of place. What they did glean from their repeated encounters with neoliberal ideas was that the market was the information conveyance device par excellence in human endeavors, and they proceeded to embed that conviction in all manner of theories, from the "efficient markets hypothesis" in finance to the "rational expectations revolution" in macroeconomics to principal/agent stories in the theory of the firm to the theory of "common knowledge" in Nash game theory. Once this conviction became entrenched throughout the discipline, it began to dawn on a number of economists that something like it must be true for the natural sciences as well.[23]

I need to be clear about why this intellectual move, now virtually ubiquitous in the economics profession, constitutes a contradiction in terms of the neoclassical model of the economic agent. It has long been understood among economic methodologists that the status of the agent has always been asymmetrical to that of the economist/analyst in modern mainstream economic theory.[24] The economist *qua* "scientist" acts as though she can stand above and outside of the economy and its agents, looking down on them benevolently and explaining why their activities and plans are thwarted or realized. But what gives economists this Godlike ability? The short answer is that the economist arrogates to herself a constitutional capacity that she denies to her little agent offspring: the ability to survey the rules and institutions imposed by the "model," to critically engage in self-reflexivity, and to decide whether or not the agent (it?) will accept the terms and conditions dictated by the model. One way to put this is the agent is doomed to be a total slave to the model, a cognitive robot, a fixed nonentity rather than a person in process of becoming someone else; by construction, the agent cannot under any circumstances rebel against the scripted role imposed by the economist. But another, better way of putting the same insight is that *if the economist and the agent were on the same epistemic footing, then the cognitive acceptance by the agent of the model putatively describing their experience would be a necessary precondition for the validity of the model.* No economist would ever grant that, so instead they pretend to glare down on the world from Mount Olympus (or was that Mont Pelerin?).

This explains in a nutshell why any neoclassical "economics of science" is a bald contradiction in terms and ends up trapped within a neoliberal double truth as astringent as anything in Hayek. Because every neoclassical economist believes in her heart of hearts that she is a scientist in good standing, to then proceed to model "science" dictates that she would have to extend to her agents the same courtesy of symmetric epistemic and cognitive status as she receives. But doing so would preclude all the standard model components of the neoclassical agent: fixed well-behaved preferences, knowledge modeled as a commodity, fixed cognitive and epistemic abilities of the agent, and the inability to survey and question the aptness and logic of the model describing his predicament. A "well-behaved neoclassical agent" would make for a lousy scientist. Yet what is science if it is not the sustained conscious alteration of previous beliefs, perceptions, existing knowledge, and, ultimately, the rules previously thought to govern reality? Isn't critical scrutiny of the received model the duty of every imaginative scientist? So if the mainstream economist wants to press ahead with her orthodox economics of science, she has one of two choices: (a) reimpose the asymmetry by making her scientist agent stupider than she is (if she is indeed smart enough to see this conceptual problem in the first place) or (b) restrict the model so that both it and the agents inside cannot comment on the quality or character of the scientific knowledge being produced (e.g., science as "black box" or disembodied "technical change"). In either case, the scientist is being portrayed in the neoclassical model as *less than fully rational.* Most natural scientists would never accept such a characterization of themselves. Hence existing models of the economics of science can't pass the snicker test: Scientists can rarely be cajoled into seeing themselves in the neoclassical speculum.

Wade Hands has touched upon the bogus character of the existing literature when he has distinguished between an "economics of science" and the possibility of an "economics of scientific knowledge," or ESK:

> [Mainstream] economics of science analyzes (explains and/or predicts) the behavior of scientists in the same way that an economist might analyze the behavior of firms or consumers. Like the Mertonian school of sociology, the economics of science almost always *presumes* that science produces products of high cognitive quality, but investigating whether it "really" does so is not considered to be the proper subject for economic analysis (it would be like an economist investigating whether the products of a firm "really" satisfy consumer wants). By contrast, ESK . . . would address the question of whether the epistemologically right stuff is being produced in the economy of science . . . [Were they to entertain the latter,] Economists doing ESK will certainly run into a similar problem. If scientists are pursuing their own self-interest (reputation, promotion, etc.) then economic scientists must be pursuing their own self-interest as well. (2001, 360, 390)[25]

Many economists, oblivious to the internal contradictions, nevertheless proceeded to apply neoclassical models to science, starting, not coincidently, with the advent of the globalized privatization regime. The chickens came home to roost when various economists then began to base policy prescriptions on the thesis that the marketplace of ideas was the appropriate framework to superimpose on any number of questions involving science and the universities. I have been taken aback in the course of the research for this volume to repeatedly stumble over evidence of how famous economists have intervened at key junctures to help institute bulwarks of the modern regime of globalized privatization of science and to undermine the Cold War university. For instance, it seems George Stigler played a critical role in mobilizing intellectual resources and justifying the shift in stance of the pharmaceutical sector toward research and pricing policies in the early 1970s.[26] Stigler, following up on his own neoliberal theory of knowledge production, convened a Pharmaceuticals Project (with funding from Big Pharma), which argued that FDA regulation of drugs was stifling pharmaceutical innovation. The solution was to privatize more aspects of drug research; one of the offshoots of the project was the Center for the Study of Drug Development, headed by Louis Lasagna. Members of this latter organization were among the earliest entrepreneurs conjuring the CRO industry covered in Chapter 5 (Petryna 2009). This was the very same George Stigler, Nobel Prize winner, who argued in discussion of Arrow's (1962) paper on the economics of science that "the paradox is that information is expensive to produce and cheap to distribute raises serious problems. In fact Arrow might have pointed out that the optimum incentive to invent presumably would require an infinite patent period, very much as the privilege of living in a house for only 17 years would lead to a suboptimal amount of building."[27] Following this logic, the pharmaceutical sector, in conjunction with some key neoliberal economist allies,[28] were instrumental in pushing the agenda at the Uruguay Round trade talks that bequeathed us TRIPs and the globalized strengthening of intellectual property (Drahos and Braithwaite 2002; Sell 2003).

There were many other examples of neoliberal interventions in the "knowledge economy." I have already mentioned Milton Friedman's proposal (1981) to abolish the National Science Foundation. As the economist George Shultz told William Simon during the energy crisis of the 1970s, "I'm so glad it's you who's heading up the energy bureaucracy. That way it will go out of business, and you'll be able to keep the damage in check" (M. Jacobs 2008, 208). Richard Levin was installed as president of Yale with a mandate to privatize its biotechnology research portfolio (Geiger and Sá 2008, 137). But interventions like these did not happen simply or solely at the level of legal infrastructure: It has also happened at the level of encouraging the junk science/sound science movement. Naomi Oreskes has recently documented the role of such economists as Nobel winner Thomas Schelling and Yale economist William Nordhaus in providing intellectual support for those who sought to contest and deny global warming (Oreskes et al. 2008; Oreskes and Conway 2010). Indeed, one of the recent books advocating "environmental skepticism" and promoting "sound science," which garnished substantial attention critiquing the physical science, was written by a specialist in game theory, not a trained climate scientist (Lomborg 2001).

Of course, mainstream economists have not been omnipresent shadow Svengalis lurking behind every important development in the new regime of science management; my point is that they have served as major promoters of the new production of ignorance, *whether they were aware of it or not.* Models of the economics of science render the scientist just as ignorant as the neoliberal is content to render the average citizen ignorant. They wind up serving as apologists for the new regime, witting or no, to the extent that they attempt to make the world conform to their image of the Market as information processor. It happens when economists who work for think tanks attempt to intervene in scientific debates in the natural sciences; it happens when economists argue that litigation science is not harmful to the research enterprise, because it maximizes competition among points of view;[29] it even happens when economists like Paul David, who pride themselves in opposing neoliberalism, argue that the best way to defend the integrity of the modern university is simply to calculate and impose the optimal mix of public- and privately funded science (David and Dasgupta 1994). One can sympathize when David argues,

> For university administrators to encourage (or even permit) political leaders to entertain the hope that the energies of their faculties and students could be harnessed to yield accelerated productivity growth, showers of better-quality products, enlarged export earnings, and local job creation—all within the brief time frame that will make a difference in the coming elections—is not merely deceptive. It is quite reckless in risking the almost certain disappointment of unrealistic expectations, and so may bring in its train public disaffection and damage to the university. (2007b, 263)

Yet sympathy turns to despair when we realize that he does not apparently notice two things: first, the provocation of disaffection of the public with their universities is *part of the design of the entire neoliberal agenda;* and second, the neoliberal production of ignorance drives a wedge between scientific research and economic growth, such that any promises of payoffs, however delayed and unpredictable, are no longer grounded in the current system of research. The neoliberals are aware of that but are unperturbed; they don't want science to be funded based on any such reasoning. They look forward to the rapturous day when all knowledge (and not just science) is comprehensively funded and coordinated by the market, and state-organized research is reduced to a pitiful insignificant remnant. Paul David, by making his benchmark some ideal neoclassical marketplace of ideas, himself ends up undermining "public" science in the name of subjecting it to a balanced optimization calculus. One particular flaw in that argument is the conflation of the state university system with the locus of publicly funded science.

Indeed, it could be suggested that the modern economics profession, insofar as it was incubated in the postwar university, has fouled its own nest by becoming highly complicit in the dissolution of the Cold War university. It started out with the hostility of neoliberals like Milton Friedman and George Stigler to the state provision of higher education. Friedman devoted much of his accumulated fortune to the privatization of state-supported education,[30] which he regarded as the largest residual sector of state socialism in the West. The more economists shifted the image of education from preparation for citizenship to accumulation of personal "human capital," the harder it became to maintain public support for state-subsidized higher education (Apple 2003, 2006). One implication of human capital theory was to shift individual student support from scholarship grants to student loans—another neoliberal innovation. In a stratagem notable for its Machiavellian brilliance, individual universities were then encouraged to solicit more private funds to offset cuts in state subsidy, but the more the university consequently became increasingly embroiled in market activities, the more it lost any political justification for state support, resulting in a downward spiral of appropriations and the de facto privatization of the American public university system. *Pace* Paul David, there is no longer any option to find that elusive optimal public/private mix in research and education, because the state system of higher education has been irreversibly privatized (Rizzo 2004). Most people seem unaware of the degree to which (in America) flagship state research universities no longer depend upon any substantial state subsidy, as dramatized in Figure 7.1.

This dynamic is exemplary of how the spread of certain neoliberal beliefs concerning how knowledge production works eventually leads to a transformation in the range of possibilities whereby research and learning can proceed. In the neoclassical economics of science one supposedly enjoys infinite choice over the shape and extent of commercialization of scientific research, as well as of scientific training and dissemination of results. But in reality, range of options and extent of commercialization have been intimately connected, and not in the liberating direction. The privatization of research funding has been followed by the privatization of the American university and the fencing off of open modes of access to knowledge; the 2008 economic contraction only exacerbated this trend. Choice over modes of conduct of science is shrinking, rather than expanding. Conveniently, intellectual rationalization by mainstream economists has smoothed the path nearly every step of the way.

It is difficult to escape the conclusion that, for most economists, much of what is happening to the conduct of natural science on the ground at the university and within the lab doesn't actually seem to raise many qualms. One fact germane to their apparent equanimity is that economics, alone among the academic professions, has enjoyed almost unchecked expansion within the university in the twentieth century *throughout the world*. For instance, in one large sample of universities sited within the British commonwealth, the only field that met or exceeded the growth of total faculty numbers in economics from 1915 to 1995 was in chemistry (Frank and Gabler

2006, 133, 160). As they suggest, "Although it is good to understand why university economics prospered more than psychology over the twentieth century, it would also be good to know why at the end of the century economists were ten times more common on average in the world's universities than psychologists" (2006, 202–203). Clearly economists think their untrammeled success within the university was entirely warranted, because the

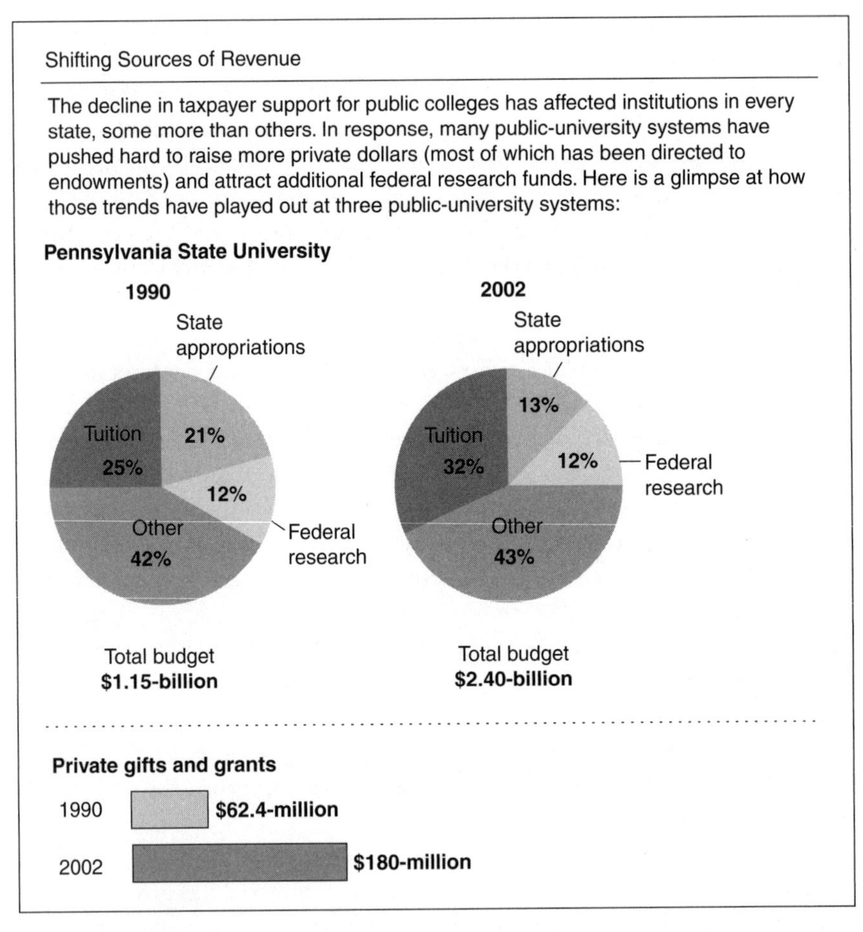

Figure 7.1. Budgets of some major American state universities. More recent (2006) approximate data on individual flagship universities: the University of California system gets 18 percent of its budget from the state; Pennsylvania State gets 13 percent; University of Colorado gets 9 percent; Michigan State University gets 18 percent; University of Wisconsin gets 19 percent; University of Massachusetts gets 26 percent. *Sources:* Selingo (2003), graphic; Rauchway (2007); Lyall and Sell (2006); and university Web sites.

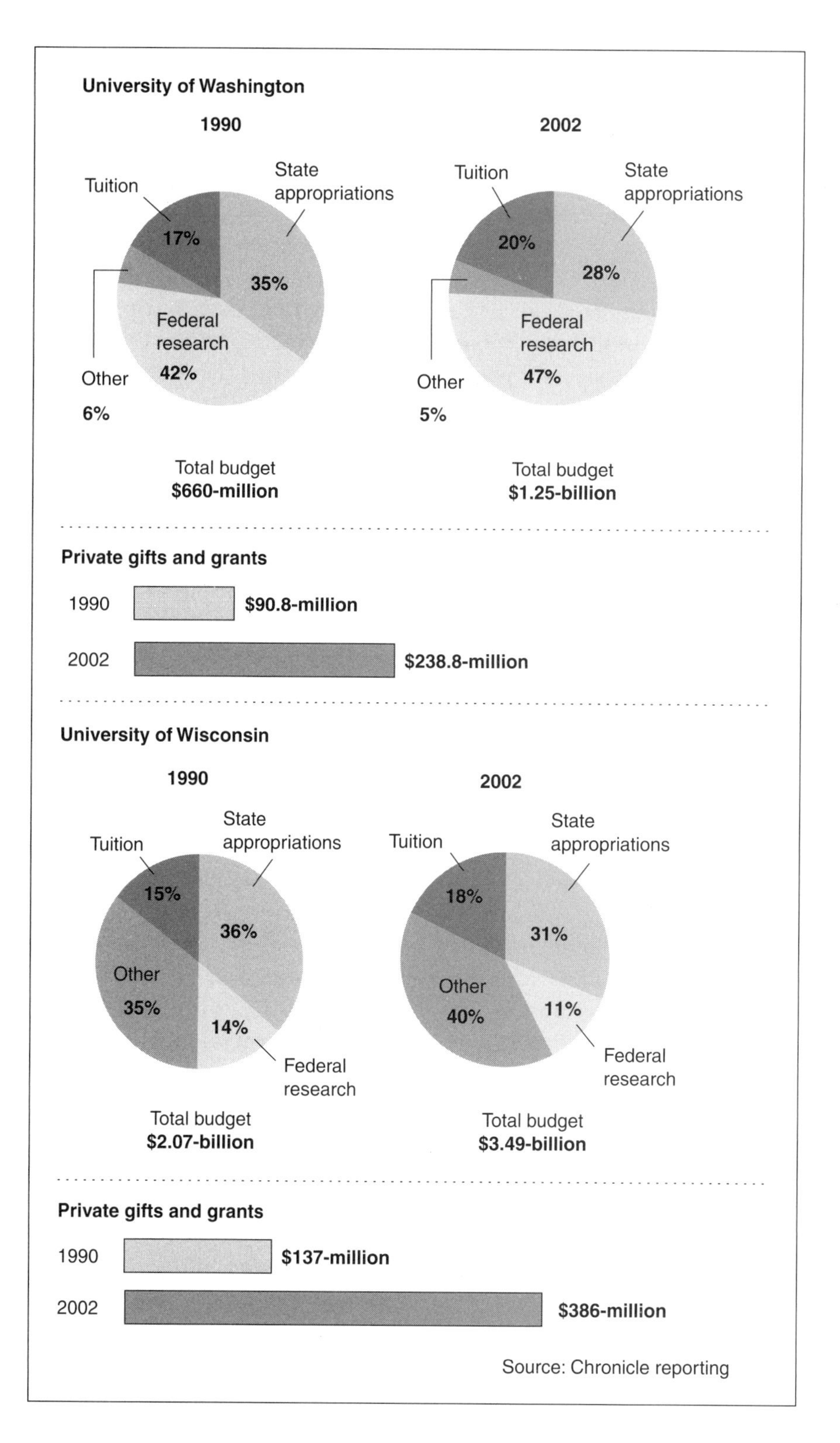

Figure 7.1. (continued)

stamp of approval of the marketplace of ideas was on their foreheads. It is just an unfortunate fact of life that if some fields expand, then others must shrink. The ebb and flow of the sciences is for believers in the marketplace of ideas just part of the natural order.

Other academics might not agree, and instead they strive for more causal accounts. One is provided by Marion Fourcade (2006), who estimates that economics has displaced other fields by growing from roughly 1 percent of all university faculty at the beginning of the last century to about 4 percent at the end, and she links the phenomenon to the role of the American economy in the globalization of world trade and discourse. As she points out, "Western companies investing abroad . . . carry with them scores of lawyers and consultants, who then find themselves in a powerful position to penetrate local markets, and in the process impose their own definitions of reality" (2006, 150). Neoliberal economics lavishly benefits from this dynamic, because it deals in the seemingly universalist rhetoric of abstract economies and efficiencies, apparently unfettered by any geographic, cultural, or intellectual specificities, and economics is thus transformed into a generic technology of bureaucratic and political power by corporations, central banks, the IMF, the World Bank, the WTO, and so on. Very few natural sciences have enjoyed a comparable strategic role in the American Century.

So it seems economists may not so intently feel the pain of their brethren in the natural sciences who are undergoing downsizing and restructuring, and they may be deficient in the moral sympathy that would allow them to focus their attention more concertedly on the drawbacks of commercialization. But that is not the only way that economists stand out in the contemporary intellectual ecology. Another is that, in the annals of the American university, the *economists,* not biomedical researchers, pioneered the practice of forming start-up firms while remaining employed as faculty members during the Cold War, by arguing that there was no conflict between their responsibilities to their shareholders and to their students. Indeed, it was instances of entrepreneurship like the Harvard neoliberal economists Martin Feldstein and Otto Eckstein in the 1960s luring venture capitalists to help them create for-profit start-up firms (in this particular instance, Data Resources, Inc., selling econometric forecasts)[31] that pressed the envelope on holding down full-time major faculty positions while simultaneously serving as CEOs of corporations (Parker 2005, 509). By the 1980s, once the biotechnology boom began, that was no longer such a rare occurrence. It would therefore naturally seem to many in the economics profession that what had been good for the economists in the past should be presumptively equated with what should be good for science as a whole in the future. Rather than imposing symmetry on the agents in their models of the economics of science, neoclassicals surreptitiously impose symmetry in real life by positing that natural scientists in vivo *should* want what the economists want. As I have demonstrated, there is growing evidence that this synecdoche is based on faulty data and flawed premises.

The Paradox of the Cretan Liar

There is one last conceptual problem that needs to be confronted in this dark view of the role of economists in the current regime of privatized science. My argument in this volume, and more explicitly of this chapter, has been that scientists should be wary of neoliberal arguments concerning the virtues of the supposed marketplace of ideas, and even more specifically, to beware economists bearing gifts of an economics of science. But then, avers Viridiana, how about you, Philip Mirowski? Aren't you an economist? Don't you yourself come bearing the promise of a sound and valid economics of science?

There are both a simple answer and a more complex answer to this objection. The simple answer is to take this book to anyone in your local economics department and ask him or her if it was written by a "real" economist. The more complex answer is that, if the neoliberal project collapses in the wake of the economic crisis of 2008 (although the chances of that happening now seem substantially less than 50/50), then a comprehensive and sophisticated appreciation for the way that science and the economy interact will not generally be conceded as forthcoming from economists any more. Past approaches to the economics of science will come to be discredited. Scientists will, by that time, have lost patience with neoliberal ripping tales concerning their natural expertise romancing the market and general omnicompetence in the all-powerful marketplace of ideas.

The Bonfire of the Vanities

Viridiana harkens back to the newspaper quote from the current president of Harvard concerning her distress: "Have universities become captive to the immediate and worldly purposes they serve? Has the market model become the fundamental and defining identity of higher education?" (Faust 2009). The implication seems to be that Harvard is willing to confront and reconsider the trends we have documented in this volume. But upon learning a little more, Viridiana finds out that stirring op-eds are one thing, but actions quite another.

I argue that a major characteristic of the modern neoliberal era of the new knowledge economy is the unapologetic production of ignorance. I think most people would accept that fraud is one manifestation of the production of ignorance, and it is one of the more perilous phenomena that are encouraged by the commercialization of research. The connection to Harvard is the largest case of fraud (in dollar terms) ever settled by Harvard University that I wish to recount here. It involves two economists, Andrei Shleifer and Lawrence Summers. Most people have heard of the latter, if only as being Faust's predecessor, the erstwhile president of Harvard University, and Barack Obama's late economic policy czar.

One of the stranger phenomena of the late twentieth century has been the noticeable inclination of major research universities in the United States, and now Europe, to resort to hiring economists as captains of their institutions.[32] This in itself has also been an artifact of the neoliberal agenda, because boards of trustees are predominantly composed of businessmen and have not often sought out humanists as ideal candidates to bring about the commercialization of various aspects of the university. In any event, Summers was explicitly recruited by the Board of Overseers at Harvard to help bring what was perceived to be a relatively recalcitrant university, with an overly autarkic faculty body, closer to the market ideal (Bradley 2005, 67). Summers came aboard with strong neoliberal credentials, for instance, having served as one of the advocates for the "Washington Consensus" while at the IMF (ibid., 31).

Most people have been bequeathed the impression from the news media that Summers had been forced to resign his position as Harvard's president because he had an "abrasive" personality, or possibly because he made some unfortunate remarks about why women were underrepresented in the higher academic ranks of the natural sciences.[33] Without gainsaying those characteristics, almost no one had mentioned the settlement of the fraud case, and the ensuing scandal that erupted six months or so after Harvard agreed to pay more than $31 million to settle a government lawsuit against itself and economist Andrei Shleifer.[34]

Shleifer was one of Summers's protégés, and a famous economist in his own right. He was widely regarded as an academic star in the economics profession, winning the Clark medal from the American Economics Association, designating him the most promising economist under age forty in 1999. One of his specialties was the express application of neoliberal theory to government regulation (Shleifer 2009), and in particular, in a piquant irony that cannot pass unnoticed, on neoclassical models of optimal levels of fraud and corruption in capitalist economies. He also was an acknowledged expert in what has come to be called "behavioral finance," which, for the purposes of the present audience, simply means he was a student of the divergences from the pure "efficient markets hypothesis" in such areas as stock markets and corporate finance. Shleifer was born in Russia in 1961, and a fluent speaker of Russian, but emigrated to the United States in 1976. He attended Harvard as an undergraduate, and was taken under Summers's wing there. Shleifer was one of the youngest economists ever granted tenure at Harvard, and he worked on Russian development economics and finance at the World Bank when Summers served there as chief economist. After the collapse of the Soviet state in 1990, USAID contracted with Harvard, and in particular the Harvard Institute for International Development (HIID) in 1992, to help administer a "shock therapy" privatization program in the New Russia. Due to his Russian expertise, Shleifer was selected to run the project. Because of his expertise both in the application of neoclassical theory to government functions and in finance, the project was particularly focused upon setting up a reformed banking and financial sector in Russia.

Now, to prevent any possibility of the appearance of taking unfair advantage of their privileged access to information, not to mention their ability to shape events, both USAID and Harvard expressly forbade anyone involved with the contract to invest in the newly privatized Russian economy. "In July 1994, Shleifer . . . began investing in Russian enterprises in direct violation of his contract and the restrictions imposed by AID and Harvard" (McClintick 2006). Their subsequent attempts to profit from the privatization process later took a number of baroque twists and turns (ibid.), which we need not recount here for our present purposes. However, it is not irrelevant that the Harvard Corporation had itself been investing up to 2 percent of its endowment in Russian securities involved in the privatization in the mid-1990s, so it was not simply an innocent bystander.[35]

It is important here to pull back from these minutiae of personalities and historical events in order to observe that this situation constituted a problem of conflict of interest, a standard pathology found throughout the sciences in the modern era of the privatized university (Slaughter et al. 2009). The economist Shleifer was supposed to both produce and convey technical academic information, in this case concerning the best way to set up a financial system in a former Soviet transitional economy, and simultaneously to commodify that information, which in this case took the form of engaging in corporate activities that would profit from that information. In a pure marketplace-of-ideas scenario, supposedly there would be no problem here: If his consulting activities made money, then that would be *prima facie* evidence that it was indeed the correct way to organize the system and propagate the knowledge. But not everyone would accept the legitimacy of the marketplace-of-ideas scenario in every instance, particularly when a substantial contingent remains skeptical that there exists a single best method to construct anything as complex as a financial system in a country that had little prior experience with one. Worse, in a fluid geopolitical situation like that after the fall of the Berlin Wall, the project was hedged round with mutual suspicions of crude taking advantage of a former rival, with the attendant degradation of the quality of the "knowledge" on offer.[36] From an Olympian perspective, this situation closely resembles a report on a clinical trial of a drug, where the principal investigator holds a substantial share of the corporation that stands to earn millions if the drug is found effective, and was paid by the corporation to carry out the clinical trial in the first place; moreover, the investigator's university also buys a stake in the company.[37] Conflicts of interest are pervasive and inevitable wherever and whenever research is fully privatized, because making money is never comprehensively isomorphic to making truth (Cook et al. 2007). Neoliberals simply deny that proposition: For them, making money is the only truth. The naive and the unwashed may choose to believe otherwise, but the neoliberal theorist then suggests the Invisible Hand or natural selection will themselves weed out these mendicants, depending upon the extent of their tough-mindedness. The only truth societies can absorb is the truth they are willing to pay for.

Sometimes, neoliberals deride concerns over conflicts of interest by attempting to reduce the problem to a spurious moralism: No one is more petty or self-interested than the ivory tower academic mired in penury, they sneer.[38] But this commits the fallacy of misplaced concreteness: Conflict of interest is an *institutional* problem, not an essentially personal matter of ethics. Different institutional structures foster or discourage different types of critique and discourse; and as a form of critique, the neoliberal marketplace of ideas is notoriously short term, weak on the handling of unintended consequences, and poor at the unpacking of subtle nuance. This is nicely illustrated in the Harvard case.

The problem of conflict of interest at HIID could not be confined to Andrei Shleifer or the small team at work on the Russia project. When the Inspector General's Office began an investigation into Harvard's conduct of the project in 1997, Shleifer dug in, insisting he had done nothing wrong. From a neoliberal viewpoint, one can understand how he might argue that position, but it was not the legal understanding of government lawyers, or the perception within the Russian Federation. Indeed, one can sympathize with Russians fed up with Americans lecturing them on corruption and their supposed lack of respect for market discipline. On May 19, the first deputy prime minister demanded USAID terminate the HIID program, and the Russia Project was then killed. The Russian authorities carted off the Harvard project's computers and files "for safekeeping." Harvard then fired most of the Russia cadre from HIID, but Shleifer retained his tenured professorship. Later, Harvard did rescind his named chair. After an FBI investigation, a grand jury rejected criminal prosecutions,[39] but USAID filed civil charges against Harvard in September 2000.

This, as they say, is where the plot thickens. Shleifer's mentor, Lawrence Summers, was appointed president of Harvard in March 2001. Indeed, Summers stayed with the Shleifer family when he interviewed for the president's job. Summers had been aware of Shleifer's problems, and if the major players had been cognizant of that, they might have suspected that Summers might shield his protégé.[40] To ward off such suspicions, Summers took the position that "conflict-of-interest issues should be left to the lawyers" and formally recused himself from the ongoing litigation. However, McClintick (2006) reveals that Summers intervened repeatedly at Harvard in order to keep Shleifer on the faculty, award him a chaired professorship, and promote him as a respected Russia expert. After all, Summers had been brought back to Harvard to nurture and promote the entrepreneurial spirit of faculty like Shleifer—in 1994, Shleifer had also helped found a firm called LSV Asset Management, again turning his financial expertise into market profit. Thus, Harvard elected to settle the civil suit on its own behalf in August 2005, with Harvard ordered to pay $26.2 million to the government, and Shleifer to pay $2 million in recompense. None of the defendants acknowledged any liability under the settlement, and the details of other lawsuits were sealed.[41]

The story does not end there. As news of these events leaked out, Harvard faculty began to challenge the administration as to why Harvard should pay out a settlement of more than $26 million plus legal fees of over $10 million for the actions of a professor (and some ancillary actors), when he seemed to get off relatively scot-free. When the McClintick article appeared in January 2006, Summers himself was challenged at a faculty meeting. His response was, "I have taken no role in Harvard's activities in the courts, nor . . . familiarized myself with the facts of the situation . . . I am not knowledgeable of the facts and circumstances to be able to express an opinion as a consequence of my recusal" (quoted in Ciarelli and Troianovski 2006; also Ivry 2006). This was more than informed observers could swallow. Met with a wall of silence, the faculty responded with incredulity, outrage, and a vote of no confidence in their president, the first in the history of Harvard. Before he died in 2008, Dean Jeremy Knowles ruled that the report on the matter by the Committee on Professional Responsibility must itself remain secret.

McClintick's article in the obscure outlet *Institutional Investor* (hardly a muckraking rag) was entitled "How Harvard Lost Russia" (2006). Perhaps it may seem to be venturing into hyperbole, but one can see what he meant: a little problem of "conflict of interest" had ballooned all out of proportion into an international crisis, which itself had repercussions for the future of Russia's economic infrastructure and (as we have learned to our regret) global geopolitics. And again, risking hyperbole, this sequence of events was possibly the straw that broke the camel's back, causing the Harvard Corporation to rethink their controversial president, leading to Summers's resignation on February 21, 2006. Chains of causality are always notoriously slippery, but isn't that just what unintended consequences are like? Under a different institutional setup, one less tolerant to a little entrepreneurialism between consenting adults, wouldn't many components of this sequence of events have been rendered less likely? A little bit of commercial action on the side for a true believer in the marketplace of ideas, trading on his academic status, led (inadvertently) to corrupted science, a crippled socialist transition to capitalism, massive court costs, and the fall of the president of the supposedly most prestigious university in America. For lack of a horse . . . An active PR campaign kept this from becoming a major topic of conversation on all the chat shows, let alone being analyzed as a paradigm instance of the fallout from the progressive privatization of the modern university.[42] Yet, "behind the scenes, the Harvard Russia scandal continues to fester" (Warsh 2009a).

The denouement to this tale is as indicative of life in the modern university as it is of the role of the contemporary economics profession. Summers did resign, but he has been rehabilitated as a political insider, becoming one of the top economic advisors for the Obama administration. He is now the subject of numerous complimentary journalistic portraits. He was the person behind the financial bailouts of Wall Street firms and the arbiter of the correct public policy to deal with the economic crisis of 2008. He did not suffer financially either, becoming part-time managing director of the D. E. Shaw Group, a hedge fund, while serving as an editor of the *Brookings Papers.*[43] Andrei Shleifer did not suffer unduly, serving as editor of the *Journal of Economic Perspectives* all during the period of his backroom battles with government attorneys, and he has continued to publish numerous papers on Russia, the role of corruption in development, and the "right" way to impose a transition from socialism to capitalism. He has proclaimed, "It is natural to refer to the last quarter century as the Age of Milton Friedman" (Shleifer 2009, 123). In what may be a damning piece of evidence concerning the current state of the economics profession, he was in 2008, according to ISI Thomson, the top-cited scientist in economics and business in the world.[44] Shleifer later sold his share of LSV Asset Management for an undisclosed sum, undoubtedly mitigating the pain of paying the court fine. In the interregnum, he was treated with kid gloves by Summers's temporary replacement, Derek Bok, so did not have to relinquish any of his Harvard perquisites.[45] He has never publicly defended any of his activities, and no one in the economics profession has ever urged him to do so. As Warsh wrote in 2008, "He has become, as he might say, a normal professor, or perhaps as another MIT-trained economist put it the other day, a normal Harvard professor."

This is not the story of "a few bad apples," or of overweening hubris, or pushy East Coast glitterati, or even unjustified government meddling in the marketplace of ideas. (I know some of you think that.) It is instead the story of the neoliberal corruption of knowledge. It is the story of the paladins of the economics profession preaching the commercialization of science, profiting from its installment, all the while denying any of its unsavory aspects. It is the story of a university that publicly seeks to assuage fears of the corruption attendant upon the commercialization of knowledge, while privately it pays to cover up the consequences. It is simply one more incident in the ongoing project to reengineer American universities to become the cash cows of the knowledge economy. It is the story of Viridiana Jones in the Temple of Mammon.

Notes

1. Viridiana Jones and the Temple of Mammon; Or, Adventures in Neoliberal Science Studies

1. Viridiana has given a copy to me, just in case I didn't believe her story. She scoffed when I asked if she had ever considered moonlighting.
2. See, for instance, Mooney (2005), and the discussion of his book on his Web site, www.waronscience.com, and the report on scientific integrity of the Union of Concerned Scientists (2004). Exclamations that a change in administration renders these worries obsolete are themselves premature, as one can glean from reading McGarity and Wagner (2008).
3. The *Times* story is by Revkin (2006). Viridiana is vaguely aware that various authors have documented systematic suppression of inconvenient results in the field of clinical research. See Callahan (2003); Mirowski and van Horn (2005); Sismondo (2008). She is also aware that one colleague in physics was hauled before a Star Chamber–style inquisition because he had included an author who "did not substantially contribute" to his paper. This reprimand was used to justify recourse to his government funding agency to try and get his grant revoked (Associated Press 2008).
4. This argument is made in detail for twentieth-century philosophy of science in Mirowski (2004b) and in a different manner by Sarkar (2007).
5. See Latour (2005); Latour and Woolgar (1979); Etzkowitz (2002, 2003); Etzkowitz et al. (2000); Leydesdorff and Etzkowitz (2003); Woolgar (2004); Shapin (2003, 2008b).
6. See Fallis (2007); Koertge (1998); Sokal and Bricmont (1999); Sokal (2008). One antidote to this opinion is Newfield (2008).
7. On the flagging quality of science journals, see Bosman (2006) and Corbyn (2009). On the thankless task of journal editorship in the biomedical sciences, see McNeil (2006). On ghost authorship in science, see Chapter 6, and in particular, the work of Sergio Sismondo.
8. http://www.sourcewatch.org/index.php?title=American_Chemical_Society (last visited August 13, 2008).
9. As the coeditor of an anthology of writings by economists on the issue (Mirowski and Sent 2002), I feel fairly confident in this generalization. For other instances,

see Mowery, Nelson, Sampat, and Ziedonis (2004); Etzkowitz (2003); Geiger (2004); Kirp (2003); Washburn (2005); Fallis (2007); Geiger and Sá (2008); McKelvey and Holmen (2009); and Andersson and Beckmann (2009). The work of Philip Kitcher (2001) has been especially unfortunate in this regard, as I have argued (Mirowski 2004a, 2004b).

10. See Murray and Stern (2007); Sampat (2005). This, of course, has been a prelude to subjecting the corruption of science to a cost-benefit analysis. I dissect this problem in Chapter 6.
11. Of course, I am nowhere near the first to notice this. "Neoliberal campaigns to downsize public education aimed to largely abolish the public character of support for broad-based access to knowledge and information, and to define education as a matter of personal responsibility" (Duggan 2003, 42). See also Apple (2003, 2006) and Slaughter, Archerd, and Campbell (2004). Where I add my modest contribution is to explore how this has worked in the case of science itself.
12. For some economic examples, see Stiglitz (1999); Powell and Snellman (2004); Feldman and Link (2001); Porat (1977); Castells (2000); May (2002); and Warsh (2006). As Stiglitz (2008) wrote, "We live in a knowledge economy, an information economy. Because of our ideas, we have all the food we can possibly eat." A Google search of the term *knowledge economy* in October 2005 produced an amazing 1,690,000 hits. By August 2008 it was up to 1,950,000. Google (www.google.com/trends) reveals a fairly even interest in the search term since 2004.
13. How many faculty members are aware if their university administrations have brought in paid consultants from McKinsey (for example) to turn their supposedly "dumb organizations" into something resembling "smart organizations" such as McDonald's or WalMart? See Fuller (2002). Viridiana stumbled across this when she found a consultant had suggested that her university install real-time audience response meters in its classrooms.
14. Some recent attempts have been by Benkler (2006); Castells (2000); Foray (2004); David and Foray (2003); Bowker (2004); and Warsh (2006). McCormack (2009) is a bracing antidote to this tendency.
15. But see Mirowski (2008, 2009). A popularization of one aspect of this development, the supposed incorporation of "knowledge" into neoclassical growth theory, can be found in Warsh (2006). I revisit the theoretical underpinnings of this incident in Chapter 2.
16. For the evolution of the computer, see Edwards (1996) and Ceruzzi (1998). For the history of the Internet, see Abbate (1999) and Gillespie (2006). A nice history of the information concept is Capuro and Hjorland (2002). I have written a history of the influence of the computer on American economics (Mirowski 2002).
17. This information is covered in detail in Chapter 6.
18. See, for instance, Drahos and Braithwaite (2002); Eisenberg (2001); Landes and Posner (2003); Nelson (2004); Mowery et al. (2004); Washburn (2005); Slaughter and Rhoades (2004); Jaffe and Lerner (2004); and Boyle (2008). The details with regard to patents are covered in Chapter 4.

19. Some of the authors that address the globalized character of the contemporary transformation of the economy of knowledge include Drahos and Braithwaite (2002); Drori, Meyer, Ramirez, and Schofer (2003); Apple (2003); Marginson (2007); and Cooper (2008).
20. The argument that universities did bring it upon themselves can be found in E. Berman (2008). This argument is refuted in Chapters 3 and 4.
21. This is the explanation offered by Landes and Posner (2003, 403–422). However, a second pass at the same issue (Landes and Posner, 2004) offers another implicit explanation, one closely linked to neoliberal doctrines, as I discuss later in this chapter and also in Chapter 4.
22. For economic historians, see Lamoreaux, Raff, and Temin (2003, 405); for legal scholars, see Boyle (2000); Lessig (2001, 2004); and McSherry (2001); for historians of education, see Geiger (2004, 3); Matkin (1990, 22); Slaughter and Rhoades (2004); Kirp (2003); and Apple (2005); for politics, see Krimsky (2003, 30–31; and Mirowski and van Horn (2005). Washburn (2005, Chap. 3) documents the political maneuvers leading up to the passage of the Bayh-Dole Act.
23. This refers to Alfred Chandler's *Visible Hand* (1977), which we take as shorthand for a historical and analytical description of the rise of the large multidivisional (M-form) firm over the course of the twentieth century. The main proponents of the thesis that this model began to break down in the 1980s have been Lamoreaux et al. (2003, 2004) and Langlois (2004).
24. I will return to this historical watershed in Chapter 3, because it explains what seem to otherwise be puzzling trends noted by some authors (e.g., Geiger and Sá 2008, 24).
25. There is some indication from news accounts in June 2009 that Congress will restore funding for collection of these accounts (Ginsbach 2009), but past history does not tend to instill confidence. Many empirical papers in economics journals using BEA data never mention its corruption over the last decade. Adding insult to injury, BEA stopped collecting this data just when it was tasked with a major revision of the national income accounts, which among other "improvements"—get ready—is slated to incorporate "investment in R&D" to the core accounts by 2013 (Aizcorbe, Moylan, and Robbins 2009). I cannot think of a better example of why almost all econometric macroeconomic analyses of the relationship of science to economic growth are literally gibberish.
26. Examples of this narrative can be found in Bok (2003); Geiger (2004); Krimsky (2003); Nelson (2001); Owen-Smith and Powell (2003); Thursby and Thursby (2003); and Kirp (2003).
27. "Innovation's Golden Goose," *Economist*, December 14, 2002, 3. After only eight more years, we observe instead that it is the *Economist* that deserves the Golden Fleece Award.
28. For the run-up to the legislation, see Washburn (2005, 60–69); for the version that universities had already been able to patent, see Mowery et al. (2004); and for an argument that universities wanted a standardized patent policy, see E. Berman (2008). The line that Stanford and Harvard/MIT had special capacities

for commercializing research that could not easily be imitated elsewhere can be found in Powell et al. (2007).

29. Consult Slaughter and Rhoades (2004, 51–57) for a roster of legislation during the 1980s and early 1990s.
30. This is the major drawback in the premises governing the work of Mowery et al. (2004); Krimsky (2003); and Geiger (2004), and the literature that restricts itself narrowly to considerations of intellectual property in academe such as McSherry (2001) and E. Berman (2008).
31. I discuss real figures from real states in Chapter 7.
32. Some of the best analyses have been by Slaughter and Rhoades (2004); Apple (2005); Duggan (2003); Kirp (2003); Marginson (2007); Frank and Gabler (2006); Washburn (2010); and Douglass (2008).
33. *Madey v. Duke University*, 307 F.3d 1351 Fed Circuit (2002). This ruling is discussed in detail in Chapter 4.
34. See Dillon, (2006); Breneman, Pusser, and Turner, (2006).
35. See Jaschik (2007). The experiment at Illinois collapsed in 2009 (Kolowich 2009), but that doesn't mean the end of that particular business model for state universities. The 2008 economic crisis has been a boon for distance education in general.
36. The most striking contemporary example of a nonacademic research unit outcompeting the academic contractor for the research dollar is the "contract research organization," which has progressively displaced the academic hospital as the primary contractor for pharmaceutical clinical trials. See Chapter 5.
37. See Nedeva and Boden (2006); Fallis (2007); Pestre (2003b, 2005); Michaels (2008); Goldman (2005); J. Fisher (2008); Foucault (2008); and Lave, Randalls, and Mirowski (2011). This section is inspired by my work with an exceptional group of scholars, gathered together in Mirowski and Plehwe (2009).
38. See, however, Backhouse (2005); Capuro and Hjorland (2002); Edwards (1996); Foucault (2008); Burchell, Gordon, and Miller (1991); Johns (2006); Mirowski (2002, 2008); Napoli (1999); Peters (2004); Porat (1977); Shinn (2003); and Starr (2004).
39. There is a historical question as to the extent to which other schools of economics, such as the American Institutionalists, possessed a distinct position with regard to cognition before World War II. This issue will have to be postponed to another venue, given our present concerns. The attempt by Giocoli (2003) to attribute cognitive content to prewar European economists also needs further attention.
40. See, for instance, Director (1964); Coase (1974b); Ingber (1984); and McCloskey (1994, 368). As far as I am aware, no one has adequately traced these early locutions back to the pragmatist school of philosophy and explicated the role that they played there. If and when this happens, we may find that it had very little resemblance to the postwar neoliberal conception. A good start on this project can be found in Peters (2004).
41. Some instances of this revaluation are Caldwell (2004); Mirowski (2007b); Mirowski and Van Horn (2009); Dixon (1998); and Schulman and Zelizer (2008).

42. The history of neoliberalism is a burgeoning topic in its own right and would require a separate survey. The interested reader might consult Hartwell (1995); Foucault (2008); Mirowski and Plehwe (2009); Harvey (2005); and Plehwe and Walpen (2005).
43. While Hayek never openly attacked public education in this immediate postwar period, his comrade Milton Friedman did; see Friedman (1962, Chap. 6; 1978). Indeed, Friedman went so far as to praise a highly skewed distribution of income as a prerequisite for a free market in ideas (1962, 17).
44. The modern attitude of neoliberals toward the *necessary* role of ignorance in society will be central to my concluding evaluation of the modern economics of science in Chapter 7.
45. More careful discriminations (too numerous to cover here) between the various factions within neoliberalism with regard to numerous political and theoretical questions are contained in Mirowski and Plehwe (2009).
46. This is a variant of copyright invented by a group of lawyers intended to place downstream restrictions on the ability of users to privatize information used. See http://creativecommons.org; Benkler (2006); Lessig (2004).
47. Washburn (2005, Chap. 4) gives some striking anecdotes that illustrate just how twisted the entire teacher-student relationship can become under the modern order.
48. For the roles of Hayek, Polanyi, Popper, and other neoliberals in opposing the state organization and patronage of science, see Johns (2006); Mirowski (2004a, chaps 2–3); and McGucken (1984).
49. This Russian doll organizational structure of the neoliberal thought collective is discussed in Walpen (2004); Plehwe and Walpen (2005); Mirowski and van Horn (2009); Mirowski and Plehwe (2009); and Cockett (1995). It was intended to provide a rival model to the postwar university.
50. Those who have followed the "Sokal hoax" and the so-called Science Wars will perhaps recognize the basic fallacy that weakens their crusade to rescue science from its supposed detractors.
51. This point was raised in an interesting post by Jacob Levy on the discussion thread of Chris Mooney's book on http://crookedtimber.org/2006/03/27 (last visited August 16, 2008). As he puts it with admirable candor, "We need to create a Hoover Institution or Federalist Society of climate science!" See also Mooney (2005, 82) regarding the API memo on how to produce a contrarian scientist.
52. A little digging reveals the presence of numerous high-profile neoliberals and neoconservatives on the board of the Ethics and Public Policy Center, including William Barr, Jeane Kirkpatrick, Father Richard Neuhaus, William Kristol, and Michael Novak. Their Web site http://www.eppc.org/ (last visited September 11, 2006) also displayed an endorsement from Nathan Hatch (the former provost of Notre Dame).
53. Consult http://www.trumpuniversity.com/ (last visited August 16, 2008). For a skeptical overview, see Von Hoffman (2006).

2. The "Economics of Science" as Repeat Offender

1. Although these assertions might sound impossibly grand and dramatic, historians have spent many pages documenting the evidence for each claim. The main source on early neoclassical economics is Mirowski (1989); for the American Institutionalists, see Hodgson (2004); for postwar American economics, see Mirowski (2002); and for Carnegie, see Sent (2001, 2005) and Crowther-Heyck (2005, 2006).
2. I am well aware there exist various analytical moves to try and get around the paradoxes of reflexivity. However, I agree with Hands (2001, 390–392) that these merely postpone the logical regress. The major implication of reflexivity for the neoliberal program is postponed until our discussion in Chapter 7 of the role of ignorance in the new economics of science.
3. Even more incongruously, one finds contemporary philosophers such as Philip Kitcher (1993) upbraiding the sociologists for their logical incoherence while at the very same time appropriating models from the neoclassical economists and claiming them as demonstrations of how science ought to work (Mirowski 2004a, intro and Chapter 5; 2004b). The more neoliberal the analyst, the more obtuse they are about reflexivity: "[George] Stigler seemingly applied the results of his research work on the economics of information to [his own] daily life" (van Overtveldt 2007, 213)."[W]hat justifies drawing an analogy between market and science is exactly that both can be viewed as evolutionary processes of knowledge creation" (Vanberg 2008).
4. But see Mirowski (2008); Zamora Bonilla (2008); Camerer, Loewenstein, and Prelec (2005); Bruni and Sugden (2007); and L. Samuelson (2004).
5. Both of these papers are reprinted, along with commentary, in Mirowski and Sent (2002). This earlier attempt to make sense of the literature is now obsolete, in my opinion.
6. It is especially instructive to observe the way that science journalists, caught between neoliberal stories about the utter incompetence of government and the knee-jerk presumption that their target audience will always come down on the side of increased public subsidy of science, stumble upon the notion that the causal connection between science and economic growth may not be all that straightforward. See Grens (2007); L. Schultz (2007); and Minkel (2003).
7. The ideas in this paragraph are primarily taken from Kline (1995); Dennis (1987, 1997); Godin (2006a); and Forman (2007).
8. There are a couple of ironies lingering here, which we must decline to explore further in this venue. The first is that the earliest school of economic thought to insist that science was the prime mover behind capitalist expansion was Marxism in the later nineteenth century. Neoclassical economics merely expropriated this doctrine without acknowledging its provenance in the twentieth century. The second irony is that Bernal claimed to have been won over to the stress on the social relations of science upon hearing the famous paper by Boris Hessen on the technological roots of Newton's physical theories delivered in London in 1931(see McGucken [1984]; Werskey [1978, 2007]; and Mayer [2004]). It is

ironic, because Hessen was denying the standard precedence of basic science over application, a doctrine that Bernal came to champion.

9. For the history of science policy, see Elzinga and Jamison in Jasanoff, Markle, Peterson, and Pinch (1995, 573); and H. Brooks (1986). For the birth of STS, see David Edge in Jasanoff et al. (1995, 7).
10. Polanyi's curious image of scientists as a monastic order deserving unrestricted funding, and a comparison to Hayek's position, is discussed at greater length in Mirowski (2004a, Chaps. 2, 3).
11. This was recently admitted by Paul Samuelson (2004, 531). Since Samuelson (1954) was also the progenitor of the analytical concept of a "public good," this helps explain how the bridge was constructed at MIT in the later 1950s between public good theory and the linear model, largely through that other MIT product, the Solow growth model, covered below.
12. Hence we must dispute the interpretation offered in Edgerton (2004, 36) that economists did not endorse the linear model in the postwar period. Three clarifications render his statement otiose: first, he does not distinguish between neoclassical economists and heterodox traditions like that associated with Nelson and Winter (1982). The latter were indeed motivated to challenge the linear model, because they were challenging the neoclassical approach to technological innovation. Second, when he cites authors like Mowery and Rosenberg (1989), he does notice that they repudiate the model early in their text (22), but, like so many other economists, then go ahead and work with the categories of "basic" and "applied" science as if they had unproblematic consequences (e.g., 159). Third, he leaves out of consideration growth theory and the theory of public goods, which constitutes roughly 80 percent of the appearance of the linear model in the economics of science.
13. For discussion of their resistance, see Hong (1999); Scranton (2006); and the authors in Grandin, Wormbs, and Widmalm (2004).
14. As per the insistence of Misa (2004, 261). This inability to understand the linear model as an adjunct of 1950s neoclassical economic theory is one of the flaws of contemporary attempts to revive the linear model, such as by Balconi, Brusoni, and Orsenigo (2009).
15. On Hindsight and its history, see Sherwin and Isenson (1967); Kreilkamp (1971); Rosenberg (1982, 207–210); and Hong (1999). Beyond Traces, the NSF also commissioned the Battelle Institute to conduct a follow-up study (National Science Foundation 1973).
16. See Griliches (1958); Mansfield (1965, 1998); for examples, and Grens (2007) for some jaundiced observations, such as "The bottom line: No one knows what the actual returns of science are."
17. "The final nail in the coffin of the linear model was its inability to explain how Japan could be so successful despite lacking a world-class science base" (Martin and Nightingale 2000, xvii). Of course, things may look somewhat different a decade later.
18. Kealey even managed to merit an interview in *Scientific American* (Minkel2003), although the neoliberal agenda of that magazine dates back at least to the early 1990s. I can personally testify to that, because *Scientific American* approached me to write an article about Hayek on science for a popular audience

in the early 1990s. When I produced a relatively critical text, they refused to publish it. Other neoliberal assaults on the linear model include Butos and McQuade (2006) and Rottenberg (1981).

19. See, for instance, Rosenberg (1976); Faulkner (1994); Pinch and Bijker (1987); and MacKenzie (1996).
20. For an example of the former, see Rammert (1997); for the latter, see the special issue of *Technology and Culture,* 1976, vol.17, no.4.
21. So much so that sometimes they get carried away, straining to see it expressed as far back as "the founding of the Republic"(Guston and Kenniston 1994, 6). Often, this is documented by a quote from a letter written by Thomas Jefferson in 1813: "That ideas should freely spread from one to another over the globe for the moral and mutual instruction of man and improvement of his condition, seems to have been peculiarly and benevolently designed by nature when she made them like fire, expansible over all space, without lessening their density in any point, and like the air which we breathe, move, and have our physical being, incapable of confinement or exclusive appropriation." The misrepresentation of the history comes in realizing Jefferson was making an argument *against* Lockean natural rights accounts of property, in service of the thesis that one should *not* treat knowledge as a thing.
22. The issue of the seemingly harmless but in practice rather crucial role of the commodity space in the neoclassical model has been addressed in Mirowski (2006, 2007a).
23. The close connections between RAND and the economists in the Cowles Commission is discussed in Mirowski (2002, Chaps. 5, 6) and Jardini (1996). The story of the genesis of the economics of technical change at RAND is related by Hounshell (2000) and Asner (2006).
24. The story of Johnson's foray into science policy advice is related in Moggridge (2008, 314–320). Johnson ended up where so many others did by the 1970s:"it appears virtually impossible to establish any empirical basis for science policy decisions" (ibid., 317).
25. This has been frequently dubbed the "Coase Theorem"; but nothing in any of his writings even remotely resembled a mathematical theorem. Indeed, there could not have been any such thing, since mostly it involved playing around with definitions. These problems are described by Medema (2011). But when economists banished history of thought from their departments, the Orwellian practice of saying anything one pleased about the history of economics grew endemic.
26. Examples of the public choice approach to the economics of science are Lütge (2001); Zamora Bonilla (2003, 2008); and Vanberg (2008).The reader less than besotted with the ins and outs of the economics literature can get a flavor of Buchanan's attitudes from the BBC documentary *The Trap* (2007), written and produced by Adam Curtis.
27. See, for instance, Garcia-Quevedo (2004) and Gorg and Strobl (2007).
28. See Mowery (2005, 44). For other rejections of Jensen and Thursby, consult Stein (2004, 108).
29. This has been the subject of extensive commentary in the economic methodology literature. See Hands (2001, 207–212); Mäki (1992, 1999, 2005); McClellan (1996); and Pels (2005).

30. This is discussed a bit further in Mirowski and Nik-Khah (2007) Mirowski (2004b); and Newfield (2008).
31. Joseph Schumpeter (1976) wrote that "the fundamental impulse that sets and keeps the capitalist engine in motion comes from the new consumers' goods, the new methods of production or transportation, the new markets, the new forms of industrial organization that capitalist enterprise creates." Godin (2008) dates many empirical attempts to demonstrate this connection to the work of the National Bureau of Economic Research in the 1930s.
32. This doesn't even take into account that they have become corrupted over time, a point already explained in Chapter 1, and explored further in Chapter 6.
33. See, for instance, H. Jones (1976); Harcourt (1972); Snowdon and Vane (2005); Warsh (2006); Solow (1994); Foss (1998); Aghion and Durlauf (2005). Godin (2007) briefly covers the prehistory of the production function as defining what technology means for the science policy community.
34. "One of the defining characteristics of growth theory as a branch of macroeconomics is that it tends to ignore all the difficult economics that is papered over by that sentence" (Solow 1994, 46). One thing it does is simply identify growth of the economy with growth of the capital stock. Another is that it banishes all the macroeconomic problems Keynesians generally used to worry about.
35. See, for instance, Foss (1998) and Park (2007). But Romer knew enough to realize where his true allegiances lay: "Solow and Samuelson had to engage in vicious trench warfare about this time with Cambridge England to make the world safe for those of us who wanted to use the concept of the production function" (Snowdon and Vane 2005, 681).
36. Quoted in Snowdon and Vane (2005, 669). Steedman (2001) wrote a scathing critique of the slapdash way that knowledge has been treated in the empirical new growth theory literature.
37. See Jones (1995a, 1995b); P. Evans (1996); Thore (1996); Kocherlakota and Yi (1997); Dinopolous and Thompson (1999); commentary and overview in Cavusoglu and Tebaldi (2006).
38. See, for instance, Rosenberg (1982); Georgescu-Roegen (1976); and Robinson (1953/1954).
39. See Dosi, Malerba, Ramello, and Silva (2006); Dosi, Llerena, and Sylos-Labini (2006); Verspagen and Werker (2003); Sharif (2006); Amin and Cohendet (2004, Chap. 2); and Albert and Laberge (2007).
40. I am particularly thinking here of the work of Paul David, Richard Nelson, David Mowery, and Sidney Winter. Another example of this ingrained conservatism can be found in Balconi et al. (2009). I discuss these problems in detail in Mirowski (2009).

3. Regimes of American Science Organization

1. Dennis (1997, 1). See also Dennis (2004) and Dickson (1988).
2. See, e.g., Brown (2000); Hollinger (2000); Miyoshi (2000); Newfield (2003, 2008); Krimsky (2003); Monbiot (2003); Washburn (2005).
3. Nelson (2001, 14). For similar assessments: the "growing commercial engagement has not, thus far, altered the research culture of universities, so as to privilege

applied orientations at the expense of basic science" (Owen-Smith and Powell 2003, 1696); "Science today is only a short way down the path to becoming a toady of corporate power" (Greenberg 2001, 3); do we see "universities compromising their core values [?] . . . at least at the major research universities, their revenue-enhancing activities have not seriously distorted such values" (Baltimore 2003, 1050); "There is evidence to suggest that university licensing facilitates technology transfer with minimal effects on the research environment" (Thursby and Thursby 2003, 1052).

4. The first two quotes are from the National Bureau of Economic Research (NBER 1962, 452) and the third is from NBER (1962, 454–455).
5. Named after Russian historian Boris Hessen, this was an argument that the content of Newtonian mechanics owed much to the artisan traditions and economic structures of the seventeenth century. For modern research on the subsequent suppression of Hessen's work and attendant economic considerations, see McGuckin (1984); Chilvers (2003); and Mayer (2004).
6. This problem of this blind spot of science studies approaches to the laboratory has recently been noticed by Kleinman (2003, Chap. 5). There is the further complication that early works such as Latour and Woolgar (1979) and Bourdieu (2004) made extensive use of economic metaphors while essentially ignoring any substantive economic structures. The *curiosum* of sociologists gleefully resorting to economic metaphors while avoiding examination of the economy is nicely discussed in Pels (2005); Hands (2001); and Mäki (2005). Nevertheless, the range of reasons behind the neglect of economic factors by STS in the 1970s to 1990s is itself too complex and cannot be covered in this venue.
7. It is of paramount importance to counter claims such as "the weakening of university-industry research linkages during a significant proportion of the postwar period was the real departure from the historical trend" (Mowery et al. 2004, 195 note 15) and "the so-called Mode 2 is not new; it is the original of science before its academic institutionalization in the 19th century" (Leydesdorff and Etzkowitz 1998, 116).
8. This literature would include Coriat and Orsi (2002); Coriat and Dosi (1998b); Coriat (2002); Coriat et al. (2003); Asner (2006); Tyfield (2006); Nedeva and Boden (2006); A. Johnson (2004); and Pestre (2003b, 2004, 2005, 2007). It is, as the discerning reader will notice, almost exclusively European in provenance.
9. However, see Chapter 6 on clinical trials; and Warwick (2003) on mathematics in Britain. The extension of the regimes analysis to other arenas of science, such as its conduct in government settings, would start with work such as Westwick (2003); and McGarity and Wagner (2008).
10. "The organizational approach to understanding big business developed by Alfred Chandler Jr. has given historians a framework within which to place the research laboratory" (Smith 1990, 121). The Chandlerian approach to business history is best represented in Chandler (1977, 2005b) and is acknowledged in Mowery (1981, 3).
11. "The postwar R&D system, with its large well-funded universities and Federal contracts with industry, had little or no precedent in the pre-1940 era and contrasted with the structure of research systems in other postwar industrial coun-

tries. In a very real sense, the US developed a postwar R&D system that was internationally unique" (Mowery and Rosenberg 1998, 12).

12. See, e.g., Fox and Guagnini (1999); Shinn (2003); Hounshell (1996); Buderi (2000, Chap. 2); Mowery (1981, 1983, 1990); Swann (1988); Smith (1990); Pickering (2005); and Grandin et al. (2004). Yet perhaps it needs to become better known in the science policy community, so as to forestall statements like "During the late 19th and early 20th centuries, practically all research was conducted outside the firm in stand-alone research organizations" (OECD 2008, 19).
13. According to Smith (1990, 124), "broad-ranging research in German dye companies began in 1890 . . . yet these research programs do not appear to have been a model that American companies emulated." As regards the German dye industry, Pickering (2005, 389) suggests that "from the late 1870s, scientific research was itself uprooted from its academic mooring and, for the first time, set down in the middle of industry."
14. A partial exception to this generalization is the agricultural experiment station attached to state land-grant universities from the 1870s onward. Their indeterminate status as scientific research poles is discussed by Charles Rosenberg in Reingold (1979). An attempt to project the extension service onto the industrial sector in the form of a bill to establish "engineering experimental stations" at land-grant colleges in 1916 was easily defeated (Tobey 1971, 40; Noble 1979, 132).
15. See, e.g., Stanley Guralnick in Reingold (1979); and Weart (1979).
16. For some examples, see Mowery and Rosenberg (1989); Smith (1990); and Lamoreaux et al. (2003, 2004).
17. See Wise (1980); Reich (1985); Dennis (1987); Hounshell (1996); Grandin et al. (2004); and Asner (2006). For instance, at the research-intensive DuPont, the first viable commercial products developed within its own labs only materialized in the 1930s (Chandler 2005b, 45).
18. For background, see Mowery (1981, 52) and Steen (2001). This incident was far more consequential for the history of American science than simply jump-starting American chemical and pharmaceutical industries. In 1919 most of the confiscated patents were bequeathed to the Chemical Foundation, which both used its revenues from licenses to support American science journal publications, and later on, helped formed the nucleus of the American Institute of Physics in 1931. This model of the nonprofit semipublic entity that performs some critical science management functions was a peculiarly idiosyncratic American phenomenon, and it will be the subject of some future joint research with Tommy Scheiding. Another example of this hybrid model was the Research Corporation, discussed later in this chapter.
19. The early twentieth-century American stress on suppression of interfirm governance of markets may account for the substantial differences in the German and American climates, which were treated as a puzzle by Mowery (1990, 346). See also Murmann (2004). The banishment of cartels resulted in a different species of corporate lab in America: "IG Farben had no corporate central laboratory comparable to that of DuPont" (Chandler 2005b, 119).
20. See, e.g., the 1921 *Wireless Specialty* case (Fisk 1998, 1176). Furthermore, the metaphor that compared exploration of the laws of nature to exploration for

minerals or gas deposits, which would dominate Kenneth Arrow's Cold War "economics of innovation" (1962), was already present in the 1911 *National Wire* decision (Fisk 1998, 1194). The legal acceptance of the ideal of collective science in the 1920s then dovetails quite nicely with recent claims that the "theoretical" treatments of the scientific community as a distinctly social entity in philosophy and sociology find their origins in the 1930s (S. Jacobs 2002; Mirowski 2004b).

21. See Mowery (1981, 1983); Mowery and Rosenberg (1989, 84–90); and the claim by Mowery (1990, 347) that "independent research organizations do not appear to have substituted for in-house research." This does not mean we endorse his explanation of the phenomenon having to do with Arrovian "shortcomings of market institutions as the mechanism for the conduct and distribution of research and development" (1983, 372). The stunted character of contract research in the first half of the twentieth century is not appreciated by historians like Shapin (2008b, 95), who then make erroneous generalizations about the historical continuity of the outsourcing of scientific research.
22. The American Institute of Chemical Engineers was chartered in 1908 to set themselves apart from the analytical chemists on the one hand and from factory hands and mechanical engineers on the other. It has often been observed that a similar specialization did not develop in Germany until the late twentieth century, possibly because of the initial industrial stress on complex organic processes in Germany, and the absence of an advanced petrochemicals industry (Divall and Johnston 2000, 81). Sharp differences in higher educational institutions must also have played a role.
23. A good example is the way that various philosophical/cultural attempts to reconcile science with democracy or science with industry tended to struggle with Einstein's theory of relativity. For an account of these struggles, see Tobey (1971, Chap. 4).
24. See, for instance, Mowery and Rosenberg (1998); Kevles (1995); Leslie (1993); Kleinman (1995); Lowen (1997); Morin (1993); and Krige (2006).
25. Examples of these sorts of arguments can be found in Capshew and Rader (1992) and Galison and Hevly (1992) and in the otherwise insufficiently appreciated pioneer researcher in economic themes of science organization, Jerry Ravetz (1971).
26. This story has been retold numerous times, usually elevating Vannevar Bush to the status of central protagonist as disappointed midwife. See Zachary (1997); Kevles (1995); Reingold (1995); Owens (1994), and the commentary in Mirowski (2002, Chap. 4) and Mirowski and Sent (2002).
27. See, e.g., Forman (1987); Graham (1985); Hounshell (1996, 47–50); and Asner (2004). As Colonel Norair Lulejian was reported to have said in a 1962 speech, "Can we for example plan and actually schedule inventions? I believe this can be done in most instances, provided we are willing to pay the price, and make no mistake about it, the price is high" (quoted in G. Johnson 2002, 19).
28. IBM stands out from these other corporations as forming its in-house research capacity rather late, creating a "pure science" department only in 1945. On this unusual history, see Akera (2002). The Yorktown Heights facility was opened only in 1960.

29. See Westwick (2003, 51). The policy was not uniformly applied to industrial contractors, but the second-source rule often mitigated any commercial advantage that the firm might enjoy from keeping patent rights. Interestingly, the supposedly public-spirited University of California resisted the AEC rule and was defeated.
30. This is not to say that foundations had no lasting effects in the Cold War regime. The Ford Foundation more or less single-handedly established the dominant model for American business schools in this era (Khurana 2007), while the Rockefeller Foundation played a crucial role in the academic development of molecular biology (Kay 1993).
31. These issues are discussed further in Hollinger (1990); Mirowski (2004b); and Thorpe (2007).
32. See, for economic historians, Lamoreaux et al. (2003, 405). For legal scholars, see Boyle (2000); Lessig (2001, 2004); McSherry (2001). For historians of education, see Geiger (2004, 3); Matkin (1990, 22); Slaughter and Rhoades (2004); Kirp (2003); Apple (2005). For politics, see Krimsky (2003: 30–31); Mirowski and van Horn (2005). Washburn (2005, Chap. 3) documents the political maneuvers leading up to the passage of the Bayh-Dole Act.
33. A good example was the economist Zvi Griliches. Starting from the dubious tradition of the empirical aggregate production function, he asserted that "it is unlikely that the recent productivity slowdown can be blamed primarily on the R&D slowdown" (1980, 347). Later, he became a convert to the neoliberal position that "privately financed R&D expenditures have a significantly larger effect on private productivity and profitability than federally financed R&D" (1986, 153).
34. Interestingly, Chandler himself did not entirely agree with this assessment (see Chandler 2005a). As we have already suggested, the market/hierarchy dichotomy had been stressed in previous neoliberal theories of the firm.
35. See Drahos and Braithwaite (2002); Sell (2003); and Chapter 4 for further description. The identity of these vanguard industries is crucial for our narrative because it is widely recognized that "in most industries, university research results play little if any role in triggering new industrial R&D projects" (Mowery et al. 2004, 31). The activist firms were located in that small group of industries that did make extensive use of academic research.
36. This crisis of confidence was the subject of a whole slew of books starting in the 1990s, the best of which were Readings (1997); Sommer (1995); Rhoads and Torres (2006); Khurana (2007); Kirp (2003); and Newfield (2008).
37. See, e.g., Slaughter and Rhoades (2002); Sampat (2006); Geiger (2004); Miyoshi (2000); Krimsky (2003); Washburn (2005); and Mowery et al. (2004). Mowery et al. (2004, 94ff.) point out the fallacies behind the rhetoric one encounters in European contexts, to the effect that if those countries would only institute their own versions of Bayh-Dole, they would automatically reap untold benefits of escalated technology transfer. Sampat (2006) is a good prophylactic against the wilder claims made for the Bayh-Dole Act, but he is in error when he asserts that "there is no systematic evidence that the growth of the private parts of academic research is negatively effecting the conduct of or returns from public science" (773), as I shall argue in Chapter 6.

38. This sea change is documented and discussed in Anderson (2004); Anderson, Butler, and Juris (2008); Buderi (2000, 2002); Economist Intelligence Unit (2004); Reddy (2000); Chesbrough (2001); D. Berman (2003); OECD (2008); and Markoff (2003).
39. "It was on the morning of Jan. 22, 2007, that Pfizer announced plans to close the Ann Arbor site by 2008, part of a larger company downsizing move. Little other explanation was given. A year later, the site that created Lipitor—the best-selling prescription drug in history—as well as Accupril, Neurontin, Caduet and Lyrica is on track to wind down operations by fall." See: http://blog.mlive.com/annarbornews/2008/01/one_year_later_pfizer_labs_emp.html (last visited August 29, 2008).
40. This includes the cancellation of the collection of multinational corporate R&D data by the BEA described in Chapter 1. The general tactic of buying time by spreading misinformation and obstructing research, both of which prevent any political mobilization against the questionable behavior, has been associated with the "sound science" movement, which is discussed in detail in Chapter 6.
41. The closest we get to complaint is the OECD's meek demurral concerning a "lack of systematic data on MNE's foreign R&D investments" (OECD 2008, 31).
42. Examples of this narrative can be found in Bok (2003); Geiger (2004); Krimsky (2003); R. Nelson (2001); Owen-Smith and Powell (2003); Thursby and Thursby (2003); Kirp (2003); and E. Berman (2008). Economists have been especially prominent in spreading this narrative.
43. See, for instance, Dosi et al. (2006); Abraham (2007b); David (1997, 2006); Douglass (2008); European Commission (2003–2004); OECD (2008); Marginson and Van der Wende (2007); and Vincent-Lancrin (2006).
44. Although national differences in corporate governance structures may still be a factor see Djelic (1998); Djelic and Quack (2003); and Guena, Salter, and Steinmueller (2003).
45. For discussion, see Narula (2003, Chap. 5); Thursby and Thursby (2006b); OECD (2008); Hegde and Hicks (2008); and Anderson et al. (2008).
46. Some of these aspects are considered in Chapter 6; see also Light (2009). "The US enjoys a global role in terms of institutional power that far exceeds its share of scientific output" (Marginson and Van der Wende 2007, 37).
47. For updates, see http://eit.europa.eu/about-eit/at-a-glance/milestones.html and http://www.euractiv.com/en/science/european-institute-innovation-technology/article-164275 (last accessed November 2008).
48. This issue has only begun to receive scholarly attention. See Drori et al. (2003); Guilhot (2005); Marginson and Rhoades (2002); and Jang (2000).

4. Lovin' Intellectual Property and Livin' with the MTA

1. See, for instance, Kapczynski (2008); Boyle (2008); Bessen and Meurer (2008); Kahin (2001); and Orsi and Coriat (2005). It is not as though people have only just realized this. There is a story that the Clinton White House had commissioned a report on the failures of the current patent system back in 1998, but

that it was suppressed due to lobbyist pressures. See http://eupat.ffii.org/papri/whitehouse98/index.en.html (last visited September 4 2008).

2. Some evidence of unhappiness in the tech transfer community can be found in the Stanford TTO as represented by Ku and Henderson (2007, 721): "Emphasis on legal contractual arrangements for the exchange of scientific materials and reagents has now gone too far."
3. The canonical citation is Levin, Klevorick, Nelson, and Winter (1987). See also Cohen, Nelson, and Walsh (2003).
4. The history of this about-face within the Mont Pelerin thought collective is the subject of some work by Rob van Horn and Matthias Klaes (forthcoming). For an example of the new-model neoliberalism, see Kahn and Sokoloff (2001).
5. See for instance Markoff (2008), where the misnamed Patent Reform Act of 2007 is criticized as harming "the little guy" because it sought to limit damage awards in patent infringement suits. Not to fret, though: the last I looked, the bill was essentially stalled in the Senate in 2010. See http://www.patentlyo.com/patent/2010/03/patent-reform-act-of-2010-an-overview.html (last consulted September 2010). The dissemination of disinformation of the above ilk concerning the relationship of IP to ideas can be considered a major instance of a neoliberal "double truth" doctrine, discussed in Chapter 7.
6. The volume of commentary on this case is huge. See Boyle (1996, 2003b, 2008); Kevles (1994, 1998, 2001); Sally Hughes (2001); Dutfield (2002). It is sometimes mistakenly stated that this was the first validation of the patenting of live organisms, but some classes of patents on plants had been granted since the Plant Patent Act of 1930.
7. 126 S. Ct. 2921 (2006). This is discussed in detail in Andrews, Paradise, Holbrook, and Bochneak (2006); Lieberwitz (2007); and Ho (2007).
8. A wonderful example of a neoliberal reform proposal has been to start up a Web site called Peer-to-Patent, a Wiki for "interested parties" where it is claimed the public would voluntarily help vet the quality of patent applications (Schrock 2007). Who needs those feckless bureaucrats, the salaried examiners?
9. See, for instance, Orsi and Coriat (2005); Bessen and Meurer (2008); Boyle (2000, 2008); Lessig (2001, 2004); Dosi et al. (2006); Eisenberg (2003, 2006); Dutfield (2004b); May and Sell (2006); Macdonald (2004). Kapczynski (2008, 837) asserts that even more recently, the Supreme Court has handed down some decisions that reverse the decades-long tide of strengthening patent holders; but I see nothing that fundamentally alters the modern regime. Perhaps she means the 2007 Supreme Court decision *KSR International v. Teleflex Inc.*; see the commentary by John Trembath (2008). I believe a case could be made that this attempt to toughen up the nonobviousness test for patents just makes things worse.
10. 307 F.3rd 1351 (Fed. Cir. 2002). For commentaries on the case see Boettinger and Bennett (2006) and Guttag (2004). For a summary of the decision, see http://cyber.law.harvard.edu/people/tfisher/2002Madeyedit.html (last visited November 2009).
11. 545 U.S. (2005); 125 S. Ct. 2372 (2005). For commentary, see Lieberwitz (2007); S. Rubin (2006); Angrist and Cook-Deegan (2006); Jaquette (2007).

12. The Act: 108th Cong., 21 July, S.15. The only commentary I have seen is Loepky (2005b). The U.S. Government Accountability Office issued a critique of the operation of the first few years of the program in 2007.
13. This history has been cobbled together from a wide range of far-flung sources, primarily because there is no history of the MTA as such. Key texts are Eisenberg (2001); OECD (2006a); Rai (2004); Rodriguez (2005, 2008). It is also based on a series of interviews I conducted with a number of TTOs at major universities, in particular with Joyce Brinton, director of the Harvard Office of Technology Transfer from 1984 to 2003.
14. For example, Gilbert Smith at Duke suggested that he only saw roughly 75 percent of outgoing MTAs at Duke; Catherine Ives at Boston University estimated their office saw less than half of outgoing MTAs; the MIT officer (in 2006) insisted they didn't even monitor the MTAs signed by their faculty. By definition, they could not say with certainty that explicit regulations had been violated. This has dire implications for the statistics reported in Table 4.1.
15. For instance, it has been reported that the Stanford TTO (generally regarded as the pacesetter) only took over responsibility for negotiations for IP rights, including MTAs, in 1987 (Matkin 1990, 131–132). When I asked Joyce Brinton when responsibility for MTAs was shifted to the TTO at Harvard, she said she could not remember.
16. Science studies have long demonstrated that this image of "gift exchange" never corresponded to the behavior of real scientists in real labs (Collins 1992; Mirowski 2004a, Chap. 10).
17. This list is extracted from examples provided in interviews with TTOs conducted in 2006.
18. For some fine examples, see the Columbia unit's annual report (called "Science and Technology Ventures") at www.techventures.columbia.edu (last visited August 25, 2010).
19. The following is based on Marshall (1997b); Kevles (2002); Murray (2006); Blaug, Chien, and Schuster (2004); Rader (2004); Dutfield (2004b). The question of how to respond specifically to the widespread moral outrage about the patenting of living things has provoked much debate and soul-searching among the IP community and within science studies. See Boyle (2003b) and Jasanoff (2005).
20. Rader mentions it only once (2004, 258), and she inaccurately describes it as engaged simply in low-end deskilled mouse production. The only dedicated source on Charles River I have found is at www.referenceforbusiness.com/history2/62 (last visited August 20, 2007). I would direct the reader to this site for a vivid description of what it entailed to produce a "disease-free" animal; see also Birke (2003).
21. Quoted from U.S. Patent 4, 736,866.
22. There was also the further consideration that (at least in the early days) Oncomice, by their very nature, were pretty fragile, and so what you were really selling were skills in building and breeding Oncomice. For more on this, see Murray (2006).
23. Knockouts allow researchers to silence targeted genes by inserting foreign DNA sequences within critical exons by homologous recombination. Among other

uses, this aids in mimicking diseases and can uncover previously unknown gene functions.

24. See http://ott.od.nih.gov/textonly/oncomous.htm (last visited November 2008) for the full document. For commentary, see Blaug et al. (2004). Maria Freire was the signatory for the NIH; Dr. Randolf Guschl was the signatory for DuPont.
25. Quoted in Murray (2006), speaker unidentified. It is interesting that Murray ends up endorsing much of the TTO account of what happened, even though she has uncovered a substantial body of the contrary evidence.
26. The group included Maria Freire (NIH), Joyce Brinton (Harvard), Sandy Shotwell (Oregon Health Sciences), Kathy Ku (Stanford), Fred Reinhardt (Wayne State), and Marjorie Forrester (Maryland).
27. See 60 Federal Register 12771 (1995); also at http://www.autm.net/aboutTT/aboutTT_fedReg.cfm (last visited August 25, 2007).
28. This was recognized early on by Rebecca Eisenberg (2001, 226fn) and later confirmed, for example, on the Science Commons Web site http://sciencecommons.org/projects/licensing/background-briefing/ and by Baca (2006). Again, there are no actual data on this practice, and TTOs are loath to provide breakdowns of UBMTA use.
29. See for a sampling of the earlier vintage J. Cohen (1995); Abbott (2000); Marshall (1997b); Cyranoski (2002). For the latter, see Cukier (2006); Rounsley (2003); and Streitz and Bennett (2003).
30. The incident was more disturbing than it initially sounds, because it became all wrapped up in the war on terror, as explained in the docudrama *Strange Culture* at http://www.strangeculture.net/. See also Park (2004).
31. See Angrist and Cook-Deegan (2006). *The New Atlantis* is the stealth neoliberal publication described in Chapter 1. The endorsement of the neoliberal line on patents originating with Kitch (1977) can be found in Thursby and Thursby (2007b). Wesley Cohen's sentiments were expressed at the Harvard Berkman Center Conference on "Patent Law and innovation in the Life Sciences" in December 2005, where I was in attendance; his endorsement of market-based incentives can be found in Cohen and Sauermann (2007). Richard Jensen has personally informed me of his convictions. Cook-Deegan's and Walsh's echoing of this chorus can be found in Caulfield, Cook-Deegan, Kieff, and Walsh (2006).
32. For evidence, consult their Web site http://www.kauffman.org/ and one of their publications, Gulbranson and Audretsch (2008).
33. *Laboratory Corporation of America Holdings v. Metabolite Laboratories Inc.* et al., No. 04–607, 2006 WL 1699360 (U.S. June 22, 2006). This is described on page 146 above.
34. This point has been made repeatedly, directed at the principal authors of the surveys that assert little harm to the research process, at various conferences I am aware of, but I have yet to see it in print. Jerome Reichman made the point at the December 2005 Harvard conference cited in note 19 above; Rebecca Eisenberg made it at the May 2006 OECD Conference on the Research Use of Patented Inventions (OECD 2006a, 20).

35. This is admitted, to a greater or lesser degree of serious documentation, in Powell et al. (2007); Greenberg (2007); Geiger and Sá (2008); and Newfield (2008, Chap. 12). The AUTM has the bad habit of only reporting gross revenues in its various survey results, as do many individual university TTO reports.
36. Here I explicitly dispute the story retailed by Paul David (2007b, 268) and others that universities ramped up their patent activity due to the rise of biotechnology, or, to put it another way, because "the science made them do it." It was not the scientists who were responsible for promoting the marketplace of ideas. The specific innovation of the biotech model is covered in the next chapter.
37. One starting point would be to consider the composition of those chosen to be members of the National Academy or officers of the AAAS. The level of commercial participation among these groups is bound to be much higher than the mean level of participation of the whole community of scientific researchers, given the effects of decades of commercialization of the university covered in this volume. Another would be to explore the neoliberal connections and convictions of certain officers of the NAS, following up on the work of Oreskes and Conway (2008). This bears directly on the neoliberal production of ignorance described in Chapter 7.
38. See, for instance, Freeman (2007) and Stephan and Ehrenberg (2007). One seldom noted consequence of the disappearance of a majority of Americans from advanced training in the natural sciences is a weakening of the political support for science graduate education in the United States.
39. The U.S. National Science Board has just come round to this orientation (as witnessed in NSB 2008c). Perhaps the only sector with a strongly organized political machine dedicated to opposition to the international strengthening of IP rights is a subset of the NGOs (Dutfield 2003, 211).
40. This is a question that has been shamefully neglected in both economics and science studies. The honorable exceptions to this generalization are Drori et al. (2003); Schofer (2004); and Goldman (2005).
41. The main sources for the following paragraphs are Drahos and Braithwaite (2002); Sell (2003); May and Sell (2006); May (2006); Dutfield (2003); and Bassett (2006). One of the major protagonists of this movement was the economist Jacques Gorlin, now head of the Gorlin Group and the American BioIndustry Alliance. See http://www.gorlingroup.com/whatsnew.html (last visited September 6, 2008).
42. Most people need to be reminded that the patenting of drugs was prohibited in the majority of developed countries until very recently. For instance, drug patents were prohibited in West Germany until 1968 and in Italy until 1978 (Dutfield 2002, 127).
43. Sell (2003, 96). Among the twelve were Pfizer, Merck, Bristol Meyers, Johnson & Johnson, Monsanto, IBM, Hewlett-Packard, CBS, GE, and GM. The predominance of Big Pharma in this list is characteristic of their wider activist role in shaping the new regime of globalized privatization of science. For the delayed growth of resistance in the developing world, see Kapczynski (2008).
44. The latter phenomenon has nicely been covered by Tamar Lewin in a series of articles for the *New York Times* in 2008, and on the outstanding Web site http://

globalhighered.wordpress.com. See also Altbach (2004); Marginson (2007); and Marginson and Van der Wende (2007).

45. Although an interesting literature in science studies denies even that level of unity, see Dupré (1993) and Galison and Stump (1996).
46. This is, of course, a hypothetical postulate, because the last time I checked, patent reform had been stifled in the Congress in 2008. See note 5 above.
47. The extent to which we might attribute this habit to the way modern thought has been mesmerized by the utterly disastrous political philosophy of John Rawls in this regard is something that we must unfortunately pass by in this volume. For some observations on Rawls as a neoliberal theorist, see Amadae (2003).
48. This is the fundamental premise behind the work of most contemporary economists on IP—for instance Scotchmer (2004); Guena et al. (2003); Guston and Kenniston (1994); Huber (2006); and Posner (2005). It is also the main theme of the work of Paul David, and in particular, the model in David and Dasgupta (1994).
49. For instance, see Metlay (2006); Biagioli and Galison (2002); Sherman (1996); McSherry (2001); and Cooper (2008).
50. For instance, the patent as a prospective promise rather than a retrospective engrossment is a key to the rise of the biotech model of scientific research, as discussed in the next chapter. It is also central to the degradation of patent quality discussed in Chapter 6.
51. A fact proudly listed on his curriculum vitae: see www.law.uchicago.edu/files/cv/Posner,%20Richard%20CV.pdf (last visited September 22, 2010). Landes and Posner (2004, 24) quote an early work of Hayek to suggest neoliberals generally opposed strong IP, but this ignores the postwar history of the Mont Pelerin Society, something that Posner himself should be intimately familiar with. See note 4 above.

5. Pharma's Market

1. See, for instance, www.biocompare.com. When I last consulted the site in February 2008, it had a click-through entry: "Look Ma, no hands! Benchtop tasks go automatic." Can full automation and technological unemployment of the scientist class be far behind? Their promotional material states, "Across the board, lab procedures suffer from the phenomenon called protocol creep. That is, different people performing the exact same protocols will have different results. In addition to circumventing the problem, automation helps to overcome that natural human tendency of boredom, which leads to shortcuts. Best of all, it says, 'Machines never get tired.'" They omit to mention that machines also never get curious.
2. Resemblances to some earlier doctrines in the philosophy of science and, in particular, the episode of "logical empiricism," are a bit too close for comfort. On this, see Reisch (2005); Mirowski (2004b); O'Neill (2003, 2006).
3. Of course, initially it is presented not as downsizing and rationalization but rather as an altruistic desire to *help* the scientists (Singer 2009b). It is then

combined with a little neoliberal outsourcing of the marketplace of ideas: "A group from the Sloan School at MIT and Harvard Business School has created *Pharmer's Market,* an online prediction market that uses crowd-sourcing to forecast the likelihood of a drug's success."

4. See, for instance, Kenney (1986); Kevles (1998, 2001); Orsenigo (1989); Robbins-Roth (2000); Owen-Smith and Powell (2003); Powell and Owen-Smith (1998); Cockburn (2004); Quéré (2003); McKelvey, Richne, and Hellman (2004); Owen-Smith, Riccaboni, Pammoli, and Powell (2002); and David (2007b). Some of the worst of this literature celebrates the thoroughly misleading notion that "drug companies began to look and behave more like universities" (Cockburn 2004, 16; Kleinman and Vallas 2006), when in fact, it was the whole sector that was being reengineered from top to bottom, as indeed were the universities.
5. The major texts are by Coriat et al. (2003); Pisano (2006a, 2006b); Mazzucato and Dosi (2006); Nightingale (2000); Tyfield (2006); Hopkins et al. (2007); Loeppky (2005a); Cooper (2008b); and Dutfield (2002).
6. As one might imagine, this serves as a red flag for modern neoliberals. Cockburn and Henderson (1997) attempted to identify the twenty-one "most important" drugs introduced between 1965 and 1992, and they argued that public funds were used to develop fourteen out of the twenty-one. The NIH itself claimed that public research was responsible for the development of the five best-selling drugs of 1995: www.nih.gov/news/070101wyden.htm (last visited September 2010). Neoliberals have sought to argue otherwise (Landau 1973).
7. For more on the economic history of the industry, see Chandler (2005b); Mazzucato and Dosi (2006); and Gottinger and Umali (2008).
8. Sharp cuts in the American pharma workforce were reported in Berenson (2007). Pisano (2006b, 121) reports a sharp drop in Big Pharma R&D spending per new drug by nearly 75 percent from 1985 to 1991. The closure of Pfizer's main research facility in Ann Arbor, Michigan, in 2008 was discussed in Chapter 3, and in http://blog.mlive.com/annarbornews/2008/01. Merck closed its San Diego research facility in 2005: see http://www.signonsandiego.com/uniontrib/20050114/news_1b14merck.html. Alza, a Johnson & Johnson subsidiary company, cut six hundred jobs in drug research, development, and support and closed its office in Mountain View, California, in 2007. Bristol Myers closed two research labs in the Philadelphia area in 2002. The situation has only further deteriorated during the economic contraction of 2008–2009.
9. The minor exception to this generalization was the formation of a few contract research firms in chemistry before World War II. These were discussed in Chapter 3 as having insignificant impact on actual research.
10. See Eisenberg (2001, 229) for a description of various attempts by the NIH panel in 1998 to define the meaning of a "research tool." As she points out, each player was "eager to establish that the term 'research tools' means something other than their own institution's crown jewels."
11. In one estimate provided by the trade group Biotechnology Industry Organization, of the 370 publicly traded American biotechs, 125 had less than six months' worth of cash on hand (Pollack 2008). Because this organization exists to pres-

ent the model in its most flattering light, the situation was probably more extreme than that. The 2008 crisis has exposed the sector's vulnerabilities: "Historically the biotech world has depended on the public markets to be there, and if there is no appetite for life sciences IPOs, that will result in a lack of capital to keep moving new drugs toward FDA approval." http://www.boston.com/business/technology/innoeco/2010/06/(last visited September 2010).

12. This situation of the biotech firm resembles the plight of the curious fly Cecidomyiidae Miastor. In what must be an uncomfortable state of parturition, the larvae hatch and consume their mother from within, before they are even "born." James Costa (2007) calls this the "ultimate Head Start Program." For those less impressed with biological metaphor, it anticipates a regularity discussed in Chapter 6: due to the fat-tailed distribution of scientific outcomes, it is nearly impossible to rationally hedge your bets on the success of a portfolio of research projects.
13. I discuss this in the following section. The main sources on this worrying trend are: Nightingale and Martin (2004); Hopkins et al. (2007); Angell (2004); Aggarwal (2007). Also, here we must insist upon highlighting the distinction between the biotech model and the CRO model in part 2, which has generally been profitable in the same time frame.
14. See Dewan (2009). The implications of the failure of the biotech model are so inflammatory, and so opposed to the interests of so many "new knowledge economy" actors, that the neoliberal think tanks have recently began to mount a counterinsurgency to argue there is no real problem. See Buckley (2007); Caulfield, Cook-Deegan, Kieff, and Walsh (2006); Adelman and de Angelis (2007). The misguided attempt of the OECD to promote the notion of a "bioeconomy" is discussed in Cooper (2008b) and Parry (2007).
15. "Aggregate biotech returns have historically been strikingly mediocre . . . Furthermore, there are large numbers of biotech walking dead—companies that survive without tangible returns to investors . . . Against this backdrop, it might seem puzzling that the capital markets continue to fund the biotech sector" (Booth 2007, 854). Only if you don't compare it to something like, say, the 2008 mortgage-backed securities financial crisis?
16. See, for instance, Shapin (2008, Chap. 8). The notion that neoliberal economists and 1980s-vintage pharma entrepreneurs were fairly intimately connected is explored in Nik-Khah (2009) and Chapter 7.
17. Pisano (2006b, 144) points out that generally accepted accounting principles do not require any disclosure of R&D projects, so basically the investor has to take the entire business plan of a biotech-style firm on faith.
18. This argument is pushed even further in Chapter 7, where I entertain the possibility that the marketplace of ideas can also produce ignorance, often in the guise of wrong or corrupted information.
19. See, for instance, Angell (2004); Avorn (2004); Hawthorne (2003). I review a number of these books in Mirowski (2007c).
20. There is even the possibility that, if one takes a more global view, including Europe, then the appearance of new chemical entities may have been declining for a much longer period (Dutfield 2002, 96). However, another author (Light 2009)

argues that if you normalize new drugs by total amounts of money spent on pharmaceutical R&D, Europe has a much better track record of success from 1982 to 2003 than the United States had.

21. An insightful comparison of the U.S. and German experience (the German pharmaceutical industry predated the U.S. one) can be found in Daemmrich (2004). The attacks by neoliberal economists such as Sam Peltzman and George Stigler are mentioned in Chapter 7. The pivotal role of the FDA standards in the global drug industry is attested in interview excerpts in Getz and de Bruin (2000, 732).
22. However, the International Conference on Harmonization of Technical Requirements for Registration of Pharmaceuticals for Human Use has sought to reduce these requirements. See Abraham and Smith (2003, 88–90).
23. There is another consideration that compromises the figures: namely, the inclusion of imaging reagents for use in diagnostic kits—something we cannot "correct" the figures for here. They do, however, also tend to overstate pharmaceutical innovation over time.
24. An active moiety designates the molecule or ion, excluding those appended portions of the molecule that cause the drug to be an ester, salt (including a salt with hydrogen or coordination bonds), or other noncovalent derivative of the molecule, responsible for the physiological or pharmacological action of the drug.
25. Of course, biotechs also originate small-molecule drugs, as well as a complement of non-NMEs. For data on total FDA approvals of products from biotech firms, see Hopkins et al. (2007, 577). Those figures do not seem to conform exactly with the data I took directly from the FDA site. NMEs are described in detail in the FDA Orange Book http://www.accessdata.fda.gov/scripts/cder/ob/default.cfm. It appears that the FDA does not deign to distinguish "me-too drugs" from novel therapies in the biologics category, further bedeviling my tables (Hopkins et al. 2007, 578).
26. Annual (Priority + Standard) in Table 4.4 = reported (NMEs + biologics) in Figure 5.2.
27. See Buckley (2007); Caulfield et al. (2006); Adelman and de Angelis (2007); Epstein (2006); Huber (2006); Owen-Smith et al. (2002); Agarwal and Searls (2009).
28. See Carey (2008). For a recall timeline, see *Drug Recall News* at http://www.consumerjusticegroup.com/drugrecall/ (last visited September 20, 2008).
29. This literature was kicked off by Heller and Eisenberg (1998) and has since reached imposing proportions: see Rai and Eisenberg (2003); Boyle (2003a, 2008); Eisenberg and Nelson (2002); Eisenberg (2006); Ostrom and Hess (2007). The main problem with this literature is it attempts to turn the concepts of "commons" and "public good" into an orthodox economics defense of open noncommercial science, something I argued was futile in Chapters 2 and 3. Because this default ends up ushering the argument down the primrose path, one subsequently tends to find dismissals of its importance, such as "the evidence for a growing biomedical research anticommons that can stifle biomedical research is almost non-existent" (Bailey 2007) or "existing empirical studies find few clear signs that the patenting of biotechnology inventions is adversely affecting biomedi-

cal innovation" (Adelman and de Angelis 2007, 1681). This chapter stands as refutation of that latter statement.

30. Big Pharma seems to have cooled its ardor for collaborative research projects of late with biotechs and seems more inclined to buy up the few demonstrably profitable biotechs left. In 2008, Roche snapped up Genentech, and Bristol-Myers bought ImClone (Saul and Pollack 2008; Huggett 2008).
31. One of the great all-time cautionary tales was the disaster at Boston University with its erstwhile president John Silber and his pet biotech startup Seragen (Barboza 1998). Some historian is still needed to do this story justice. Until then, suffice it to say that Silber not only sunk a large proportion of his institution's modest endowment into equity in the firm, but he sucked trustees and other colleagues into risky personal investments as well. When the firm went bust, as most biotechs do, recriminations ran wild, and BU lost more than 90 percent of its investment. Silber's legacy is thus a compromised BU. Where are such moral tales in the current go-go climate of inexperienced universities lusting after equity in their professors' start-ups?
32. There is as yet no serious history of the CRO phenomenon. As Shuchman (2007, 1365) says, it has attracted surprisingly little attention. Some interviews with various early figures appear in Petryna (2009), but there is little attempt to weave them into a coherent narrative. Actually, I shall use the term "CRO" to refer to what is in reality an entire ecosystem of partially differentiated for-profit firms that carve up the clinical trial into many possible subsets: site management organizations, central patient recruitment companies, study brokers, bioinformatics firms, medical writing firms, and so on. An elaborate business history awaits its (for-profit?) historian.
33. The main source of statistics is CenterWatch (2008), the product of a for-profit firm, with an unapologetic pro-CRO bias. Secondary sources are BioPharm Publishing and the trade organization ACRO. This situation is yet another illustration of the current regime's restrictions on sources of knowledge concerning the operation of its science organization and management to those willing and able to pay. It also brings to mind the similar opacity of the shadow banking system, brought into the twilight by the crisis of 2008–2009. Some indication of the inability of the FDA to monitor clinical trials can be found in the Inspector General's report (OIG 2010).
34. Enthusiasm for this approach has spread throughout much of clinical R&D but has yet to demonstrate substantial benefits. See Hedgecoe and Martin (2003); Hopkins et al. (2007, 573).
35. Tufts Center for the Study of Drug Development (2006). Because there is no such thing as a global record of clinical trials, estimates do vary. By comparison, CenterWatch (2008, 92) estimates 53,000 Phase I–III trials conducted in 2004, and 59,000 in 2006. One thing is clear: Trials conducted in the United States have been roughly flat since 2001, while those overseas have grown dramatically. Nevertheless, it is entirely likely no one really knows how many clinical trials are being conducted at any one time (Petryna 2007, 31; OIG 2010). The legal loopholes to avoid disclosure are just too numerous (Lurie and Zieve 2006).

36. AHCs are predominantly university teaching hospitals that conduct clinical drug trials. Here, we presume that CROs and AHCs are discrete mutually exclusive categories, as they were in the 1980s. The evidence presented later in this chapter raises a disturbing possibility that, as commercial pressures at universities have rendered them more similar, they also are becoming less distinct as separate entities.
37. With regard to an earlier version of this chapter, Nick Rasmussen asked me in private communication, "Why regard the decade or two of post-Kefauver vigilance, ending with Reagan in 1980, as representing the baseline natural relationship between drug companies, academia, and regulators? What if one chose to regard the mid-60s through 1970s as aberrant, and the preceding two decades as the natural baseline?" After all, it has been estimated that approximately 90 percent of all drugs licensed prior to 1970 were first tested on prison populations (Petryna, Lakoff, and Kleinman 2006, 38). The FDA only outlawed this practice in 1981, so the AHC cannot be imagined to have enjoyed a long reign dominating clinical trials. The short answer is that we treat the AHC as an important institutional component of the Cold War regime, and we approach the CRO as a harbinger of the globalized privatization regime. Neither relationship was "natural" in any coherent sense.
38. The preeminent advocate of this position is Nicholas Rasmussen. See, in particular, Rasmussen (2002, 2004, 2005). Another representative would be Harry Marks, who nevertheless has pointed out that up until our modern period, those whom he calls "therapeutic reformers" stigmatized "those who operate in profit-making institutions . . . as operating on the edges of, if not outside, the boundaries of science" (1997, 234).
39. See Getz and de Bruin (2000); Kaitin and Healy (2000); CenterWatch (2008). The FDA now posts mean durations on their Web site (www.fda.gov).
40. See, in particular, Rettig (2000), Pichaud (2002), Davies (2001), Gelijns and Their (2002), and Azoulay (2002).
41. "For such drugs as antibiotics for acute infections, large populations and long timelines are seldom needed to establish efficacy and safety. With the new emphasis on prevention and treatment of chronic diseases, however, clinical drug research has changed. Many people take antihypertensive drugs and lipid-lowering drugs for many years in order to prevent relatively few undesired clinical endpoints" (Bodenheimer 2000, 1539).
42. See, for instance, Stephens (2000), Lemonick and Goldstein (2000), and Shah (2003, 2006). More scholarly works in this vein are Petryna (2007, 2009); Petryna et al. (2006); J. Fisher (2006, 2008).
43. "The varieties of roles that bioethicists inhabit . . . complicates the question of corruption, because it is not clear what duties and loyalties are expected of them . . . Until recently, students studying the ethics of stem cell research would not have suspected that their teacher was a consultant for Geron; scholars criticizing industry-sponsored clinical trials would not have imagined the editor evaluating their manuscript was working for Eli Lilly; newspaper readers would not have thought that the ethicist commenting on genetic engineering was drawing a pay cheque from Celera" (Elliott 2002, 36). See also the papers by Foster and Corrigan in Abraham and Smith (2003).

44. The anthropologist Jill Fisher has managed to get a CRO coordinator to admit this directly: "It's just getting that thing in your head that it's *not* a patient-doctor or nurse relationship. It's a *participant-research* relationship . . . We're doing medical tests and they're still expecting medical treatment *appropriate* for their conditions" (2006, 684).
45. This issue received wide publicity with the death in 1999 of Jesse Gelsinger in a gene therapy trial at the University of Pennsylvania. James Wilson, the principal investigator in the study, held a 30 percent stake in Genovo, which owned the rights to the drug Wilson was testing. Investigations into patient deaths in clinical trials in other cities also raised the issue of whether experimental subjects realized that they were risking their lives, given the confidentiality arrangements imposed by the commercial ties of their physicians (see Wilson and Heath 2001).
46. For instance, Petryna (2009, 18) reports that when Poland started to implement stricter EU guidelines concerning human subjects in clinical trials, her unidentified CRO just shifted its activities to Russia and the Balkans.
47. See Stephens (2000), Shah (2003, 2006), Flaherty, Nelson, and Stephan (2000), DuBois (2003), Stone (2003), Cooper (2008a) and especially Petryna (2007) and Petryna et al. (2006).
48. See DeYoung and Nelson (2000); Pomfret and Nelson (2000); Stephens (2000). Foreign trials offer the freedom to adjust the protocols to achieve the preconceived answer one seeks, as was admitted by a CRO scientific officer: "In my recruitment strategy, I can use subject inclusion criteria that are so selective I can 'engineer out' the possibility of adverse events being seen. Or, I can demonstrate that my new drug is better by 'engineering up' a side effect in another drug . . . That is the big game of clinical trials" (Petryna 2007, 27).
49. The variables that regularly enter into the "auxiliary hypotheses" in drug trials are discussed in Bero and Rennie (1996) and Morgan et al. (2000). A nice online lecture by Dr. Lisa Bero explaining the ways this works is: http://fora.tv/2009/02/04/Dr_Lisa_Bero_Bias_in_Drug_Trials (last visited September 2010). Those familiar with the philosophy of science literature will recognize the general problems with falsification often discussed under the rubric of "Duhem's Thesis."
50. See Davidoff et al. (2001) and www.icmje.org for the actual guidelines. A reflection on their motivation is Frank Davidoff, "Between the Lines," at http://content.healthaffairs.org/cgi/content/full/21/2/235 (last visited September 2010). These incidents are also described by Krimsky (2003).
51. There is a small literature in the philosophy of science that attempts to maintain that marketplace models of science can demonstrate that self-interested biases need not or will not impugn the quality and integrity of the knowledge produced under such circumstances. Undoubtedly much of this literature has been prompted by phenomena such as described in this section. Some of these arguments may be sampled in Mirowski and Sent (2002); this literature is criticized in Mirowski (2004b).
52. This innovation was directly relevant to the later disavowal of the "research exemption" in patent law for universities in the case of *Madey v. Duke University,*

discussed in Chapter 4. For further insight, one might consult www.dcri.duke.edu.

53. As described by Azoulay (2003), curiosity leading to appropriable IP is consciously engineered out of the system in a CRO. This was put in a colorful fashion by a financial officer of a pharmaceutical firm: "Our purchasing department uses a matrix where relationships can be described anywhere from a continuum that goes from 'used-car salesman' to 'we're married' kind of thing. For the used-car salesmen, we try to squeeze whatever we can out of the price, and we don't care if they go out of business, we do not care if they lose money, we are just trying to get the best deal we can . . . And we are more on the used-car salesman end of the spectrum. I think that's the case for most sponsors. I think that Merck and Pfizer are even tougher with the CROs than we are" (16).
54. Moffat and Elliott (2007) quote a 2000 Pfizer sales document: "What is the purpose of publication? . . . High quality and timely publications optimize our ability to sell Zoloft most effectively." The impact of the monopolization of scientific publication by a small number of companies is another topic in the new economics of science that has languished, in the sense that economists seem incurious about how it might alter the character of the science produced. Unfortunately, we must postpone its sustained examination to another day. But for a convenient timeline of the concentration of scientific publishers, consult http://www.ulib.niu.edu/publishers (last visited September 19, 2008).
55. The chronology of events can be found at www.councilscienceeditors.org (last visited September 2010). For further chronological developments, see http://www.icmje.org/update.html (last visited September 2010).
56. See Giombetti (2002); Healy (2003). Perhaps the first revelations of medical ghostwriters surfaced in court cases related to the diet drug "fen-phen" (actually a combination of fenfluramine, dexfenfluramine, and phentermine). Company documents subpoenaed from the producer Wyeth-Ayerst Laboratories revealed that it had commissioned Exerpta Medica, Inc., to write ten papers concerning the drug, two of which were subsequently published in refereed medical journals under the names of prominent researchers, one of whom claimed in testimony he had no idea that Wyeth had commissioned the paper (Zuckerman 2002). But each new round of plaintiff suits in pharma cases brings new evidence of ghostwriters. See the recent imbroglio concerning Vioxx (Ross et al. 2008).
57. See the links on http://post.queensu.ca/~sismondo/page1/page1.html (last visited September 2010). The circumstances of the case, and a diagrammatic guide as to how to ghostwrite a paper, were presented in Singer (2009b).
58. CROs had openly advertised these services on their websites, as in, for instance, the site of Parexel at www.parexel.com/products_and_services (last consulted May 2003), where it was stated, "PAREXEL medical writers provide ghost writing services both as stand-alone projects and as part of a larger scope of PAREXEL services." CROs have since become a bit more subtle in how they organize and advertise these kinds of services, as explained further in the chapter.
59. As I write this chapter in 2008, we in America have been living through a Red-Queen style race between disclosure and secrecy, where both the ICJME and Congress have been trying to move away from the industry's preferred approach

to a system of "voluntary" registrations at the government's site www.clinicaltrials.gov, and to make registration mandatory. The ICJME attempted this through an editorial policy statement in 2005 that it would only publish studies based upon trials which had been previously registered, and Congress through the FDA Amendments Act of 2007, which expanded the scope of the registration mandate. On these developments, see Lurie and Zieve (2006) and Zarin and Tee (2008). The industry and especially the CROs have responded with enhanced innovation in data control and offshoring of trials aimed at evading or otherwise neutralizing these mandates, usually through appeals to multiple jurisdictions and conflicting IP law. The gross number of registrations have indeed increased in the last year or two, but we still have no idea of how many studies evade these regulations. This is just one more instance of how globalization serves to undermine nationalist conceptions of "reform" of science and thwart knowledge about the operation of the new regime.

60. As one might expect, representatives of the industry take umbrage at anyone pointing out that CROs have utterly undermined older notions of "science." Douglas Peddicord, the director of the Association of Clinical Research Organizations, reacted to earlier versions of these arguments by stating that "CROs were found to have acted appropriately and well within regulatory limits . . . One of the things we certainly reacted to . . . is the assertion that this kind of specialized expertise has begun to kill clinical research. It was absolutely unwarranted and there is no empirical data provided" (DeSantis 2007). The way in which the CRO industry falls back on the state to define "legitimate behavior" of the lowest common denominator (such as "good clinical practice") in dodgy circumstances illustrates another key tenet of the neoliberal playbook.
61. Again, Rasmussen (2004, 2005) suggests that an indeterminate amount of this disguised advertising in the form of "seeding trials" has occurred for more than a century. The point here is simply that the CRO facilitates and stabilizes a practice that might be more difficult to prosecute on an industrial scale in AHCs.
62. See, for instance, the launch of 3Sbio, http://bbs.3sbio.com/en, in February 2007, on the NASDAQ. WuXi PharmaTech and Simcere Pharmaceutical were also launched on the NYSE in 2007. See Frew et al. (2008). Currently, the three largest CROs in China are reported to be WuXi PharmaTech, ChemExplorer, and Pharmaron. Other Chinese CROs familiar to Western firms are Chemizon, BioDuro, and Medicilon. Founded in 2000, Shanghai-based WuXi PharmaTech is apparently the leading China-based pharmaceutical and biotechnology CRO. In 2006, WuXi PharmaTech provided services to seventy pharmaceutical and biotechnology customers, including nine of the top ten pharmaceutical companies in the world, as measured by 2006 total revenues, as in Table 5.1 (from its IPO prospectus in the United States, 2007). A track record of working for Merck, Pfizer, and AstraZeneca, among others, has enhanced its international credibility, in an environment that is often suspected of cutting costs by cutting corners on the quality of the data produced, and having a relaxed attitude to the protection of IP. Other firms, such as Shanghai Genomics, offer intercession to help gain regulatory approval for drugs from the Chinese government.

6. Has Science Been "Harmed" by the Modern Commercial Regime?

1. *Science* magazine, which likes to present itself as an advocate of open publication, in fact has repeatedly condoned the practice of publishing papers without disclosure of the underlying proprietary data, first in the case of the human genome, as explained in this chapter in note 25, and then in the case of the *indica* rice genome reported in the April 5, 2002, issue, where Syngenta withheld disclosure of the actual DNA sequence. Daniel Koshland, the editor, had himself been one of the earliest and staunchest advocates for the biotech model at the University of California—Berkeley, one of the first public universities to take the tech transfer plunge (Jong 2008, 1272–73).
2. The first salvo in this dispute was Mooney (2005), who has been followed by Wagner and Steinzor (2006); Shulman (2006); Steinzor, Wagner and Shudtz (2008) and the Union of Concerned Scientists (2007; 2008). The Web site http://www.waronscience.com/new.php tracks subsequent developments. An attempt by the Democratic Party to claim the high ground before the 2008 elections was the report by the Committee on Oversight and Government Reform, *Political Interference with Climate Change Science under the Bush Administration*. December 2007, http://oversight.house.gov/documents/20071210101633.pdf (last visited November 2009).
3. Senator Charles Grassley has recently gone on the warpath in an attempt to embarrass the NIH about its nonchalance about conflicts of interest. See Harris (2008).
4. For those warning of decline, see Broad (2004); Washburn (2005); Hicks (2005); Lemonick (2006); Guess (2007); Lederman (2008b); S. Moore (2009); Mooney and Kirshenbaum (2009). Among those who have sought to quell our fears of the worst are Greenberg (2007); D. King (2004); Bok (2003); S. Levin (2005); Owen-Smith (2006); Baltimore (2003); and Shapin (2008b).
5. See Godin (2006b) on the birth of bibliometrics, where we learn that it was the psychology profession that first propounded their salience and carried out these kinds of exercises. As one might suspect, these were motivated by a desire to accrue legitimacy for their own fledgling science. The use of bibliometrics to discuss the economics of the natural sciences was greatly popularized by Derek de Solla Price 1963, 1976) More recently, it has been the subject of a dedicated issue of *Nature* (June 17, 2010).
6. Thomson Scientific's flagship products and services include the *Derwent World Patents Index*, *Medistat*, *Current Drugs and Micromedex*, as well as its *Web of Science*, which in 2005 provided access to information from roughly 8,700 research journals. Thomson Scientific enjoyed a 2005 turnover of just over $500 million, of which 83 percent came from electronic software and services. The *Web of Science* derives its depth of historical coverage from what was formerly Garfield's ISI and has given Thomson Scientific a headstart in the innovation of commercialization of the online world of scientific citations across different sectors and client groups. The pharmaceutical sector is hugely important to Thomson Scientific; for instance, GlaxoSmithKline is its largest customer. A great deal

of Thomson Scientific's recent emphasis on developing new products and services has occurred with its pharma customers foremost in mind. As for universities, a subscription to *Web of Science* is often more useful to the administration for retrospective surveillance than it is for the individual faculty for prospective research purposes, especially in its 2008 configuration more attuned to blocking access to articles from for-profit scientific journals without an institutional subscription. Google Scholar delivers a wider range and more nuanced network of citation data (and in certain respects with more accuracy) in a more convenient and timely fashion. The pharma connection may explain why, only in 2008, Thomson opted to incorporate a freestanding non-English language citation database. The *Chinese Science Citation Database* (*CSCD*) is the first non-English content to appear on the *Web of Knowledge* platform. *CSCD* contains more than 1,000 scholarly science publications, more than 1.3 million records, and is searchable from within the Web of Knowledge.

7. See, for instance, http://info.scopus.com/scopus-in-detail/facts/ (last accessed November 2009). Not unexpectedly, Scopus is skewed toward Elsevier and Springer/Kluwer titles but also seems less biased toward Anglo literature sources. Of course, some of those are Elsevier-sponsored ghost journals, as described in Chapter 5.
8. The incorporation of the *CSCD* [see note 6 above] in 2008 inflates this number further. This has rendered the data on the explosion of Chinese publications in Adams, King, and Ma (2009) somewhat ambiguous. Indeed, since 2001 Thomson has roughly doubled the size of its core set, partly in response to competition from Scopus.
9. Much of this information was provided by Professor Diana Hicks in personal communication, who used to work for CHI, Inc., but who now is an academic at Georgia State. In the CHI days, it seems Francis Narin was the primary figure who insisted that NSF produce indicators using a fixed journal set to avoid artifacts of this nature. However, NSF indicators dropped this method when he was out of the picture, and CHI seemingly now counts in its time series whatever Thomson provides.
10. One of the few attempts to explore this profound aspect of the economics of science at an analytical level is Scheiding (2006); another is European Commission (2006).
11. The economic history of these companies has been conveniently summarized by the librarian Mary Munroe at the Web site http://www.ulib.niu.edu/publishers/. As she writes there, "A collection development colleague of mine once said that he had a recurring dream. In his dream he awoke to find that there was only one journal and it cost $5 million. A look at the shrinking number of companies that are displayed here could cause one to think that my friend is not too far from the mark."
12. The journal *Physical Review Letters* has a lower acceptance rate than the other *Physical Review* journals cited in the text, but the "rapid telegraphed" publication model does not have a serious counterpart in the economics profession in terms of status, and therefore I have opted for the comparison displayed above.

13. The way that this curious journal structure, combined with a mania for precise journal rankings over a large expanding universe, serves to keep heterodoxy out in the cold is discussed in Davis (1998) and Hodgson and Rothman (2001). From the vantage point of a new economics of science, the influence of the dominant neoliberal notion of a marketplace of ideas within the economics profession upon scientific publication behavior is nicely illustrated in this differential "market structure" of its journals.
14. The self-referential irony here feels like it requires some sort of comment. The recent initiative by the NSF to revive the old dream of a collective "science of science and information policy" and bring collective authorship to study of the modern predicament of science is just one more symptom of the dissolution of the scientific author. See www.nsf.gov/pubs/2008/nsf08520/nsf08520.htm (last visited September 2010). The science studies literature on the modern transformation of scientific authorship is nicely represented in Biagioli and Galison (2002); McSherry (2001); and Birnholtz (2006).
15. In an amazing exercise in the sociology of science, all the more striking because of its appearance in an incongruous journal, Birnholtz (2006) points out that at CERN one exceptionally diligent physicist claimed to have read all but two of the 250 papers that listed him as an author. The very idea you would *not even have read what you putatively wrote* shows how authorship in the new regime bears almost no relationship to old-fashioned notions of responsibility, ownership as related to effort, or personality as displayed by expression.
16. The NSF now hires another private firm, ipIQ, Inc. (formerly CHI), to perform these database manipulations on the Thomson database. It has occurred to me that one reason the median scientist has remained relatively unaware of the trends recounted in this section is that academic tenure and promotion cases are generally based on the whole-count method, which, of course, grossly overstates the general expansion of the science literature. One wonders whether a movement for "evidence-based tenure and promotion decisions" might take hold if the biases in whole-count bibliometrics ever arose in academic circles.
17. See, for instance, Mervis (2007); NSF SRS (2007); NSB (2008a, 5–36); Guess (2007). One example of how the NSB companion (2008b) broke ranks with this consensus was its willingness to discuss "declines" in article output.
18. We have already encountered the politics of for-profit CROs in telling their pharma clients what they want to hear as a major theme in Chapter 5. Why shouldn't it happen as well with bibliometrics?
19. This analysis has been seconded by a few others, like Richard Freeman (2007). Professor Hicks has also suggested to me in private correspondence that declines in scientific output and share have been noted by analysts for the United Kingdom and Australia without their respective national science policy agencies trying to mask or otherwise hide the phenomenon.
20. Another recent book (Cross and Goldenberg 2009), while stressing the ambiguities in definitions of job categories, does document the expansion of the resort to adjunct faculty at elite universities as well from 1995 to 2005.
21. This is a fascinating question, which deserves its own chapter. In lieu of that, one could begin by consulting Altbach and Levy (2005) and the informative Web site http://globalhighered.wordpress.com/ (last visited September 2010).

22. Perhaps defenders of the economics profession would insist that they can easily handle quality variations in commodities as well; but I side with that unsung genius the late Nicholas Georgescu-Roegen, who, in his amazing book *The Entropy Law and the Economic Process* (1971), explained why their presumptions were groundless.
23. The modern advocates of "social epistemology" might feel this critique was misplaced, but I have my doubts. These issues were aired in the symposium surrounding Mirowski (2004b). One exception to this generalization is Kourany (2010).
24. Gary Becker (1971, 129ff.) praises the salutary effects of the "race to the double helix," without knowing much about the actual circumstances of that curious episode.
25. The available accounts of the contest between Celera and the public HGP are strikingly biased in favor of the neoliberal version of events, which praises the incursion of for-profit activity into the public project. The bias may be due to the fascination of the outsiders with the swashbuckling public persona of Craig Venter. See Shreeve (2004); Shapin (2008a, 2008b); Cook-Deegan (2003); Levin (2005). For more balanced accounts, see Olson (2002); Bell (2003).
26. This information comes from the business press; in particular, the *Washington Post* of March 1, 2008, page D01. Further problems from the breakup were admitted: "Celera confirmed in its filing that it would trade on the NASDAQ under its current symbol, CRA, with its current tracking stock becoming delisted from the New York Stock Exchange. Celera President Kathy Ordonez is expected to continue to use Alameda, California as the company's main base. While the separation makes business sense in terms of allowing each company to form specialised niches that play to their productive strengths, the allocation of intellectual property rights between the firms could potentially harm Celera's business." http://www.labtechnologist.com/Industry-Drivers/Celera-turns-up-heat-on-intended-Applera-split (last visited October 3, 2008). See also https://www.celera.com/celera/history (last visited September 2010).
27. Insiders had their doubts from the get-go: "conditions were ripe for a cream-skimming effort by a new-economy company seeking to harvest IP from the human genome . . . Less clear, was how such an initiative could be embedded in a viable business plan. But this was the late 1990s, and business plans did not need to be viable to attract billions of dollars in private investment" Olson, (2002, 933). Yet credulous writers in the generalist press continue to write as though Celera were some kind of glowing success story: "Venter is a hugely ingenious scientist, but his greatest originality has probably been in the design of the new arrangements for doing genomic research . . . turning Celera Genomics into the Bloomberg of biology. (On that, Venter turned out to be mainly right.)" Shapin, (2008a, 7). One reads this, and weeps for the future of science studies.
27. See, for instance Cheung et al. (2003); Waterston, Lander, and Sulston (2002); She et al. (2004). Admittedly, subsequent innovations in bioinformatics have reduced this disadvantage, but there is as yet no real solution to the problem of errors introduced due to long repetitive stretches in the human genome.
28. This asymmetry existed even up to the so-called publication of the genome (Venter et al. 2001). When *Science* magazine agreed to give Celera publication

credit without actually providing the full sequence to researchers, the HGP withdrew its paper and published in *Nature* instead. Some scientists accused *Science* of sinking to merely selling ad space to Celera (Shreeve 2004, 361): yet another instance of how the AAAS has actually acted as facilitator of the privatization of modern science. I should also warn the reader about the outrageous bias also evident in Shreeve (2004, 367), who argues the quality of Celera's genome was "better" than that of the HGP when the tie was declared. Because there was no independent comparative evaluation possible at that time, his statement is misleading at best.

29. See Quackenbush (2001). A detailed comparative retrospective evaluation was carried out by academic biologists, and the Celera draft was then pronounced inferior (Waterston, Lander, and Sulston 2003; Li et al. 2003; She et al. 2004). In 2004, Francis Collins was quoted in *New Scientist* as insisting, "If we want to finish genome sequences for organisms besides the human, one cannot just count on the shotgun method to do that correctly, at least in its current form."
30. Bob Waterston, quoted in "Celera Defends Human Genome Sequence," *Genome Biology,* February 20, 2004.
31. Craig Venter, interview in *Bio-It World,* http://www.bio-itworld.com/archive/11202/horizons_venter.html (last visited November 2009). The publication of Venter's autobiography, *A Life Decoded,* just provided the pretense for another round of such blather in the public press: "Commercial goals of some kind had to be achieved if genomic knowledge was going to cure sick people and extend life. Bioscience needed capital as much as capital needed bioscience" (Shapin 2008a, 7).
32. http://www.aps.org/publications/apsnews/200301/guidelines.cfm. (last visited September 2010). For some background on the case, see Monastersky (2002); G. Johnson (2002). Indeed, the commercialized regime has prompted many professional scientific societies to adopt ethical codes for the first time in their histories (Kourany 2008, 770).
33. See also the work of Evans (forthcoming) on this issue.
34. The documents can be found at http://tobaccodocuments.org, (last visited September 2010) one of the few truly useful outcomes of the settlement of the lawsuit. Some of the best sources commenting on these documents and events are Glantz et al. (1996); L. Friedman, Daynard, and Banthin (2005); Michaels (2006, 2008); Baba, Cook, McGarity, and Bero (2005); Cook and Bero (2006); Union of Concerned Scientists (2007); McGarity and Wagner (2008). Background on Hill and Knowlton can be found in K. Miller (1999).
35. Brown and Williamson memo, "Smoking and Health Proposal" (snapshot_bw 0000332501), dated 1969, http://tobaccodocuments.org/bw/332501.html (last visited September 2010).
36. The Manhattan Institute was founded in 1978 by Anthony Fisher (the man also responsible for the IEA) and William Casey. Some historical background can be found in O'Connor (2008).
37. http://web.archive.org/web/19980112135500/www.tassc.org/about.html.
38. The main sources for this paragraph are http://tobaccodcuments.org, but see also Thacker (2005); L. Friedman et al. (2005); Mooney (2005); Union of Con-

cerned Scientists (2007). The journalist Paul Thacker was fired by the American Chemical Society for looking too closely into the relationship between mirror-world promoters like Milloy and the AAAS, and the role of the chemical industry in supporting astroturfed political interest organizations (Thacker 2007)—so much for our vigilant professional scientific societies assiduously protecting academic science.

39. For further discussion on these legal developments, see Wagner and Steinzor (2006); Mooney (2005); Jasanoff (2008); L. Friedman et al. (2005); Cook and Bero (2006); Steinzor et al. (2008); McGarity and Wagner (2008).
40. One pertinent example can be found in the article by Kalman Applbaum in Petryna, Lakoff, and Kleinman (2006), which documents how American drug representatives brutally disparage Japanese approaches to pharmaceuticals as "junk science" because they do not conform to U.S. drug regulatory practices, particularly with regard to SSRIs like Prozac and Zoloft.
41. The truly disturbing theoretical implications of this statement are explored in Chapter 7.
42. See Baba et al. (2005); Cook and Bero (2006); Mooney (2005); Shulman (2006); Wagner and Steinzor (2006); Steinzor et al. (2008); McGarity and Wagner (2008).
43. See Raloff (2008); Boden and Ozonoff (2008); Jasanoff (2008); Haack (2008).
44. See, for instance, Boldrin and Levine (2005, 2008); David (2006); Murray and Stern (2007).
45. See, for instance, the following sites: the Lessig blog, http://www.lessig.org/blog/; Against Monopoly, http://www.againstmonopoly.org/index.php?limit=&chunk=0&topic=Patents%20(General); Jim Bessen's Patent Failure, http://researchoninnovation.org/dopatentswork/; and my personal favorite, Patently Silly, http://www.patentlysilly.com/ (last visited November 2009). The movies have yet to catch up, as usual, still mired in portraying intellectual property as the last hope of the rugged individual inventor, as in *Flash of Genius* (2008).
46. This point is made with great care by Daines (2007). It never ceases to amaze (me, anyway) the extent to which neoliberal economists just cannot grok the fundamental contradiction in their curious catechism that science and the economy are in some fundamental way independent, but yet that market prices represent the relevant gist of all knowledge.
47. Unlike in the European patent system, where a special procedure exists for external challenges to patents before they are awarded and is built into the examiner system.
48. Mowery was a Stanford PhD, and is now a faculty member at the Berkeley business school. Ziedonis was a Berkeley PhD. Both based their case on proprietary data not available to other researchers. This is another manifestation of the creeping conflict of interest that extends well beyond biomedical research in the modern commercial regime. These authors' defense of Berkeley is essentially uncheckable. By the way, one hastens to point out we are talking about the very same eminently vigilant and competent University of California, Berkeley, campus, which was the site of the Ninov incident, not to mention the Novartis debacle (described in Rudy et al. 2007).

49. Similar exercises have also been performed for some European countries, although with smaller data samples. See Sapsalis and Navon (2006) for Belgium, which ignores the rather different structure of patent examination at the EPO.
50. See, for instance, Mowery et al. (2004); Mowery and Rosenberg (1998). It is important to note that the authors could not *replicate* the original Henderson-Jaffe regressions, due to their different benchmark random sample; this inability to replicate statistics is one of the banes of econometric research. On this general point concerning replication, see Mirowski (2004a, Chap. 10).
51. Indeed, in the midst of the confusion of the economic crisis of the fall of 2008, neoliberals proposed and Congress quietly passed the strongest legal fortification of IP yet, in the form of the "Prioritizing Resources and Organization for Intellectual Property Act" (Hess 2008). Among other things, it creates a cabinet-level post for IP coordination. Rarely has bald retrogression passed so thoroughly unchallenged under the oxymoronic rubric of "reform."
52. Examples of this "new wisdom" concerning patents are Farrell and Shapiro (2008); Shapiro (2004); Mills and Tereskerz (2007). The height of neoliberal presumption came from Mark Lemley and Bhaven Sampat, who suggested it was futile for the government to "throw money" at the problem, and they proposed to *weaken* the formal standards of patent acceptance even further and then offer applicants a commercial service that would supposedly seriously review and sanction their validity. See "What to Do about Bad Patents" (2006) at http://repositories.cdlib.org/cgi/viewcontent.cgi?article=1017&context=bclt (last visited October 6, 2008). Talk about "science for sale"!
53. One example would be Derwent Innovation, which links Thomson ISI data with U.S. and European patent data, and is owned by Thomson-Reuters. Other for-profit services are Delphion and Patent Café. See http://www.researchinformation.info/riaut02patents.html (last visited September 2010). Some of this is discussed in Leydesdorff (2004). It goes without saying I have been unable to access any of these services.
54. Some meditations upon this phenomenon can be found in Scherer and Hartoff (2000) and Scherer, Hartoff, and Kukies (2000).
55. Consult Chatterjee, Yarlagadda, and Chakrabarti (2005) and Dragulescu and Yakovenko (2001) for some examples.
56. More directly, it is also an argument against the advocacy of the venture capital model found in Scherer and Hartoff (2000, 562).
57. This was written before the current economic crisis, with the failure of "rocket science" finance staffed by former physicists. Risk management gets less respect these days (Power (2007). Orthodox risk management is refuted in finance from a Minskyan perspective in Mirowski (2010).

7. The New Production of Ignorance

1. The latest work to notice this most plangent of ironies is Geiger and Sá (2008, 120 et seq.). They report upon the real neoliberal answer later in their book, when they quote a policy paper from the Ewing Marion Kauffman Foundation, which

exhorts administrators to stop worrying about disappointing revenues from IP but pull out all the stops and patent like mad anyway (140).

2. On the role of Lippmann in the genesis of neoliberalism, see Mirowski and Plehwe (2009, 13–16).
3. I have unexpectedly encountered this accusation of nostalgia on the road numerous times when I have lectured on portions of this book. Independent instances have happened in person, in e-mail correspondence after the fact, and even in print (B. Miller 2007, 132). What is perhaps more distressing about what is becoming a predictable reaction is that the objection tends to emanate from younger scholars—older academics like Viridiana don't seem quite as bothered by it.
4. For one version of that exercise, see Mirowski (2008).
5. The poor showing of American students vis-à-vis their poorer Chinese contemporaries with regard to knowledge of physics has even been the subject of research in *Science* (Bao et al. 2009). A cute survey done by George Mason University showed 27 percent of American broadcast meteorologists reported that climate change is a hoax: http://www.physorg.com/news196769149.html (last visited September 2010).
6. I hope my publisher does not take umbrage if I insert a personal note here. When I first got a contract to write this book, both the referees and my editor suggested that, relative to my previous books, I should make some effort to write shorter sentences with a more restrained vocabulary, and, indeed, make the whole thing more truncated in length. As one said, "People don't buy long books anymore." I did try to accede to their demands in this book (though with only partial success). Did you notice, dear reader?
7. And hence explain why Stephen Colbert's notorious neologism "truthiness" enjoyed such cultural resonance with the chattering classes. A nice meditation on this phenomenon of modern life is Hesse (2008).
8. The work of a few brave souls seeking to call the entire construct of the "knowledge economy" into question, while instructive (Carlaw, Oxley, and Walker 2006); Perraton 2006), tends to miss out on the crucial role of ignorance in a theory of the knowledge society. The one paper I have found that raises the issue is by the sadly neglected Louis Schneider (1962).
9. Hayek (1948, 290–291). What is regarded as wickedly radical in science studies is propounded as eminently conservative by the Hayek wing of neoliberalism.
10. Hayek (1960, 378), my italics. This explains why neoliberals tend to diverge from their predecessors: "most nineteenth century liberals were guided by a naive overconfidence in what mere communication of knowledge could achieve" (377).
11. Hayek reveled in the tough-minded stance of the scholar who disparaged any recourse to the Third Way, most notoriously in his denunciation of the welfare state as just the slippery *Road to Serfdom* (1944). This sets him apart from his contemporary figures like Walter Lippmann, or Keynes, or their modern epigones like Cass Sunstein (2006) or Joseph Stiglitz.
12. George Stigler Papers, University of Chicago, Box 20, File: Schools in Science. Thanks to Eddie Nik-Khah (2008) for these and other quotes from the Stigler Papers.

13. George Stigler, Address to the Mont Pelerin Society 10th Anniversary Meeting, George Stigler Papers, University of Chicago, Box 26.
14. Hayek entertained the possibility of blaming "the engineers" for the frustration of the neoliberal project in his *Counter-Revolution of Science* (1952), but he subsequently came round to the position that it was politically unwise to demonize such a powerful constituency in the twentieth-century economy. See Mirowski (2007b).
15. The actual history is discussed at great length in Mirowski and Plehwe (2009). A concise statement of this position is Blundell (2005).
16. See http://en.wikipedia.org/wiki/Jimmy_Wales (last visited November 2009).
17. Hayek (1967, 178). Attacks upon "intellectuals" were a common refrain in the history of Mont Pelerin and were not restricted to Hayek. See, for instance, Hartwell (1995, 161); Friedman (1962, 8; 1978). But of course the neoliberals don't renounce *all* expertise—just the stuff they don't like.
18. I explain this in some detail in the "postface" to Mirowski and Plehwe (2009). Recent developments have only accentuated the oligarchy; see http://www.time.com/time/magazine/article/0,9171,1924492,00.html (last visited September 2010).
19. One example of the double truth doctrine is that Hayek does admit that "spontaneous order and organization will always coexist" (1973, 48). The codicil for the elect then comes with a rather tendentious rationalization concerning when and how organizations like Mont Pelerin are legitimate within the doctrine concerning the evolution of natural orders, like, say, its mandate for the construction of market forms that do not already exist. The double truth doctrine was explicated along different lines by one of Hayek's fellow exiles at the University of Chicago, Leo Strauss.
20. For more on this see Mirowski (2008). However, Arnsperger (2008) explores the possibility that both the Walrasian program and the more modern agent-based complexity approach are driven to treat the individuals in their models as sociological and political dopes as part of the specification of equilibrium.
21. As I have already signaled, I believe the previous attempt at diagnosis in Mirowski and Sent (2002) falls well short of doing justice to the topic. Mirowski (2009) does a little better.
22. Blundell (2005) makes this assertion about the economics profession; it also corresponds to my own personal experience. Skeptics will, of course, want quantitative data. I suggest they compare the responses in the original *Making of an Economist* (Colander and Klamer 1990) and *The Making of an Economist Redux* (Colander 2007).
23. The literature here is quite large and includes Lutge (2001); Zamora Bonilla (2003, 2008); Shi (2001); Stiglitz (1999, 2000); Thomsen (1992); Wible (1998); Scotchmer (2004); Brock and Durlauf (1999); Durlauf (1997, 2005); Kitcher (1993); and a number of papers gathered together in Mirowski and Sent (2002).
24. The best general explanation of this predicament is Arnsperger (2008), but see also J. Davis (2003) and Sent (2001). Other, more minor instances of contradiction are explored in Mäki (1999).

25. Although he does not mention it there, Hands is also discussing the symmetry/asymmetry problem of the economist and her agents. It should be obvious by now that this current book is almost entirely devoted to something resembling Hands's ESK, and therefore takes it as a stipulation of consistency that standard neoclassical models are nowhere used herein to explain economics or science.
26. The evidence is contained in the Stigler Papers, University of Chicago Regenstein Library, Box 3: Walgreen Conferences. The incident is discussed in Nik-Khah (2009).
27. Stigler, "Discussion of K. Arrow," Box 23, Stigler Papers, University of Chicago Regenstein Library.
28. See, for instance, Gary Becker's proposals (2002) for gutting the FDA. The neoliberal economist Jacques Gorlin played an important role in mobilizing intellectual rationales for TRIPs.
29. See Boden and Ozonoff (2008) .Unfortunately, a similar position has been voiced in the science studies community: see Jasanoff (2008).
30. See the public face of "The Rose and Milton Friedman Foundation for Educational Choice" at http://www.edchoice.org/friedmans/friedmansbio.jsp. (last visited September 2010).
31. This incident is described in some detail in Marron (1984), leaving out the controversy it created at Harvard regarding the role of faculty in the postwar university. Later, Feldstein was a member of the board of directors of AIG, Inc, while at the same time providing expert analysis of the 2008 crisis in numerous public outlets. Early insensitivity to possible conflicts of interest tends to then encourage testing the envelope later in life.
32. These include Summers at Harvard, Hugo Sonnenschein at Chicago, Harold Shapiro at Princeton, Elizabeth Hoffman at University of Colorado, Lars-Hendrik Roller at the European School of Management, and Richard Levin at Yale. A few of these university presidents have even had the temerity to write about the effects of marketization upon higher education, although this literature has proven next to useless when it comes to trying to understand how the modern privatization regime arose, or how it operates.
33. This impression has now become ingrained in the secondary literature: see Jacoby (2008, 232–233); Fallis (2007, p. 258); Rich (2008). One possible excuse might be the only popular book on Summers's stint as president was written before the major event recounted here became public Bradley (2005).
34. The following account is based primarily on McClintick (2006); Ciarelli and Troianovski (2006); and David Warsh's blog www.economicprincipals.com, especially the entry for June 14, 2009. It is noteworthy that one lone article in the *New York Times* acknowledged this incident, but then that paper of record never mentioned it ever again (Ivry 2006). Shleifer was levied a $2 million dollar fine, while Harvard paid $31 million in fines for breach of contract with the government.
35. David Warsh writes:

> Meyer and the Harvard endowment are part of the story simply because Harvard Management was an aggressive investor in Russia in the mid-1990s, during the period that Shleifer was advising the Russian government

> and illicitly investing on his own behalf. For example, Euromoney magazine at the time described Harvard as "bolder than most," because it bought stakes in companies themselves, instead of relying on intermediaries such as hedge fund managers. When government attorneys deposed Meyer in the course of their suit against Harvard, they learned that as much as 1.8 percent of Harvard's endowment had been invested in Russia in the years before the US government fired Shleifer—some $200 million of a portfolio then worth around $11 billion—not the 10 percent that gossip had it at the time.

Column of Jan 15 2005. Available at: http://www.economicprincipals.com/issues/2005.01.16/131.html (last visited September 2010).

36. Indeed, there had been very little formal argument from neoclassical bases that "shock therapy" was the correct way to initiate a transition to a market economy, and the widespread attitude toward events in Russia nowadays is that the shock therapy was more a political tool to destroy the Communist Party, rather than a specifically economic instrument to foster development. In any event, after 2008 it is not so clear one would want to trumpet the virtues of an American-style financial system.
37. This is not an imaginary scenario. See the history of John Silber and the faculty start-up Seragen at Boston University (Barboza 1998). Silber managed to lose $50 million of the BU endowment, not to mention the investment of a number of trustees, by attempting to use privileged proprietary information to orchestrate the testing and development of, and therefore profit from, an anticancer drug.
38. See the comments by Thomas Stossel, a doctor at the Harvard Medical School, in the *New York Times:* "Academic socialists and conflict of interest vigilantes are stifling the biotechnology revolution . . . The idea that money is evil and academia is made up of saints is nonsense" (quoted in McNeil 2006). Stossel has recently used his Harvard post to found a neoliberal astroturf organization, the Association of Clinical Researchers and Educators, to oppose those critical of the effects of commercialization on medical science (Maternowski 2009).
39. "The case was downgraded when US attorney Donald Stern concluded a perjury charge would be expensive and difficult to prove" (Warsh 2009a), possibly due to those confiscated Russian computers.
40. "In contrast, Deputy Treasury Secretary Lawrence Summers provided some measure of cover for his old friend Shleifer when the Harvard scandal broke in 1997. After becoming president of Harvard University, in 2001, he again shielded Shleifer from the consequences of an ultimately successful Justice Department law suit" (Warsh 2009b).
41. Technically, as part of the agreement, the payment obligation was entered as an order of the court, which differs from a fine imposed through trial process or verdict.
42. The stunning silence surrounding the issue has been noted by David Warsh on his Web site, www.economicprincipals.org, and by George Krasnow on http://www.raga.org/Harvard_and_Russia.html (last visited August 13, 2008). Warsh

wrote, "The aftermath of the court case has been characterized by reluctance in almost all quarters to speak frankly about the case." But many scientists do not suffer unduly from research misconduct (Guterman 2008).

43. Further potential conflicts of interest were revealed in April 2009, involving Summers receiving $2.7 million for speaking engagements in 2008 at the very financial institutions he would shortly be tasked to rescue, including Lehman Brothers, Goldman Sachs, Citigroup and Merrill Lynch (Story 2009). I mention these in passing only to illustrate Summers's demonstrated lack of concern about such matters.
44. http://www.in-cites.com/nobel/2007-eco-top100.html (last visited August 13, 2008).
45. "Bok had also served the Summers presidency by agreeing to try to mediate a dispute between Harvard and star economist Andrei Shleifer, a friend of Summers's who had been accused by the U.S. government of committing fraud" (Bradley 2007).

Bibliography

Unpublished Theses

Asner, Glen. 2006. "The Cold War and American Industrial Research." PhD diss., History, Carnegie-Mellon University.

Daines, Gregory. 2007. "Patent Citations and Licensing Value." PhD diss., Business, MIT.

Jardini, David. 1996. "Out of the Blue Yonder." PhD diss., History, Carnegie-Mellon University.

Lee, Kyu-Sang. 2004. "Rationality, Mind and Machine in the Laboratory." PhD diss., Economics, University of Notre Dame.

Lütge, Christoph. 2001. *Ökonomische Wissenschaftstheorie.* Wurtzburg: Konigshausen & Neumann.

Mowery, David. 1981. "The Emergence and Growth of Industrial Research in American Manufacturing, 1899–1945." PhD diss., Stanford University.

Raynor, Gregory. 2000. "Engineering Social Reform: The Ford Foundation and Cold War Liberalism, 1908–1959." PhD diss., History, New York University.

Rizzo, Michael. 2004. "A (Less Than) Zero Sum Game? State Funding for Public Higher Education: How Public Higher Education Institutions Have Lost." PhD diss., Economics, Cornell University.

Scheiding, Thomas. 2006. "Publish and Perish." PhD diss., Economics, University of Notre Dame.

Van Horn, Robert. 2007. *The Origins of Chicago School of Law and Economics.* PhD diss., Economics, University of Notre Dame.

Zamora Bonilla, Jesús. 2003. *La Lonja del Saber.* Madrid: UNED.

Published Works

Abbate, Janet. 1999. *Inventing the Internet.* Cambridge, Mass.: MIT Press.

Abbott, Alison. 2000. "Mouse Geneticists Call for Unified Rules of Exchange." *Nature* (403): 236.

Abraham, John. 2002. "The Pharmaceutical Industry as a Political Player." *Lancet* 360(9344): 1498–1502.

———. 2007a. "Drug Trials and Evidence Bases in International Regulatory Context." *BioSocieties* (2): 41–56.

———. 2007b. "From Evidence to Theory: Neoliberal Corporate Bias as a Framework for Understanding UK Pharmaceuticals Regulation." *Social Theory and Health* (5): 161–175.

Abraham, John; and Smith, Helen, eds. 2003. *Regulation of the Pharmaceutical Industry.* London: Palgrave.

Adams, James. 2007. "Recent Trends in US Science and Engineering." In Galama and Hosek, 2007.

Adams, James; and Griliches, Zvi. 1996. "Mapping Science: An Exploration." *Proceedings of the National Academy of Sciences* (93): 12664–12670.

Adams, Jonathan; King, Christopher; and Nan Ma. 2009. *China: Research and Collaboration in the New Geography of Science.* Leeds, U.K.: Thomson Reuters.

Adelman, David; and deAngeles, Kathryn. 2007. "Patent Metrics: The Mismeasure of Innovation in the Biotech Patent Debate." *Texas Law Review* (85): 1677–1741.

Agarwal, Pankaj; and Searls, David. 2009. "Can Literature Analysis Identify Innovation Drivers in Drug Discovery?" *Nature Reviews Drug Discovery* (8): 865–878.

Aggarwal, Saurabh. 2007. "What's Fueling the Biotech Engine?" *Nature Biotechnology* (25): 1097–1103.

Aghion, Philippe; David, Paul; and Foray, Dominique. 2006. "Linking Policy Research and Practice in STIG Systems." Paper presented to SPRU fortieth anniversary conference.

Aghion, Philippe; and Durlauf, Steven, eds. 2005. *Handbook of Economic Growth.* Amsterdam: Elsevier.

Agrawal, Ajay; and Henderson, Rebecca. 2002. "Putting Patents in Context: Exploring Knowledge Transfer from MIT." *Management Science* (48): 44–60.

Agres, Ted. 2003. "The Cost of Commercializing Academic Research." *The Scientist* 17: 58–59.

Ainsworth, Susan. 2007. "Pharma Adapts." *Chemical and Engineering News,* December 3 (85:49): 13–24.

Aizcorbe, Ana; Moylan, Carol; and Robbins, Carol. 2009. "Toward Better Measurement of Innovation and Intangibles." BEA Briefing, www.bea.gov.

Akera, Atsushi. 2002. "IBM's Early Adaptation to Cold War Markets." *Business History Review* 76: 767–802.

Akerlof, George. 1970. "Market for Lemons: Quality, Uncertainty and Market Mechanisms." *Quarterly Journal of Economics* (84): 488–500.

———. 1984. *An Economic Theorist's Book of Tales.* New York: Cambridge University Press.

———. 2002. "Behavioral Macroeconomics and Macroeconomic Behavior." *American Economic Review* (92): 411–433.

Albert, Mathieu; and Laberge, Suzanne. 2007. "The Legitimation and Dissemination Processes of the Innovation System Approach." *Science, Technology and Human Values* (32): 221–249.

Alchian, Armen. 1950. "Uncertainty, Evolution and Economic Theory." *Journal of Political Economy* (58): 211–221.

———. 1953a. "Biological Analogies in the Theory of the Firm: Comment." *American Economic Review* (43): 600–603.

———. 1953b. "Systems Analysis—Friend or Foe?" RAND D-1778.
———. 2006. *The Collected Works of Armen Alchian, vol. 1.* Indianapolis: Liberty Fund.
Alchian, Armen; and Kessell, Reuben. 1954. "A Proper Role for Systems Analysis." RAND D-2057.
Allison Commission. 1884. 49th Congress, 1st sess., Senate Misc. Doc. 82 (Series 2345).
Altbach, Philip. 2004. "Higher Education Crosses Borders." *Change,* March–April.
Altbach, Philip; and Levy, Daniel. 2005. *Private Higher Education: A Global Revolution.* Rotterdam: Sense Publishers.
Amadae, S. M. 2003. *Rationalizing Capitalist Democracy.* Chicago: University of Chicago Press.
Amin, Ash; and Cohendet, Patrick. 2004. *Architectures of Knowledge: Firms, Capabilities and Communities.* Oxford: Oxford University Press.
Ancori, Bernard; Bureth, Antoine; and Cohendet, Patrick. 2000. "The Economics of Knowledge: The Debate about Codification and Tacit Knowledge." *Industrial and Corporate Change* (9): 255–287.
Anderson, Howard. 2004. "Why Big Companies Can't Invent." *Technology Review* May: 56–59.
Anderson, R. A.; Butler, O.; and Juris, M. 2008. *History of Physicists in Industry.* www.aip.org/history.
Andersson, Åke; and Beckmann, Martin. 2009. *Economics of Knowledge.* Cheltenham, U.K.: Elgar.
Andrews, Lori; Paradise, Jordan; Holbrook, Timothy; and Bochneak, Danielle. 2006. "When Patents Threaten Science." *Science* (314): 1395–1396.
Angell, Marcia. 2004. *The Truth about the Drug Companies.* New York: Random House.
———. 2010. "FDA: This Agency Can Be Dangerous." *New York Review of Books,* (September 30) (57:14):66–68.
Angell, Marcia; and Relman, Arnold. 2002. "Patents, Profits and American Medicine." *Daedalus* (Spring): 102–111.
Angrist, Misha; and Cook-Deegan, Robert. 2006. "Who Owns the Genome?" *New Atlantis,* (Winter): 87–96.
Anon. 2008. "State of the Biotech Sector—2007." *Nature Biotechnology,* July.
Antonelli, Christiano. 2005. "Models of Knowledge and Systems of Governance." *Journal of Institutional Economics* (1): 51–73.
Appel, Tobey. 2000. *Shaping Biology.* Baltimore: Johns Hopkins University Press.
Apple, Michael, ed. 2003. *The State and the Politics of Knowledge.* New York: RoutledgeFalmer.
———. 2005. "Education, Markets and an Audit Culture." *Critical Quarterly* (47): 11–29.
———. 2006. *Educating the Right Way.* 2nd ed. London: Routledge.
Arnsperger, Christian. 2008. *Critical Political Economy.* London: Routledge.
Arrow, Kenneth. 1962. "Economic Welfare and the Allocation of Resources for Invention." In Richard Nelson, ed., *The Rate and Direction of Inventive Activity.* Princeton: Princeton University Press. Reprinted in Mirowski and Sent 2002.

———. 1984. *The Economics of Information*. Vol. 4 of his Collected Papers. Cambridge, Mass.: Harvard University Press.

———. 1985. "The Organization of Economic Activity: Issues Pertinent to the Choice of Market vs. Nonmarket Allocation." In Edwin Mansfield, *Microeconomics,* 5th ed., New York: Norton.

———. 1996. "The Economics of Information: An Exposition." *Empirica* (23): 119–128.

Asner, Glen. 2004. "The Linear Model, the US Department of Defense, and Golden Age of Industrial Research." In Karl Grandin, Nina Wormbs, and Sven Widmalm, eds. *The Science-Industry Nexus.* Sagamore Beach, Mass.: Science History Publications.

Aspray, William; Mayadas, Frank; and Varda, Moshe, eds. 2006. *Globalization and Offshoring of Software.* ACM Publications, www.acm.org.

Associated Press. 2008. "Purdue Panel Finds Misconduct by Fusion Scientist." *New York Times,* July 18.

Athreye, Suma; and Godley, Andrew. 2009. "Internationalization and Technological Leapfrogging in the Pharmaceutical Industry." *Industrial and Corporate Change* (18): 295–323.

Auerbach, Lewis. 1965. "Scientists in the New Deal." *Minerva* (3): 457–482.

Augier, Mie. 2005. "Why Is Management an Evolutionary Science? An Interview with Sidney G. Winter." *Journal of Management Inquiry* (14): 344–354.

Aumann, Robert. 1976. "Agreeing to Disagree." *Annals of Statistics* (4): 1236–1239.

———. 2000. *Collected Papers, vol. 1.* Cambridge, Mass.: MIT Press.

———. 2005. "Musings on Information and Knowledge." *Economics Journal Watch* (2): 88–96.

Avorn, Jerry. 2004. *Powerful Medicines.* New York: Knopf.

Azoulay, Pierre. 2002. "Acquiring Knowledge Within and Across Firm Boundaries." Columbia University School of Business, Working Paper.

———. 2003. "Agents of Embeddedness." Columbia University School of Business, Working Paper.

Baars, Bernard. 1986. *The Cognitive Revolution in Psychology.* New York: Guilford.

Baba, Annamaria; Cook, Daniel M.; McGarity, Thomas O.; and Bero, Lisa. 2005. "Legislating Sound Science: The Role of the Tobacco Industry." *American Journal of Public Health,* June (95 Suppl.): S20–S27.

Babe, Robert. 1994. "Information as an Economic Commodity." In Babe, ed., *Information and Communication in Economics.* Boston: Kluwer.

Baca, Megan. 2006. "Barriers to Innovation: Intellectual Property Transaction Costs in Scientific Collaboration." *Duke Law & Technology Review,* no. 4: 1011, 1026–1030.

Bachelard, Gaston. [1928] 1987. *Essai sur la connaissance approchée.* Paris: Vrin.

Backhouse, Roger. 2005. "The Rise of Free Market Economics: Economists and the Role of the State since 1970." In Peter Boettke and Steven Medema, eds. *The Role of Government in the History of Economic Thought.* Durham, N.C.: Duke University Press.

Bailey, Ronald. 2007. "The Tragedy of the Anticommons." *Reason Magazine Online,* October 2. www.reason.com/news/show/122785.html.

Balconi, Margherita; Brusoni, Stefano; and Orsenigo, Luigi. 2009. "In Defense of the Linear Model." *Research Policy,* (39): 1–13.

Baldini, Nicola. 2006. "University Patenting and Licensing Activity: A Review of the Literature." *Research Evaluation* (15:3): 197–207.

Baltimore, David. 2003. "On Over-Weighting the Bottom Line." *Science* 301: 1050–1051.

Barboza, David. 1998. "Loving a Stock, Not Wisely But Too Well." *New York Times,* September 20.

Barry, Andrew; Osborne, T.; and Rose, Nikolas, eds. 1996. *Foucault and Political Reason—Liberalism, Neoliberalism and the Rationalities of Government.* London: UCL Press.

Barry, Andrew; and Slater, Daniel. 2003. "Technology, Politics and the Market: An Interview with Michel Callon." *Economy and Society* 31: 285–306.

Bartley, W. W. 1990. *Unfathomed Knowledge, Unmeasurable Wealth.* La Salle, Ill.: Open Court.

Bassett, Roberta. 2006. *The WTO and the University.* London: Routledge.

Bassett, Ross. 2003. "Review of Etzkowitz: *MIT and the Rise of Entrepreneurial Science.*" *Isis* 94(4): 768–769.

Bao, Lei; Cai, Tianfan; Koenig, Kathy, et al. 2009. "Learning and Scientific Reasoning." *Science,* January 30 (323): 586–587.

Bauerlein, Mark. 2008. *The Dumbest Generation.* New York: Tarcher.

Becker, Gary. 1971. *Economic Theory.* New York: Knopf.

———. 2000. "Cracking the Genetic Code: Competition Was the Catalyst." *Business Week,* August 14 (3694): 26.

———. 2002. "Get the FDA out of the Way." *Business Week,* September 16.

Bekelman, Justin; Yan Li; and Gross, Cary. 2003. "Scope and Impact of Financial Conflicts of Interest in Biomedical Research." *Journal of the American Medical Association* 284: 454–465.

Bell, J. William. 2003. "Our Genome in Common: Genomic Data Release Policies." *Libraries and the Academy* (3): 293–306.

Bell, Robert; Hill, Derek; and Lehming, Rolf. 2007. "The Changing Research and Publication Environment in American Research Universities." NSF Working Paper SRS 07–204.

Ben-David, Joseph. 1991. *Scientific Growth.* Berkeley: University of California Press.

Bender, Thomas; and Schorske, Carl, eds. 1997. *American Academic Culture in Transition.* Princeton: Princeton University Press.

Benkler, Yochai. 2006. *The Wealth of Networks.* www.benkler.org/wealth_of_net works.

Berenson, Alex. 2004. "An Industry in Poor Health." *New York Times,* December 18, p. A1.

———. 2007. "Tax Break Used by Drug Makers Failed to Add Jobs as Advertised." *New York Times,* July 24, pp. A1, A16.

———. 2008. "Study Reveals Doubt on Drug for Cholesterol." *New York Times,* January 15.

Berlin, Isaiah. [1958] 1969. "Two Concepts of Liberty." In *Four Essays on Liberty.* Oxford: Oxford University Press.

Berman, Dennis. 2003. "At Bell Labs, Hard Times Take Toll on Pure Science." *Wall Street Journal,* May 23, p. A1.

Berman, Elizabeth. 2008. "Why Did Universities Start Patenting?" *Social Studies of Science* (38): 835–872.

Bernal, J. D. 1939. *The Social Function of Science.* London: Routledge.

Bernstein, Elizabeth. 2009. "Madoff Scandal's Deep Impact on Funding for Health, Science." *Wall Street Journal,* February 12.

Bernstein, Richard. 2004. "Germany's Halls of Ivy Are Needing Miracle-Gro." *New York Times,* May 9.

Bero, Lisa; and Rennie, Drummond. 1996. "Influences on the Quality of Public Drug Studies." *International Journal of Technological Assessment in Health Care* 12(2): 209–237.

Bertrand, Elodie. 2005. "Two Complex Lighthouse Production Systems." In John Finch and Magali Orillard, eds. *Complexity and the Economy,* 191–206. Northampton, U.K.: Elgar.

Bessen, James; and Meurer, Michael. 2008. *Patent Failure.* Princeton: Princeton University Press.

Beyler, Richard; and Low, Morris. 2003. "Science Policy in West Germany and Japan." In Mark Walker, ed., *Science and Ideology.* London: Routledge.

Bhagat, Sanjai; Shleifer, Andrei; Vishny, Robert W.; Jarrel, Gregg; and Summers, Lawrence. 1990. "Hostile Takeovers in the 1980s." *Brookings Papers on Economic Activity. Microeconomics* 1990: 1–84.

Bhopal, R.; Rankin, J.; McColl, E.; Thomas, L.; Kaner, E. et al. 1997. "The Vexed Question of Authorship." *British Medical Journal* 314: 1009–1012.

Biagioli, Mario. 2006a. *Galileo's Instruments of Credit: Telescopes, Images, Secrecy.* Chicago: University of Chicago Press.

———. 2006b. "Patent Republic: Specifying Inventions, Constructing Authors and Rights." *Social Research* (73): 1129–1172.

———. . 2007. "Bringing Peer Review to Patents," *First Monday,* (12:6) at: http://firstmonday.org/htbin/cgiwrap/bin/ojs/index.php/fm/article/viewArticle/1868/1751.

Biagioli, Mario; and Galison, Peter, eds. 2003. *Scientific Authorship: Credit and Intellectual Property in Science.* London: Routledge.

Birke, Lynda. 2003. "Who or What Are the Rats (and Mice) in the Laboratory." *Society and Animals* (11:3): 207–224.

Birnholtz, Jeremy. 2006. "What Does It Mean to Be an Author? The Intersection of Credit, Contribution and Collaboration in Science." *Journal of the American Society for Information Science and Technology* (57:13): 1758–1770.

Blaug, Sasha; Chien, Coleen; and Schuster, Michael. 2004. "Managing Innovation: University-Industry Partnerships and the Licensing of the Harvard Mouse." *Nature Biotechnology* (22:6): 761–763.

Block, Fred; and Keller, Matthew. 2009. "Where Do Innovations Come From? Transformations in the US Economy 1970–2006." http://www.longviewinstitute.org/itifinnovations.

Bloor, David. 2007. "Ideals and Monisms: Recent Criticisms of the Strong Programme in the Sociology of Knowledge." *Studies in the History and Philosophy of Science* (38): 210–234.

Blumenstyk, Goldie. 1998. "Academic Medical Centers Race to Compete in Clinical Drug Trials." *Chronicle of Higher Education,* March 20.

———. 2007a. "Scientists Can Usually Acquire Research Tools and Data, Despite Patents and Other Protections, Survey Finds." *Chronicle of Higher Education,* January 17.

———. 2007b. "Iowa University Sues over Soybeans." *Chronicle of Higher Education,* June 8.

Blumenthal, D.; Campbell, E.; Causino, N.; and Seashore Lewis, K. 1996. "Participation of Life-Science Faculty in Research Relationships with Industry." *New England Journal of Medicine* 335: 1734–1739.

Blundell, John. 2005. *Waging the War of Ideas.* London: IEA.

Boden, Leslie; and Ozonoff, David. 2008. "Litigation-Generated Science: Why Should We Care?" *Environmental Health Perspectives* (116): 117–122.

Bodenheimer, Thomas (2000) "Uneasy Alliance—Clinical Investigators and the Pharmaceutical Industry." *New England Journal of Medicine* 342: 1539–1544.

Boersma, Kees. 2002. *Inventing Structures for Industrial Research: A History of the Philips Natlab, 1914–1946.* Amsterdam: Askant.

Boettinger, Sara; and Bennett, Alan. 2006. "Bayh-Dole: If We Knew Then What We Know Now." *Nature Biotechnology* (24): 320–324.

Bogdanich, Walt; and Petersen, Melody. 2002. "Science for Sale—Ad Agencies and Drug Companies Hand in Hand?" *Bill Moyers Now,* www.pbs.org/now/transcript/transcript144_full.html (last visited March 18, 2005).

Boisot, Max; and Canals, Agusti. 2004. "Data, Information and Knowledge: Have We Got It Right?" *Journal of Evolutionary Economics* (14): 43–67.

Bok, Derek. 2003. *Universities in the Marketplace: The Commercialization of Higher Education.* Princeton: Princeton University Press.

Boldrin, Michele; and Levine, David. 2005. "The Economics of Ideas and Intellectual Property." *Proceedings of the National Academy of Science,* January 25, (102): 1252–1256.

———. 2008. *Against Intellectual Monopoly.* New York: Cambridge University Press.

Booth, Bruce. 2007. "When Less Is More." *Nature Biotechnology* (25): 853–857.

Bork, Robert. 1963. "Civil Rights—A Challenge." *New Republic.* August 31: 21.

Borrás, Susana. 2003. *The Innovation Policy of the European Union: From Government to Governance.* Cheltenham, U.K.: Elgar.

Bosman, Julie. 2006. "Science Journals Harder to Trust." *New York Times,* February 13, p. C1.

Bouchard, Ron; and Lemmens, Trudo. 2008. "Privatizing Biomedical Research—a Third Way." *Nature Biotechnology* (26): 31–36.

Bourdieu, Pierre. 2004. *Science of Science and Reflexivity.* Cambridge: Polity Press.

Bowker, Geoffrey. 2004. "The New Knowledge Economy and Science and Technology Policy." Paper presented to 3rd MIT/UCI Knowledge Organizations Conference.

Boyd, Elizabeth; Cho, Mildred; and Bero, Lisa. 2003. "Financial Conflict-of-Interest Policies in Clinical Research: Issues for Clinical Investigators." *Academic Medicine* 78: 769–774.

Boyle, James. 1996. *Software, shamans and Spleens.* Cambridge, Mass.: Harvard University Press.

———. 2000. "Cruel, Mean or Lavish? Economic Analysis, Price Discrimination and Digital Intellectual Property." *Vanderbilt Law Review* (53): 2007–2039.

———. ed. 2003a. "The Public Domain." *Law and Contemporary Problems* 66(1/2).

———. 2003b. "Enclosing the Genome: What the Squabbles over Genetic Patents Could Teach Us." *Academic Genetics* (50): 97–122.

———. 2008. *The Public Domain.* New Haven: Yale University Press.

Bradley, Richard. 2005. *Harvard Rules.* New York: HarperCollins.

———. 2007. "The Healer." Winter.

Brainard, Jeffrey. 2008. "Senator Grassley Pressures Universities on Science Conflicts." *Chronicle of Higher Education,* July 25.

Braman, Sandra. 2006. "The Micro- and Macroeconomics of Information." *Annual Review of Information Science and Technology* (40): 3–52.

Brender, Alan. 2004. "In Japan, Radical Reform or Same Old Subservience? National Universities Wonder How Much Freedom They Will Be Given Under Looser Government Oversight." *Chronicle of Higher Education,* March 12.

Breneman, David; Pusser, Brian; and Turner, Sarah, eds. 2006. *Earnings from Learnings: The Rise of For-Profit Universities.* Albany: State University of New York Press.

Britt, Ronda. 2008. "Universities Report Continued Decline in Real Federal S&E Funding in FY 2007." NSF *Infobrief,* NSF 08–320.

Broad, William. 2004. "US Is Losing Its Dominance in the Sciences." *New York Times,* May 3.

Brock, William; and Durlauf, Steven. 1999. "A Formal Model of Theory Choice in Science." *Economic Theory* (14): 113–130.

Bronfenbrenner, Kate; and Luce, Stephanie. 2004. "The Changing Nature of Corporate Global Restructuring: The Impact of Production Shifts on Jobs in the US, China, and Around the Globe." http://digitalcommons.ilr.cornell.edu/cbpubs/16 (last visited August 30, 2008).

Brooks, Harvey. 1986. "National Science Policy and Technological Innovation." In Ralph Landau and Nathan Rosenberg, eds., *The Positive Sum Strategy.* Washington: National Academy of Sciences.

Brooks, Kristin, 2008. "CRO Asia Pacific Expansion." *Contract Pharma,* June.

Brown, James R. 2000. "Privatizing the University." *Science* 290: 1701.

Brumfiel, Geoff. 2006. "Theorists Snap over the String Pieces." *Nature* (443): 491.

———. 2008. "Bell Labs Bottoms Out." *Nature,* 21 August (454): 927.

Bruni, Luigi; and Sugden, Robert. 2007. "The Road Not Taken: How Psychology Was Removed from Economics." *Economic Journal* (117): 146–173.

Buckley, Ted. 2007. *The Myth of the Anticommons.* http://bio.org/ip/domestic/themythoftheanticommons.pdf (last visited September 26, 2010).

Buderi, Robert. 2000. *Engines of Tomorrow.* New York: Simon & Schuster.

———. 2002. "The Once and Future Industrial Research." In Albert H. Teich, Stephen D. Nelson, and Stephen J. Lita, eds., *AAAS Science and Technology Policy Yearbook:* 245–251.

Buderi, Robert; and Huang, Gregory. 2006. *Guanxi the Art of Relationships: Microsoft, China, and Bill Gates's Plan to Win the Road Ahead.* New York: Simon & Schuster.
Burchell, Graham; Gordon, Colin; and Miller, Peter, eds. 1991. *The Foucault Effect.* Chicago: University of Chicago Press.
Burke, James; Epstein, Gerald; and Minsik Choi. 2004. "Rising Foreign Outsourcing and Employment Losses in US Manufacturing, 1987–2002." University of Massachusetts Working Paper No. 89.
Busch, Lawrence, Allison, Richard; Harris, Craig; Rudy, Alan; Shaw, Bradley T.; Ten Eyck, Toby; Coppin, Dawn; Konefal, Jason; and Oliver, Christopher. 2004. *External Review of the Collaborative Research Agreement Between Novartis and the Regents of the University of California.* East Lansing Michigan State University Institute for Food and Agricultural Standards.
Bush, Vannevar. 1945. *Science—the Endless Frontier.* Washington: Government Printing Office.
Business Week. 2005. "Just the Bright Side, Thanks." October 17, www.businessweek.com/magazine/content/05_42/c3955025.htm (last visited August 30, 2008).
Business Wire (2002) "PharmaLinkFHI to Present." January 30.
Butos, William; and McQuade, Thomas. 2006. "Government and Science: Dangerous Liaison?" *Independent Review* (11): 177–208.
Caldwell, Bruce. 2004. *Hayek's Challenge.* Chicago: University of Chicago Press.
Callahan, Daniel. 2003. *What Price Better Health?* Berkeley: University of California Press.
Callon, Michel. 1994. "Is Science a Public Good?" *Science, Technology and Human Values* (19:4): 395–424.
———, ed. 1998. *The Laws of the Markets.* Oxford: Blackwell.
Caloghirou, Yannis; Ionnides, Stavres; and Vontoras, Nicholas. 2003. "Research Joint Ventures." *Journal of Economic Surveys* 17: 541–570.
Calvert, Jane. 2004. "The Idea of Basic Research in Language and Practice." *Minerva* 42: 251–268.
———. 2008. "The Commodification of Emergence." *BioSocieties (*3): 383–398.
Camerer, Colin; Loewenstein, George; and Prelec, Drazen. 2005. "Neuroeconomics: How Neuroscience Can Inform Economics." *Journal of Economic Literature* (63): 9–64.
Campbell, Eric; Clarridge, Brian; Gokhale, Manjusha; Birenbaum, Lauren; Hilgartner, Stephen; Holtzman, Neil; and Blumenthal, David. 2002. "Data Withholding in Academic Genetics: Evidence from a National Survey." *Journal of the American Medical Association* (287:4): 473–480.
Campbell, Eric; Powers, Joshua; Blumenthal, David; and Biles, Brian. 2004. "Inside the Triple Helix: Technology Transfer and the Commercialization of the Life Sciences." *Health Affairs* (January–February) 23(1): 64–76.
Campbell, Eric; Weissman, Joel S.; Moy, Ernest; and Blumenthal, David. 2001. "Status of Clinical Research in Academic Health Centers: Views from the Research Leadership." *Journal of the American Medical Association* 286: 800–806.
Capshew, James; and Rader, Karen. 1992. "Big Science: Price to the Present." *Osiris* 2nd series 7: 3–25.

Capuro, R.; and Hjorland, B. H. 2002. "The Concept of Information." *Annual Review of Information Science and Technology* (37): 343–411.

Carey, Benedict. 2008. "Antidepressant Studies Unpublished." *New York Times,* January 17.

Carlaw, Kenneth; Oxley, Les; and Walker, Paul. 2006. "Beyond the Hype: Intellectual Property and the Knowledge Economy." *Journal of Economic Surveys* (20): 633–658.

Carnap, Rudolf. 1947. *Meaning and Necessity.* Chicago: University of Chicago Press.

Carr, Nicholas. 2008. "Is Google Making Us Stupid?" *Atlantic,* July/August: 56–63, at http://www.theatlantic.com/doc/print/200807/google.

———. 2010. *The Shallows.* New York: Norton.

Castells, Manuel. 2000. *The Rise of the Network Society.* Oxford: Blackwell.

Caulfield, Timothy; Cook-Deegan, Robert; Kieff, F.; and Walsh, John. 2006. "Evidence and Anecdotes: An Analysis of Human Gene Patenting Controversies." *Nature Biotechnology* (24:9): 1091–1094.

Cavusoglu, Nevin; and Tebaldi, Edinaldo. 2006. "Evaluating Growth Theories and Their Empirical Support." *Journal of Economic Methodology* (13): 49–75.

CenterWatch. 2008. *State of the Clinical Trials Industry 2008.* Boston: CenterWatch.

Ceruzzi, Paul. 2003. *A History of Modern Computing.* Cambridge, Mass.: MIT Press.

Chait, Jonathan. 2008. "The End of an Error." *New Republic,* October 22.

Chandler, Alfred. 1977. *The Visible Hand.* Cambridge, Mass.: Harvard University Press.

———. 2005a. "Response to the Symposium." *Enterprise and Society* 6: 134–137.

———. 2005b. *Shaping the Industrial Century.* Cambridge, Mass.: Harvard University Press.

Chang, Kenneth; and Revkin, Andrew. 2008. "At a Sleek Bioenergy Lab, a Lens on a Cabinet Pick." *New York Times,* December 23.

Charbonneau, Louis. 2006. "Google Scholar Matches Thomson ISI Citation Index." *University Affairs,* March.

Charlton, Bruce. 2008. "What Has the RAE Done for Oxford University?" *Oxford Magazine* (271): 3–5.

Chatterjee, Arnab; Yarlagadda, Sudharkar; and Chakrabarti, Bikas, eds. 2005. *Econophysics of Wealth Distributions.* Milan: Springer.

Chaudhuri, Sudip. 2005. *The WTO and India's Pharmaceuticals Industry.* New Delhi: Oxford University Press.

Chesbrough, Hank. 2001. "Is the Central R&D Lab Obsolete?" *Technology Review,* April.

Cheung, J; Estivill, X.; Khaja, R; MacDonald, J.; Lau, K.; and Tsu, L. 2003. "Genome-Wide Detection of Segmental Duplications and Potential Assembly Errors in the Human Genome Sequence." *Genome Biology* (4:4): R25.

Chilvers, C. (2003) "The Dilemmas of Seditious Men: The Crowther-Hessen Correspondence." *British Journal for the History of Science* 36(4): 417–435.

Cho, Mildred K., Shohara, Roy; Schissel, Anna; and Rennie, Drummond. 2000. "Policies on Faculty Conflict of Interest at US Universities." *Journal of the American Medical Association* 284: 2203–2208.

Choi, Hyungsub. 2007. "The Boundaries of Industrial Research: Making Transistors at RCA." *Technology and Culture* (48): 758–782.

Ciarelli, Nicholas; and Troianovski, Anton. 2006. "Tawdry Shleifer Affair Stokes Faculty Anger towards Summers." *Harvard Crimson*, February 10, www.thecrimson.com.

Clark, Don; and Rhoads, Christopher. 2009. "Basic Research Loses Some Allure." *Wall Street Journal*, October 7.

Clark, William. 2006. *Academic Charisma and the Origins of the Research University*. Chicago: University of Chicago Press.

Coase, Ronald. 1960. "The Problem of Social Cost." *Journal of Law and Economics* (3): 1–44.

———. 1974a. "The Lighthouse in Economics." *Journal of Law and Economics* (17): 357–376.

———. 1974b. "The Market for Goods and the Market for Ideas." *American Economic Review, Papers and Proceedings* (64:2): 384–391.

Cockburn, Iain. 2004. "The Changing Structure of the Pharmaceutical Industry." *Health Affairs* (23): 10–22.

Cockburn, Iain; and Henderson, Rebecca. 1997. "Public-Private Interaction and the Productivity of Pharmaceutical Research." NBER Working Paper #6108.

Cockett, Richard. 1995. *Thinking the Unthinkable . . .* London: Fontana.

Cohen, Jon. 1995. "Share and Share Alike Isn't Always the Rule in Science." *Science*, June 23 (268): 1715–1718.

Cohen, Wesley. 2005. "Patents and Appropriation: Concerns and Evidence." *Journal of Technology Transfer* (30): 57–71.

Cohen, Wesley; and Merrill, Stephen, eds. 2003. *Patents in the Knowledge-Based Economy*. Washington: National Academies Press.

Cohen, Wesley; Nelson, Richard; and Walsh, J. P. 2003. "Limits and Impacts." In Guena et al. 2003.

———. 2003. "Protecting Their Intellectual Assets: Appropriability Conditions and Why U.S. Manufacturing Firms Patent (or Not)." NBER Working Paper No. W7552.

Cohen, Wesley; and Sauermann, Paul. 2007. "Schumpeter's Prophecy and Industrial Incentives." In Malerba and Brusoni 2007.

Cohen, Wesley; and Walsh, John. 2007. "Real Impediments to Academic Biomedical Research." *Innovation Policy and the Economy* (8): 1–30.

Colander, David. 2007. *The Making of an Economist Redux*. Princeton: Princeton University Press.

Colander, David; Holt, Ric; and Rosser, J. B., eds. 2004. *The Changing Face of Economics*. Ann Arbor: University of Michigan Press.

Colander, David; and Klamer, Arjo. 1990. *The Making of an Economist*. Boulder, Colo.: Westview.

Collini, Stefan. 2009. "Impact on the Humanities." *Times Literary Supplement,* November 13 (5563): 18–19.

Collins, Harry. 1992. *Changing Order.* Chicago: University of Chicago Press.

———. 2004. *Gravity's Shadow.* Chicago: University of Chicago Press.

Conlisk, John. 1996. "Why Bounded Rationality?" *Journal of Economic Literature* (34): 669–700.

Cook, Daniel; and Bero, Lisa. 2006. "Identifying Carcinogens: The Tobacco Industry and Regulatory Politics in the United States." *International Journal of Health Services* 36: 4.

Cook, Daniel; Boyd, Elizabeth; Grossmann, Claudia; and Bero, Lisa. 2007. "Reporting Science and Conflicts of Interest in the Lay Press." *PLoS One* (12): e1266.

Cook-Deegan, Robert. 2003. "The Urge to Commercialize." In *The Role of Scientific and Technical Data and Information in the Public Domain.* Washington: National Academies Press.

———. 2007. "The Science Commons in Health Research." *Journal of Technology Transfer* (32): 133–156.

Cooke, Philip. 2002. *Knowledge Economies.* London: Routledge.

Cooper, Melinda. 2008a. "Experimental Labour—Offshoring Clinical Trials to China." *East Asian Science, Technology and Society* (2): 73–92.

———. 2008b. *Life as Surplus.* Seattle: University of Washington Press.

Corbyn, Zoe. 2009. "A Threat to Scientific Communication." *Times Higher Education Supplement,* August 13.

Coriat, Benjamin. 2002. "The New Global Intellectual Property Rights Regime and Its Imperial Dimension." Paper delivered to BNDS seminar, Rio de Janeiro.

Coriat, Benjamin; and Dosi, Giovanni. 1998a. "Institutional Embeddedness of Economic Change." In K. Nielsen and B. Johnson, eds., *Institutions and Economic Change,* Cheltenham, U.K.: Elgar.

———. 1998b. "Learning to Govern and Learning How to Solve Problems." In A. Chandler, P. Hagstrom, and O. Solvell, eds. *The Dynamic Firm: The Role of Technology, Strategy and Organization,* 103–133. Oxford: Oxford University Press.

Coriat, Benjamin; and Orsi, Fabienne. 2002. "Establishing a New Intellectual Property Rights Regime in the United States." *Research Policy* (31): 1491–1507.

Coriat, Benjamin; Orsi, Fabienne; and Weinstein, Olivier. 2003. "Does Biotech Reflect a New Science-Based Innovation Regime?" *Industry and Innovation* (10): 231–253.

Costa, James. 2007. *The Other Insect Societies.* Cambridge, Mass.: Harvard University Press.

Costa, Manuel. 1998. *General Equilibrium Analysis and the Theory of Markets.* Cheltenham, U.K.: Elgar.

Couzin, Jennifer; and Miller, Greg. 2007. "Boom and Bust." *Science,* April 20 (316): 356–361.

Cowan, Robin; David, Paul; and Foray, Dominique. 2000. "The Explicit Economics of Knowledge Codification and Tacitness." *Industrial and Corporate Change* (9): 211–253.

Cowan, Robin; and Foray, Dominique. 1997. "The Economics of Codification and the Diffusion of Knowledge." *Industrial and Corporate Change* (9): 211–253.

Coyle, Diana. 2007. *The Soulful Science.* Princeton: Princeton University Press.
Crespi, Gustavo; and Guena, Aldo. 2008. "An Empirical Study of Scientific Production: A Cross-Country Analysis, 1981–2002." *Research Policy* (37): 565–579.
Croissant, Jennifer; and Restivo, Sal, eds. 2001. *Degrees of Compromise: Industrial Interests and Academic Values.* Albany, N.Y.: SUNY Press.
Cross, John; and Goldenberg, Edie. 2009. *Off-Track Profs.* Cambridge, Mass.: MIT Press.
Crotty, Shane. 2003. *Ahead of the Curve.* Berkeley: University of California Press.
Crowther-Heyck, Hunter. 2005. *Herbert A. Simon; the Bounds of Reason.* Baltimore: Johns Hopkins University Press.
———. 2006. "Patrons of the Revolution." *Isis* (97): 42–46.
Cukier, Kenneth. 2006. "Navigating the Futures of Biotech Intellectual Property." *Nature Biotechnology* (24): 249–251.
Cyranoski, David. 2002. "Share and Share Alike?" *Nature* (420): 602–604.
Daemmrich, Arthur. 2004. *Pharmacopolitics.* Chapel Hill: University of North Carolina Press.
Dasgupta, Partha. 1988. "The Welfare Economics of Knowledge Production." *Oxford Review of Economic Policy* (4:4): 1–12.
Datta, P. 2003. "Are We Ready for Drug Trials on a Large Scale?" *Financial Times,* September 24.
David, Paul. 1975. *Technical Change, Innovation and Economic Growth.* New York: Cambridge University Press.
———. 1994. "Positive Feedbacks and Research Productivity in Science." In O. Grandstrand, ed., *Economics of Technology.* Amsterdam: Elsevier.
———. 1997. "From Market Magic to Calypso Science Policy." *Research Policy* (26): 229–255.
———. 1998. "Common Agency Contracting and the Emergence of Open Science Institutions." *American Economic Review* (88:2): 15–21.
———. 2001. "Path Dependence, Its Critics and the Quest for Historical Economics." In P. Garrouste and S. Ioannides, eds., *Evolution and Path Dependence in Economic Ideas,* 15–40. Cheltenham, U.K.: Elgar.
———. 2003. "The Economic Logic of Open Science and the Balance between Private Property Rights and the Public Domain." In J. Esanau and P. Uhlir, eds., *The Role of Scientific and Technical Data in Information in the Public Domain.* Washington: National Academies Press.
———. 2004a. "Understanding the Emergence of Open Science." *Industrial and Corporate Change* (13): 571–589.
———. 2004b. "Can Open Science Be Protected?" *Journal of Institutional and Theoretical Economics* (167): 1–26.
———. 2006. "Reflections on the Patent System and IPR Protection in the Past, Present and Future." In *Interviews for the Future.* European Patent Office.
———. 2007a. "Path Dependence—a Foundational Concept." *Cliometrica* (1): 91–114.
———. 2007b. "Innovation and Europe's Academic Institutions—Second Thoughts on Embracing the Bayh-Dole Regime." In Malerba and Brusoni 2007.

David, Paul; and Dasgupta, Partha. 1994. "Toward a New Economics of Science." *Research Policy* (23): 487–521.

David, Paul; and Foray, Dominique. 2002. "Economic Fundamentals of the Knowledge Society." *Policy Futures in Education* (1): 20–49.

———. 2003. "Economic Fundamentals of the Knowledge Society." *Policy Futures in Education* (1): 20–49.

———. 2003. "General Purpose Technologies and Surges in Productivity." In Paul David and Mark Thomas, eds., *The Economic Future in Historical Perspective.* Stanford: Stanford University Press.

Davidoff, Frank, DeAngelis, C. D.; Drazen, J. M.; Hoey, J.; Hojgaard, L., et al. 2001. "Sponsorship, Authorship and Accountability." *New England Journal of Medicine* 345: 825–827.

Davidson, R. A. 1986. "Source of Funding and the Outcome of Clinical Trials." *Journal of General Internal Medicine* 1: 155–158.

Davies, Helen. 2001. "The Role of the Private Sector in Protecting Human Research Subjects: A CRO Perspective." Talk delivered to the Institute of Medicine on 21 August. Available at www.acrohealth.org/policy/pdfs/testimony_082101.pdf (last visited March 18, 2005).

———. 2002. "Role of Contract Research Organizations (CRO) in Protecting Human Subjects." Talk delivered at the DIA 2002 Annual Meeting. Available at www.quintiles.com/NR/rdonlyres/eynjegbwzxe4gsp2huilgm5p7eym4zq3d7skncf7cpxr6g5c6zalzh65fgrxbpjdn7hoephi3vwkzon7e4tgupbas5a/Helen_Davies_ DIA2002.pdf (last visited March 18, 2005).

Davis, John. 1998. "Problems in Using the Social Sciences Citation Index to Rank Economics Journals." *American Economist* (42): 59–64.

———. 2003. *The Theory of the Individual in Economics.* London: Routledge.

Day, Ronald. 2001. *The Modern Invention of Information.* Carbondale: Southern Illinois University Press.

Dean, Cornelia. 2008. "If You Have a Problem, Ask Everyone." *New York Times,* July 22.

Delanty, Gerard. 2001. *Challenging Knowledge: The University in the Knowledge Society.* Buckingham, U.K.: Open University Press.

———. 2003. "Ideologies of the Knowledge Society." *Policy Futures in Education* (1): 71–81.

Demsetz, Harold. 1969. "Information and Efficiency: Another Viewpoint." *Journal of Law and Economics* (12): 1–22.

Dennis, Michael. 1987. "Accounting for Research." *Social Studies of Science* (17): 479–518.

———. 1994. "Our First Line of Defense: Two University Laboratories in the Postwar American State." *Isis* (85): 427–455.

———. 1997. "Historiography of Science: An American Perspective." In Krige and Pestre 1997.

———. 2004. "Reconstructing Sociotechnical Order." In Sheila Jasanoff, ed., *States of Knowledge.* London: Routledge.

Denord, Francois. 2001. "Aux Origines du neo-liberalism en France." *Le Mouvement Social* (195): 9–34.

DeSantis, Stephen. 2007. "NEJM Takes Shots at CROs." *Clinical Trials Today,* October 31.

Dewan, Shaila. 2009. "Despite Odds, Cities Race to Bet on Biotech." *New York Times,* June 11.

DeYoung, Karen; and Nelson, Deborah. 2000. "Latin America Is Ripe for Trials, and Fraud." *Washington Post* December 21.

Diamond, Arthur. 1996. "The Economics of Science." *Knowledge and Policy* (9): 6–49.

Dickerstein, Kay; and Rennie, Drummond. 2003. "Registering Clinical Trials." *Journal of the American Medical Association* 290: 516–523.

Dickson, David. 1988. *The New Politics of Science.* Rev. ed. Chicago: University of Chicago Press.

Dickson, David; and Noble, David. 1981. "The Politics of Science and Technology Policy." in Tom Ferguson and Joel Rogers, eds., *The Hidden Election.* New York: Pantheon.

Dillon, Sam. 2004. "US Slips in Status as a Hub of Higher Education." *New York Times,* December 21, p. A1.

———. 2006. "Online Colleges Receive a Boost from Congress." *New York Times,* March 1, p. A1.

Dinopolous, E.; and Thompson, P. 1999. "Reassessing the Empirical Validity of the Human-Capital Augmented Neoclassical Model." *Journal of Evolutionary Economics* (9): 135–154.

Director, Aaron. 1964. "The Parity of the Economic Market Place." *Journal of Law and Economics* (7): 1–10.

Divall, Colin; and Johnston, Sean. 2000. *Scaling Up: The Institution of Chemical Engineers.* Dordrecht: Kluwer.

Dixon, Keith. 1998. *Les Évangélistes du marché.* Paris: Raisons d'agir.

Djelic, Marie-Laure. 1998. *Exporting the American Model.* New York: Oxford University Press.

Djelic, Marie-Laure; and Quack, Sigrid. eds. 2003. *Globalization and Institutions: Rewriting the Rules of the Economic Game.* Cheltenham, U.K.: Elgar.

Dorfman, Robert. 1960. "Operations Research." *American Economic Review* (50): 575–623.

Dosi, Giovanni. 1982. "Technological Paradigms and Technological Trajectories." *Research Policy* (11): 147–162.

———. 1988a. "Sources, Procedures and Microeconomic Effects of Innovation." *Journal of Economic Literature* (26): 126–171.

———. 1988b. "The Nature of the Innovative Process." In Giovanni Dosi et al, eds., *Technical Change and Economic Theory.* London: Pinter.

———. 1997. "Opportunities, Incentives and the Collective Patterns of Technological Change." *Economic Journal* (107): 1530–1547.

Dosi, Giovanni; Bassanini, A.; and Valente, M. 1999. "Norms as Emergent Properties of Adaptive Learning: The Case of Economic Routines." *Journal of Evolutionary Economics* (9): 5–26.

Dosi, Giovanni; and Castaldi, Carolina. 2004. "The Grip of History and the Scope for Novelty." LEM Working Paper.

Dosi, Giovanni; and Grazzi, Marco. 2006. "Technologies as Problem-Solving Procedures and Technologies as Input-Output Relations." *Industrial and Corporate Change* (15): 173–202.

Dosi, Giovanni; and Kaniovski, Y. 1994. "On Badly Behaved Dynamics: Some Applications of Generalized Urn Schemes to Technological Change." *Journal of Evolutionary Economics* (4): 93–123.

Dosi, Giovanni; Llerena, Patrick; and Sylos-Labini, Mauro. 2006. "The Relationships between Science, Technologies, and their Industrial Exploitation." *Research Policy* (35):1450–1464.

Dosi, Giovanni; Malerba, Franco; Ramello, Giovanni; and Silva, Francesco. 2006. "Information, Appropriability and the Generation of Innovative Knowledge Four Decades after Arrow and Nelson." *Industrial and Corporate Change* (15): 891–901.

Dosi, Giovanni; Marengo, Luigi; and Corrado, Pasquali. 2006. "How Much Should Society Fuel the Greed of Innovators?" *Research Policy* (35): 1110–1121.

Dosi, Giovanni; and Nelson, Richard. 1994. "An Introduction to Evolutionary Theories in Economics." *Journal of Evolutionary Economics* (4): 153–172.

Douglass, John. 2006. "Universities and the Entrepreneurial State: A New Wave of State-Based Economic Initiatives." CSHE Research Paper Series 14.06.

———. 2008. "Universities, the US High Tech Advantage, and the Process of Globalization." CSHE Research Paper 8.2008.

Dragulescu, A. A.; and Yakovenko, Victor. 2001. "Exponential and Power Law Probability Distributions of Wealth and Income." *Physica A* (299): 213–231.

Drahos, Peter; and Braithwaite, John. 2002. *Information Feudalism: Who Owns the Knowledge Economy?* New York: New Press.

Drennan, Katherine. 2002. "Patient Recruitment: The Costly and Growing Bottleneck in Drug Development." *Drug Discovery Today* 7(3): 167–170.

Dreyfuss, Rachel; Zimmerman, Diane; and First, H., eds. 2001. *Expanding the Boundaries of Intellectual Property.* New York: Oxford University Press.

Drori, Gili; Meyer, John; Ramirez, Francisco; and Schofer, Eric. 2003. *Science in the Modern World Polity: Institutionalization and Globalization.* Stanford: Stanford University Press.

DuBois, William. 2003. "New Drug Research, the Extraterritorial Application of FDA Regulations, and the Need for International Cooperation." *Vanderbilt Journal of Transnational Law* 36: 161–207.

Duggan, Lisa. 2003. *Twilight of Equality.* Boston: Beacon Press.

Dupré, John. 1993. *The Disorder of Things.* Cambridge, Mass.: Harvard University Press.

Dupuy, J-P. 2000. *Mechanization of the Mind.* Princeton: Princeton University Press.

Durlauf, Steven. 1997. "Limits to Science or Limits to Epistemology?" *Complexity* (January/February 1997): 31–37.

———. 2005. "Dismal Science." *American Scientist,* http://www.americanscientist.org/bookshelf/pub/dismal-science.

Dutfield, Graham. 2002. *Intellectual Property Rights and the Life Sciences Industries.* Burlington, U.K.: Ashgate.

———. 2003. *Protecting Traditional Knowledge and Folklore.* ICTSD Issue Paper 1. Geneva: UNCTAD, http://ictsd.net/downloads/2008/06/cs_dutfield.pdf.

———. 2004a. "Does One Size Fit All? The International Patent Regime." *Harvard International Review,* Summer: 50–54.

———. 2004b. "From Mousetraps to (Onco)mice." Queen Mary IPR Institute.

Eamon, William. 1985. "From the Secrets of Nature to Public Knowledge: The Origins of the Concept of Openness in Science." *Minerva* (23): 321–347.

Economist Intelligence Unit. 2004. *Scattering the Seed of Invention: The Globalization of Research and Development.* Available at http://www.eiu.com.

Edgerton, David. 2004. "The Linear Model Did Not Exist." In Grandin et al. 2004.

———. 2007. *The Shock of the Old.* Oxford: Oxford University Press.

Edgerton, David; and Hughes, Kirsty. 1989. "The Poverty of Science: A Critical Analysis of Scientific and Industrial Policy under Mrs. Thatcher." *Public Administration* (67): 419–433.

Edwards, Paul. 1996. *The Closed World.* Cambridge, Mass.: MIT Press.

Eisenberg, Rebecca. 2001. "Bargaining over the Transfer of Research Tools." In R. Dreyfuss, H. First, and D. Zimmerman, eds., *Expanding the Boundaries of Intellectual Property,* 223–249. Oxford: Oxford University Press.

———. 2003. "Patenting Genome Research Tools and the Law." *Comptes Rendus Biologies* (326): 1115–1120.

———. 2006. "Patents and Data Sharing in Public Science." *Industrial and Corporate Change* (15): 1013–1031.

Eisenberg, Rebecca; and Nelson, Richard. 2002. "Public vs. Proprietary Science: A Fruitful Tension?" *Academic Medicine* (77): 1392–1399.

Elliott, Carl. 2002. "Diary." *London Review of Books* (November 28): 36–37.

Elzinga, Aant. 2001. "Science and Technology: Internationalization." In Neil J. Smelser and Paul B. Baltes, eds., *International Encyclopedia of the Social and Behavioral Sciences,* vol. 20, 13633–13638. Amsterdam: Elsevier.

———. 2004. "The New Production of Reductionism in Models Relating to Research Policy." In Grandin et al., 2004, 277–304.

Endlich, Lisa. 2004. *Optical Illusions: Lucent and the Crash of Telecom.* New York: Simon & Schuster.

Engardio, Peter; and Weintraub, Arlene. 2008. "Outsourcing the Drug Industry." *Business Week,* September 4.

Engelbrecht, H. 2005. "ICT Research, the New Economy, and the Evolving Discipline of Economics: Back to the Future?" *The Information Society* (21): 317–320.

Enserink, Martin. 1999. "NIH Proposes Rules for Materials Exchange." *Science* 284 (May 28): 1445.

Epstein, Richard. 2006. *Overdose: How Excessive Government Regulation Stifles Pharmaceutical Innovation.* New Haven: Yale University Press.

Etzkowitz, Henry. 2002. *MIT and the Rise of Entrepreneurial Science.* London: Routledge.

———. 2003. "Innovation in Innovation: The Triple Helix in University-Industry-Government Relations." *Social Science Information* (42:3): 293–337.

Etzkowitz, Henry; and Leydesdorff, Loet. 2000. "The Dynamics of Innovation: A Triple Helix of University-Industry-Government Relations." *Research Policy* (29:2): 109–123.

Etzkowitz, Henry; and Webster, Andrew. 1995. "Science as Intellectual Property." In Sheila Jasanoff, Gerald Markle, James Petersen, and Trevor Pinch, eds., *Handbook of Science and Technology Studies*, 480–505. Thousand Oaks, Calif.: Sage.

Etzkowitz, Henry; Webster, Andrew; Gebhardt, Christiane; and Terra, Branca. 2000. "The Future of the University and the University of the Future: Evolution of Ivory Tower to Entrepreneurial Paradigm." *Research Policy* (29): 313–330.

Etzkowitz, Henry; Webster, Andrew; and Healey, P., eds. 1998. *Capitalizing Knowledge: New Intersections of Industry and Academia*. Albany: State University of New York Press.

European Commission. 2003–2004. *Towards a European Research Area: Science, Technology, and Innovation: Key Figures 2003–2004*. Available at ec.europa.eu/research.

———. 2006. *Study on the Economic and Technical Evolution of the Scientific Publication Markets in Europe*. http://ec.europa.eu/research/science-society/pdf/scientific-publication-study_en.pdf.

European Union. 2008. *CREST Report on the Internationalization of R&D. Facing the Challenge of Globalization*. Luxembourg: Office for Official Publications of the European Communities.

Evans, James. 2008. "Electronic Publication and the Narrowing of Science and Scholarship." *Science,* 18 July (321): 395–399.

———. Forthcoming. "Industry Collaboration and Secrecy in Academic Science." *Social Studies of Science*.

Evans, Paul. 1996. "Evaluating Growth Models Using Cross-Country Variances." *Journal of Economic Dynamics and Control* (20): 1027–1049.

Ezrahi, Yaron. 1978. "The Political Contexts of Scientific Indicators." In Yahuda Elkhana et al., eds., *Towards a Metric of Science*. New York: Wiley.

———. 1990. *The Descent of Icarus*. Cambridge, Mass.: Harvard University Press.

Fagin, R.; Halpern, J.; Moses, Y.; and Vardi, M. 1995. *Reasoning about Knowledge*. Cambridge, Mass.: MIT Press.

Fallis, George. 2007. *Multiversities, Ideas, and Democracy*. Toronto: University of Toronto Press.

Farrell, Joseph; and Shapiro, Carl. 2008. "How Strong Are Weak Patents?" *American Economic Review* (98): 1347–1369.

Faulkner, Wendy. 1994. "Conceptualizing Knowledge Used in Innovation." *Science, Technology and Human Values* (19): 425–458.

Faust, Drew Galpin. 2009. "The University's Crisis of Purpose." *New York Times,* September 6.

Fazeli, Sam. 2005. "The European Biotech Sector: Could It Achieve More?" *Journal of Commercial Biotechnology* (12:1): 10–19.

Feinstein, Alvan. 2003. "Scholars, Investigators and Entrepreneurs." *Perspectives in Biology and Medicine* 46: 234–253.

Feldman, Maryann; Link, Albert; and Siegel, Donald. 2002. *The Economics of Science and Technology.* Boston: Kluwer.

Feldman, Maryann; and Link, Albert, eds. 2001. *Innovation Policy in the Knowledge-Based Economy.* Boston: Kluwer.

Feller, William. 1971. *An Introduction to Probability Theory and Applications.* New York: Wiley.

Field, Alexander. 2003. "The Most Technologically Progressive Decade of the Century." *American Economic Review* (93): 1300–1413.

Finder, Alan. 2008. "At One University, Tobacco Money Is a Secret." *New York Times,* May 22.

Fisk, Catherine. 1998. "Removing the Fuel of Interest from the Fires of Genius." *University of Chicago Law Review* 65: 1127–1198.

———. 2009. *Working Knowledge.* Chapel Hill: University of North Carolina Press.

Fisher, Jill. 2006. "Co-Ordinating 'Ethical' Clinical Trials: The Role of Research Coordinators in the Contract Research Industry." *Sociology of Health & Illness* (28): 678–694.

———. 2008. *Medical Research for Hire: The Political Economy of Pharmaceutical Clinical Trials.* New Brunswick, N.J.: Rutgers University Press.

Fisher, Morris. 2003. "Physicians and the Pharmaceutical Industry." *Perspectives in Biology and Medicine* 46: 254–272.

Flaherty, Mary P.; Nelson, Deborah; and Stephan, Joe. 2000. "The Body Hunters." *Washington Post,* December 18, p. A1.

Flanagin, A.; Carey, L.; Fontanarosa, P.; Phillips, Stephanie, G.; and Pace, Brian P. et al. 1998. "Prevalence of Articles with Honorary Authors and Ghost Authors in Peer-Reviewed Medical Journals." *Journal of the American Medical Association* 280: 222–224.

Foray, Dominique. 2004. *The Economics of Knowledge.* Cambridge, Mass.: MIT Press.

Forman, Paul. 1987. "Beyond Quantum Electronics." *Historical Studies in the Physical Sciences* 18: 149–229.

———. 2007. "The Primacy of Science in Modernity, of Technology in Postmodernity, and of Ideology in the History of Technology." *History and Technology* (23): 1–152.

Forster, David. 2002. "Independent Institutional Review Boards." *Seton Hall Law Review* 32: 513–523.

Foss, Nicolai. 1998. "The New Growth Theory: Some Intellectual Growth Accounting." *Journal of Economic Methodology* (5): 223–246.

Foucault, Michel. 2004. *Naissance de la biopolitique.* Paris: Editions Gallimard.

———. 2008. *The Birth of Biopolitics.* Basingstoke, U.K.: Palgrave.

Fourcade, Marion. 2006. "The Construction of a Global Profession." *American Journal of Sociology* (117): 145–194.

———. 2008. *Economists and Societies.* Princeton: Princeton University Press.

Fox, Justin. 2009. "In the Long Run." *New York Times Book Review,* November 1, p. 13.

Fox, Robert; and Guagnini, Anna. 1999. *Laboratories, Workshops and Sites.* Berkeley: University of California Press.

Frangioni, John. 2008. "The Impact of Greed on Academic Medicine." *Nature Biotechnology* (26): 503–507.

Frank, Arthur. 2002. "What's Wrong with Medical Consumerism?" In S. Henderson and A. Petersen, eds., *Consuming Health*. London: Routledge.
Frank, David; and Gabler, Jay. 2006. *Reconstructing the University*. Stanford: Stanford University Press.
Freeburg, Ruth. 2005. "Is It Time for Compulsory Licensing of Biotech Tools?" *Buffalo Law Review* (53): 351.
Freeman, Chris. 1996. "The Greening of Technology and Models of Innovation." *Technological Forecasting and Social Change* (53): 27–39.
Freeman, Richard. 2007. "Globalization of the Scientific/Engineering Workforce and National Security." In Galama and Hosek 2007.
Freudenberg, William. 2005. "Seeding Science, Courting Conclusions." *Sociological Forum* (20:1): 3–33.
Frew, Sarah; Sammut, Stephen; Shore, Alysha; Ramjist, Joshua; Al-Bader, Sara; Rezaie, Rahim; Daar, Abdallah S.; and Singer, Peter. 2008. "Chinese Health Biotech and the Billion-Patient Market." *Nature Biotechnology* (26): 37–53.
Frickel, Scott; and Moore, Kelly, eds. 2006. *The New Political Sociology of Science*. Madison: University of Wisconsin Press.
Friedberg, M.; Saffran, B.; Stinson, T.; Nelson, Wendy; and Bennett, Charles E. 1999. "Evaluation of Conflict of Interest in Economic Analyses of New Drugs Used in Oncology." *Journal of the American Medical Association* 282: 1453–1457.
Friedman, Lissy; Daynard, Richard; and Banthin, Christopher. 2005. "How Tobacco-Friendly Science Escapes Scrutiny in the Courtroom." *American Journal of Public Health* (95 Suppl. 1): S16-S20.
Friedman, Milton. 1951. "Neo-Liberalism and Its Prospects." *Farmand* (February 17): 89–93.
———. 1953. *Essays in Positive Economics*. Chicago: University of Chicago Press.
———. 1962. *Capitalism and Freedom*. Chicago: University of Chicago Press.
———. 1973. "Frustrating Drug Development." *Newsweek*, January 8, p. 49.
———. 1978. Introduction to William Simon, *A Time for Truth*. New York: McGraw-Hill.
———. 1981. "An Open Letter on Grants." *Newsweek*, May 18, p. 99.
Friedman, Milton; and Friedman, Rose. 1998. *Two Lucky People*. Chicago: University of Chicago Press.
Fuller, Steve. 2000. *Thomas Kuhn: A Philosophical History for Our Times*. Chicago: University of Chicago Press.
———. 2002. *Knowledge Management Foundations*. London: Butterworth.
———. 2005. "Knowledge as Product and Property." In Nico Stehr and Volker Meja, eds. *Society and Knowledge*. 2nd ed. New Brunswick, N.J.: Transaction.
Gad, Shane C. 2003. *The Selection and Use of Contract Research Organizations*. London: Taylor & Francis.
Galama, Titus; and Hosek, James, eds. 2007. *Perspectives on US Competitiveness in Science and Technology*. Santa Monica, Calif.: RAND.
———. 2008. *US Competitiveness in Science and Technology*. Santa Monica, Calif.: RAND.
Galison, Peter. 2003. *Einstein's Clocks, Poincarés Maps*. New York: Norton.

———. 2004. "Mirror Symmetry." In Norton Wise, ed., *Growing Explanations: Historical Perspectives on Recent Science*. Durham, N.C.: Duke University Press.

———. 2008. "Removing Knowledge: The Logic of Modern Censorship." In Proctor and Scheibinger 2008.

Galison, Peter; and Hevly, Bruce, eds. 1992. *Big Science*. Stanford: Stanford University Press.

Galison, Peter; and Stump, D.J., eds. 1996. *The Disunity of Science*. Stanford: Stanford University Press.

Garcia-Quevedo, Jorge. 2004. "Do Public Subsidies Complement Business R&D?" *Kyklos* (57): 87–102.

Gaudillière, Jean-Paul; and Löwy, Ilana, eds. 1998. *The Invisible Industrialist*. London: Macmillan.

Gauffriau, Marianne; Larsen, Peder; and Maye, Isabelle. 2008. "Comparisons of Different Results of Publication Counting Using Different Methods." *Scientometrics* (77): 147–176.

Geanakoplos, John. 1992. "Common Knowledge." *Journal of Economic Perspectives* (6): 53–82.

Geenens, Raf; and Tinnevelt, Ronald, eds. 2009. *Does the Truth Matter?* Berlin: Springer.

Geiger, Roger. 1997. "Science and the University: Patters from U.S. Experience." In Krige and Pestre 1997, 159–174.

———. 2004. *Knowledge and Money*. Stanford: Stanford University Press.

Geiger, Roger; and Sá, Creso. 2008. *Tapping the Riches of Science: Universities and the Promise of Economic Growth*. Cambridge, Mass.: Harvard University Press.

Gelijns, Annetine; and Their, Samuel. 2002. "Medical Innovation and Institutional Interdependence." *Journal of the American Medical Association* 287: 72–78.

Georgescu-Roegen, Nicholas. 1971. *The Entropy Law and the Economic Process*. Cambridge, Mass.: Harvard University Press.

———. 1975. "The Measure of Information—a Critique." In J.Rose and C. Bilciu, eds., *Modern Trends in Cybernetics and Systems*, 187–217. Berlin: Springer Verlag.

———. 1976. *Energy and Economic Myths*. New York: Pergamon.

Getz, Kenneth; and de Bruin, Annick. 2000. "Breaking the Development Speed Barrier: Assessing Successful Practices of the Fastest Drug Development Companies." *Drug Information Journal* 34: 725–736.

Ghersi, Davina; Campbell, E.G.; Pentz, R.; and Cox, Macpherson C. 2004. "The Future of Institutional Review Boards." *Lancet Oncology* (5): 325–329.

Gibbons, Michael. 2003. "Globalization and the Future of Higher Education." In Gilles Breton and Michel Lambert. eds., *Universities and Globalization*, 107–116. Quebec: UNESCO.

Gibbons, Michael; Limoges, Camille; Nowotny, Helga; Schwartzman, S.; Scott, Peter; and Trow, Martin. 1994. *The New Production of Knowledge*. London: Sage.

Gigerenzer, Gerd; and Murray, David. 1987. *Cognition as Intuitive Statistics*. Hillsdale, N.J.: Erlbaum.

Gillespie, Tarleton. 2006. "Engineering a Principle: End-to-End in the Design of the Internet." *Social Studies of Science* (36): 427–458.

Gingras, Yves. 2002. "Beautiful Mind, Ugly Deception: The Bank of Sweden Prize in Economics." *Post Autistic Economics Review* no. 17, December 4.

Gingras, Yves; and Gosselin, Pierre-Marc. 2007. "The Emergence and Evolution of the Expression 'Conflict of Interest' in *Science*." Unpublished manuscript.

Ginsbach, Pam. 2009. "Data Agencies' Budgets Have Some Funding Restored." *NABE News,* http://www.nabe.com.

Giocoli, Nicola. 2003. *Modeling Rational Agents.* Cheltenham, U.K.: Elgar.

Giombetti, Ric. 2002. "Suicide Science—Paxil's Friendly Ghostwriter?" April 8. Available at www.counterpunch.org (last visited March 18, 2005).

Glantz, Sheldon, Slade, John; and Bero, Lisa. 1996. *The Cigarette Papers.* Berkeley: University of California Press.

Glanzel, Wolfgang; Debackere, Konrad; and Meyer, Martin. 2008. "Triad or Tetrad? On Global Changes in a Dynamic World." *Scientometrics* (74): 71–88.

Glimcher, Paul. 2003. *Decisions, Uncertainty and the Brain: The Science of Neuroeconomics.* Cambridge, Mass.: MIT Press.

Godin, Benoît. 2003. "Measuring Science: Is There Basic Research without Statistics?" *Social Science Information* (42:1): 57–90.

———. 2004. "The New Economy: What the Concept Owes to the OECD." *Research Policy* (33): 679–690.

———. 2005. *Measurement and Statistics on Science and Technology.* London: Routledge.

———. 2006a. "The Linear Model of Innovation." *Science, Technology and Human Values* (31): 639–667.

———. 2006b. "The Birth of Bibliometrics." *Scientometrics* (68): 109–133.

———. 2008. "In the Shadow of Schumpeter." *Minerva* (46): 343–360.

Goldacre, Ben. 2009. "The Danger of Drugs . . . and Data." *Guardian,* May 9.

Goldman, Michael. 2005. *Imperial Nature.* New Haven: Yale University Press.

Goliszek, Andrew. 2003. *In the Name of Science.* New York: St. Martin's.

Goodstein, David. 2010. *On Fact and Fraud.* Princeton: Princeton University Press.

Goodyear, Michael. 2006. "Learning from the TGN1412 Trial." *British Medical Journal* (332): 677–678.

Goolsbee, Austan. 1998. "Does Government R&D Policy Mainly Benefit Scientists and Engineers?" *American Economic Review* (88:2): 298–302.

Gordon, R. 2000. "Does the New Economy Measure Up to the Great Inventions of the Past?" *Journal of Economic Perspectives* (14): 49–72.

Gore, Al. 2007. *The Assault on Reason.* New York: Penguin.

Gorg, Holger; and Strobl, Eric. 2007. "The Effect of R&D Subsidies on Private R&D." *Economica* (74): 215–234.

Gottinger, Hans-Werner; and Umali, Celia. 2008. "The Evolution of the Pharmaceutical-Biotechnology Industry." *Business History* (50): 583–601.

Gotzsche, Peter; Hrobjartsson, Asbjorn; Johansen, Helle; Haar, Mette; and Altman, Douglas. 2007. "Ghost Authorship in Industry-Initiated Randomised Trials." *PLOS Medicine* (4:1): 47–52.

Graham, Margaret. 1985. "Corporate Research and Development: The Latest Transformation." *Technology in Society* 7: 179–195.

Grandin, Karl; Wormbs, Nina; and Widmalm, Sven, eds. 2004. *The Science-Industry Nexus.* Sagamore Beach, Mass.: Science History Publishers.

Grant, Bob. 2009a. "Merck Published Fake Journal." *The Scientist,* April 30.

———. 2009b. "Elsevier Published Six Fake Journals." *The Scientist,* May 7.

Gravois, John. 2006. "Tracking the Invisible Faculty." *Chronicle of Higher Education,* December 15.

Gray, George. 1935. "Science and Profits." *Harper's Magazine* (172): 539–549.

Gray, Steven. 2009. "Can Corporate funding Save Endangered College Classes?" *Time Magazine,* July 14.

Green, Peter. 2004. "Material Transfer Agreements." *Microbiologist,* June: 36.

Greenberg, Daniel. 2001. *Science, Money and Politics.* Chicago: University of Chicago Press.

———. 2007. *Science for Sale.* Chicago: University of Chicago Press.

Grens, Kerry. 2007. "An Economic Gamble." *The Scientist* (21:7): 28.

Griliches, Zvi. 1958. "Research Costs and Social Returns: Hybrid Corn." *Journal of Political Economy* (66): 419–431.

———. 1978. "Economic Problems of Measuring Returns to Science." In Yehuda Elkhana et al., eds., *Towards a Metric of Science.* New York: Wiley.

———. 1980. "R&D and the Productivity Slowdown." *American Economic Review, Papers and Proceedings* (70:2): 343–348.

———. 1986. "Productivity, R&D, and Basic Research at the Firm Level in the 1970s." *American Economic Review* (76): 141–154.

Grossman, Gene; and Helpman, Elhanan. 1994. "Endogenous Innovation in the Theory of Growth." *Journal of Economic Perspectives* (8): 23–44.

Grossman, Sanford. 1989. *The Informational Role of Prices.* Cambridge, Mass.: MIT Press.

Grossman, Sanford; and Stiglitz, Joseph. 1980. "On the Impossibility of Informationally Efficient Markets." *American Economic Review* (70): 393–408.

Gruber, Carol. 1995. "The Overhead System in Government-Sponsored Academic Science: Origins and Early Development." *Historical Studies in the Physical and Biological Sciences* 25(2): 241–268.

Grushcow, Jeremy. 2004. "Measuring Secrecy: A Cost of the Patent System Revealed." *Journal of Legal Studies* (33): 59–84.

Guena, Aldo; Salter, Ammon; and Steinmueller, Edward, eds. 2003. *Science and Innovation: Rethinking the Rationales for Funding and Governance.* Cheltenham, U.K.: Elgar.

Guess, Andy. 2007. "American Science Plateau." *Inside Higher Ed,* July 20, www.insidehighered.com.

Guggenheim, Michael; and Nowotny, Helga. 2003. "Joy in Repetition Makes the Future Disappear." In Bernward Joerges and Helga Nowotny, eds., *Social Studies of Science and Technology: Looking Back, Looking Ahead,* 229–258. Dordrecht, Netherlands: Kluwer.

Guilhot, Nicolas. 2005. *The Democracy Makers.* New York: Columbia University Press.

Gulbranson, Christine; and Audretsch, David. 2008. *Accelerating Commercialization of University Innovation.* Kansas City, Mo.: Kauffman Foundation.

Guston, David; and Kenniston, Kenneth, eds. 1994. *The Fragile Contract: University Science and the Federal Government.* Cambridge, Mass.: MIT Press.

Guterman, Lila. 2008. "Crime and Punishment: Research Misconduct Verdicts May Not End Careers." *Chronicle of Higher Education,* August 8.

Guttag, Eric. 2004. "Immunizing University Research from Patent Infringement: Implications of *Madey v Duke University.*" *Industry and Higher Education* (18): 157–165.

Haack, Susan. 2008. "Of Truth in Science and the Law." *Brooklyn Law Review* (73: 2).

Hackett, Ed; Amsterdamska, Olga; Lynch, Michael; and Wajcman, Judy, eds. 2007. The *Handbook of Science and Technology Studies.* 3rd ed. Cambridge, Mass.: MIT Press.

Hacohen, Mordachai. 2000. *Karl Popper: The Formative Years.* New York: Cambridge University Press.

Hands, D. Wade. 1994. "The Sociology of Scientific Knowledge and Economics." In Roger Backhouse, ed., *New Perspectives in Economic Methodology,* London: Routledge.

———. 2001. *Reflection without Rules.* New York: Cambridge University Press.

———. 2006. "(Continued) Reconsideration of Individual Psychology, Rational Choice and Demand Theory: Some Remarks on Three Recent Studies." *Revue de Philosophie Economique* (13): 3–48.

Hands, Wade; and Mirowski, Philip. 1998. "Harold Hotelling and the Neoclassical Dream." In Roger Backhouse, D. Hausman, U. Maki, and A. Salanti, eds., *Economics and Methodology: Crossing Boundaries,* edited by Roger Backhouse, D. Hausman, U.Maki and A. Salanti, 322–397. London: Macmillan.

Hansen, Stephen; Brewster, Amanda; and Asher, Jane. 2005. *Intellectual Property in the AAAS Community,* http://www.juergen-ernst.de/download_swpat/studie_sippi.pdf.

Hansen, Stephen; Brewster, Amanda; Ascher, Jana; and Kisielewski, Michael. 2006. *The Effects of Patenting in the AAAS Scientific Community,* http://sippi.aaas.org.

Harcourt, Geoff. 1972. *Some Cambridge Controversies in the Theory of Capital.* Cambridge: Cambridge University Press.

Hargreaves Heap, Shaun. 2002. "Making British Universities Accountable: In the Public Interest?" In Philip Mirowski and Esther-Mirjam Sent, eds., *Science Bought and Sold,* 387–411. Chicago: University of Chicago Press.

Harris, Gardiner. 2008. "Top Psychiatrist Didn't Report Drugmaker's Pay." *New York Times,* October 4.

Hart, David. 1998a. *Forged Consensus.* Princeton: Princeton University Press.

———. 1998b. "Antitrust and Technological Innovation." *Issues in Science and Technology* (15): 75–82.

———. 2001. "Antitrust and Technological Innovation in the U.S.: Ideas, Institutions, Decisions and Impacts, 1890–2000." *Research Policy* 30: 923–936.

———. 2009. "Accounting for Change in National Systems of Innovation: A Friendly Critique." *Research Policy* (38): 647–654.

Hartwell, R. Max. 1995. *A History of the Mont Pelerin Society.* Indianapolis: Liberty Fund.

Harvey, David. 2005. *A Brief History of Neoliberalism.* Oxford: Oxford University Press.

Hather, Gregory; Haynes, Winston; Higdon, Roger et al. 2010. "The United States of America and Scientific Research." *PLoS One* (5): e12203.

Hawthorne, Fran. 2003. *The Merck Druggernaught.* New York: Wiley.

Hayek, Friedrich. 1944. *The Road to Serfdom.* Chicago: University of Chicago Press.

———. 1948. *Individualism and Economic Order.* Chicago: Regnery.

———. 1952. *Counter-Revolution of Science.* Indianapolis: Liberty Press.

———. 1960. *The Constitution of Liberty.* Chicago: University of Chicago Press.

———. 1967. *Studies in Philosophy, Politics and Political Economy.* New York: Simon and Schuster.

———. 1973. *Law, Legislation and Liberty, vol.1.* Chicago: University of Chicago Press.

Healy, David. 2003. "In the Grip of the Python: Conflicts at the University–Industry Interface." *Science and Engineering Ethics* 9: 59–71.

Healy, David; and Cattell, D. 2003. "Interface between Authorship, Industry and Science in the Domain of Therapeutics." *British Journal of Psychiatry* 183: 22–27.

Hedgecoe, Adam; and Martin, Paul. 2003. "The Drugs Don't Work." *Social Studies of Science* 33(3): 327–364.

Hedges, Chris. 2009. *Empire of Illusion.* New York: Nation.

Hegde, Deepak; and Hicks, Diana. 2008. "The Maturation of Global Corporate R&D: Evidence from Activity of US Foreign Subsidiaries." *Research Policy* (37): 390–406.

Heller, Michael; and Eisenberg, Rebecca. 1998. "Can Patents Deter Innovation? The Anticommons in Biomedical Research." *Science* (280): 698.

Hemphill, Thomas. 2003. "Role of Competition Policy in the U.S. Innovation System." *Science and Public Policy* 30: 285–294.

Henderson, Rebecca; Jaffe, Adam; and Trajtenberg, Manuel. 1998. "Universities as a Source of Commercial Technology: University Patenting 1965–1988." *Review of Economics and Statistics* (80): 119–127.

Henry, Michelle; Cho, Mildred; Weaver, Meredith; and Merz, Jon. 2003. "A Pilot Survey of the Licensing of DNA Inventions." *Journal of Law, Medicine and Ethics* (31): 442.

Hess, Glenn. 2008. "Congress Gets Tough on Intellectual Property." *Chemical and Engineering News,* September 30.

Hesse, Monica. 2008. "Truth: Can You Handle It?" *Washington Post,* April 27.

Hicks, Diana. 1995. "Published Papers, Tacit Competencies." *Industrial and Corporate Change* 4: 401–424.

———. 2005. "America's Innovative Edge at Risk?" *Research Technology and Management* (48): 8–12.

———. 2007. "Global Research Competition Affects Measured US Academic Output." In Stephan and Ehrenberg 2007, 223–242.

———. 2008. "The US Research Enterprise in a Changing Global Science System." Georgia Institute of Technology Working Paper.

Hill, Christopher T. 2007. "The Post-Scientific Society." *Issues in Science and Technology,* Winter.

Hill, Derek; Rapoport, Alan; Lehming, Rolf; and Bell, Robert. 2007. *Changing US Output of Scientific Articles 1988–2003.* Arlington, Va.: NSF Division of Science Resource Statistics.

Ho, Cynthia. 2007. "Lessons from Labcorp v. Metabolite." *Santa Clara Computer and High Technology Law Journal* (37): 463–487.

Hochstettler, Thomas John. 2004. "Aspiring to Steeples of Excellence at German Universities." *Chronicle of Higher Education,* July 30.

Hodgson, Geoffrey. 2004. *The Evolution of Institutionalist Economics.* London: Routledge.

Hodgson, Geoffrey; and Rothman, Harry. 2001. "The Editors and Authors of Economics Journals: A Case of Institutional Oligopoly?" *Economic Journal* (109): 165–186.

Hollinger, David. 1990. "Free Enterprise and Free Inquiry: The Emergence of Laissez-Faire Communitarianism in the Ideology of Science in the U.S." *New Literary History* 21: 897–919.

———. 2000. "Money and Academic Freedom a Half Century after McCarthyism: Universities amid the Force Fields of Capital." In Peggie Hollingsworth, ed., *Unfettered Expression,* 161–184. Ann Arbor: University of Michigan Press.

Hong, Sungook. 1999. "Historiographic Layers in the Relationship between Science and Technology." *History and Technology* (15): 289–311.

Hopkins, Michael; Ibaretta, Dolores; Gaisser, Sibylle; Enzing, Christien M.; Ryan, Jim; Martin, Paul A.; Lewis, Graham; Detmar, Symone; van den Akker-van Marle, M. Elske; Hedgecoe, Adam M.; Nightingale, Paul; Dreiling, Marieke; Hartig, K. Juliane; Vullings, Wieneke; and Forde, Tony. 2006. "Putting Pharmacogenetics into Practice." *Nature Biotechnology* (24): 403–410.

Hopkins, Michael; Martin, Paul; Nightingale, Paul; Kraft, Alison; and Mahdi, Surya. 2007. "The Myth of the Biotech Revolution: An Assessment of Technological, Clinical and Organizational Change." *Research Policy* (36): 566–589.

Horta, Hugo; and Veloso, Francisco. 2007. "Opening the Box: Comparing EU and US Scientific Output by Field." *Technological Forecasting and Social Change* (74): 1334–1356.

Horton, Richard. 2002. "The Hidden Research Paper." *Journal of the American Medical Association* 287: 2775–2778.

Hounshell, David. 1996. "The Evolution of Industrial Research in the U.S." In Richard Rosenbloom and William Spencer, eds., *Engines of Innovation,* 13–85. Cambridge, MA: Harvard Business School Press.

———. 2000. "The Medium Is the Message." In A. Hughes and T. Hughes, eds., *Systems, Experts and Computers,* 255–310. Cambridge, Mass.: MIT Press.

Hounshell, David; and Smith, John. 1988. *Science and Corporate Strategy.* New York: Cambridge University Press.

Huber, Peter. 1991. *Galileo's Revenge: Junk Science in the Courtroom.* New York: Basic.

———. 2006. "Of Pills and Profits: In Defense of Big Pharma." *Commentary,* July: 21–28.

Huggett, Brady. 2008. "Big Pharma Swallows Biotech's Pride." *Nature Biotechnology* (26: 9): 955–956.

Hughes, Sally. 2001. "Making Dollars out of DNA." *Isis* (92): 541–575.

Hughes, Sue. 2002. "Does the Pharmaceutical Industry Have Too Much Control Over Clinical Trials?" *Heartwire* (15 November).

Hull, David. 1988. *Science as a Process.* Chicago: University of Chicago Press.

Huws, Ursula; Dahlmann, S.; and Flecker, J. 2004. *Outsourcing of ICT and Related Services in the EU,* available at www.eurofound.eu.int.

Ingber, Stanley. 1984. "The Marketplace of Ideas: A Legitimating Myth." *Duke Law Journal* (1): 1–91.

Israel, Paul. 1992. *From Machine Shop to Industrial Laboratory.* Baltimore: Johns Hopkins University Press.

Istrail, Sorin; Sutton, Granger G.; Florea, Liliana et al. 2004. "Whole Genome Shotgun Assembly and Comparison of Human Genome Assemblies." *Proceedings of the National Academy of Sciences* (101:7): 1916–1921.

Ivry, Sara. 2006. "Did an Exposé Help Sink Harvard's President?" *New York Times,* February 27.

Jacobs, Meg. 2008. "The Conservative Struggle and the Energy Crisis." In Schulman and Zelizer 2008.

Jacobs, Struan. 2002. "The Genesis of the 'Scientific Community.'" *Social Epistemology* 16: 157–168.

Jacoby, Susan. 2008. *Age of American Unreason.* New York: Pantheon.

Jacques, Peter; Dunlap, Riley; and Freeman, Mark. 2008. "The Organization of Denial: Conservative Think Tanks and Environmental Skepticism." *Environmental Politics* (17): 349–385.

Jaffe, Adam; and Lerner, Josh. 2004. *Innovation and Its Discontents.* Princeton: Princeton University Press.

Jaffe, Sam. 2004. "Ongoing Battle over Transgenic Mice." *The Scientist* (18:14): 46–49.

Jang, Yong Suk. 2000. "The Worldwide Founding of Ministries of Science and Technology, 1950–90." *Sociological Perspectives* (43): 247–270.

Jaquette, I. 2007. "Merck vs. Integra Life Sciences." *American Journal of Law and Medicine* (33): 97–117.

Jasanoff, Sheila. 2005. *Designs on Nature.* Princeton: Princeton University Press.

———. 2006. "Transparency in Public Science: Purposes, Reasons, Limits." *Law and Contemporary Problems* (69): 21–45.

———. 2008. "Representation and Re-Presentation in Litigation Science." *Environmental Health Perspectives* (116): 123–129.

Jasanoff, Sheila; Markle, Gerald; Peterson, James; and Pinch, Trevor, eds. 1995. *Handbook of Science and Technology Studies.* Thousand Oaks, Calif.: Sage.

Jaschik, Scott. 2007. "Defeat for For-Profit Model." *Inside Higher Ed.com,* January 12.

———. 2008. "The Shrinking Professoriate." *Inside Higher Ed.com,* March 12.

Jensen, Richard; and Thursby, Marie. 2001. "Proofs and Prototypes for Sale: Licensing of University Inventions." *American Economic Review* (91): 240–259.

Jevons, William Stanley. 1905. *The Principles of Science.* 2nd ed. London: Macmillan.

Johns, Adrian. 2006. "Intellectual Property and the Nature of Science." *Cultural Studies* (20): 145–164.

Johnson, Ann. 2004. "The End of Pure Science: Science Policy from Bayh-Dole to the NNI." In D. Baird, A. Nordmann, and J. Schummer, eds., *Discovering the Nanoscale,* 217–230. Amsterdam: IOS Press.

Johnson, Erica. 2003. "Medical Ghostwriting." *CBC Marketplace* (air date March 25).

Johnson, George. 2002. "At Lawrence Berkeley, Physicists Say Colleague Took Them for a Ride." *New York Times,* October 15.

Johnson, Harry. 1972. "Some Economic Aspects of Science." *Minerva* (10): 10–18.

Johnson, Jeffrey. 1990. *The Kaiser's Chemists.* Chapel Hill: University of North Carolina Press.

Johnson, Joseph. 2009. "Less Research, More Economies of Scale." *Inside Higher Ed,* June 19.

Johnson, Stephen B. 2002. *The Secret of Apollo.* Baltimore: Johns Hopkins University Press.

Jones, Charles. 1995a. "R&D Based Models of Economic Growth." *Journal of Political Economy* (103): 759–784.

———. 1995b. "Time Series Tests of Endogenous Growth Models." *Quarterly Journal of Economics* (100): 1127–1170.

———. 2005. "Growth and Ideas." In Philippe Aghion and Steven Durlauf, eds., *Handbook of Economic Growth,* 1063–1111. Amsterdam: Elsevier.

Jones, Hywel. 1976. *An Introduction to Modern Theories of Economic Growth.* New York: McGraw-Hill.

Jong, Simcha. 2008. "Academic Organizations and New Industrial Fields: Berkeley and Stanford." *Research Policy* (37): 1267–1282.

Jordan, J. S. 1982. "The Competitive Allocation Process Is Informationally Efficient Uniquely." *Journal of Economic Theory* (28): 1–18.

Judson, Horace. 2004. *The Great Betrayal: Fraud in Science.* New York: Harcourt.

Kahin, Brian. 2001. "The Expansion of the Patent System." *First Monday* 6(1).

Kahn, Zorina; and Sokoloff, Kenneth. 2001. "The Early Development of Intellectual Property Institutions in the United States." *Journal of Economic Perspectives* 15.

Kaitin, Kenneth; and DiMasi, Joseph. 2000. "Measuring the Pace of New Drug Development in the User Fee Era." *Drug Information Journal* 34: 673–680.

Kaitin, Kenneth; and Healy, Elaine. 2000. "The New Drug Approvals of 1996, 1997 and 1998." *Drug Information Journal* 34: 1–14.

Kapczynski, Amy. 2008. "The Access to Knowledge Mobilization and the New Politics of Intellectual Property." *Yale Law Journal* (117): 804–885.

Kay, Lily. 1993. *The Molecular Vision of Life.* New York: Oxford University Press.

———. 2000. *Who Wrote the Book of Life?* Stanford: Stanford University Press.

Kealey, Terrence. 1996. *The Economic Laws of Scientific Research.* London: Macmillan.

Keen, Andrew. 2007. *The Cult of the Amateur.* New York: Doubleday.

Keim, Brandon. 2009. "Computer Program Self-Discovers Laws of Physics." *Wired,* April.

Kelderman, Eric. 2009. "Public Colleges Consider Privatization as a Cure for the Common Recession." *Chronicle of Higher Education,* May 1.

Kenney, Martin. 1986. *Biotechnology: The University-Industry Complex.* New Haven: Yale University Press.

Kesan, Jay; and Gallo, Andres. 2006. "Why Bad Patents Survive in the Market and How Should We Change?" *Emory University Law Journal* (55):61.

Kessler, David. 2001. *A Question of Intent.* New York: Public Affairs.

Kevles, Daniel. 1995. *The Physicists.* Cambridge, Mass.: Harvard University Press.

———. 1994. "Ananda Chakrabarty Wins a Patent." *Historical Studies in the Physical and Biological Sciences* 25: 111–136.

———. 1998. "Diamond v. Chakrabarty and Beyond: The Political Economy of Patenting Life." In Arnold Thackray, ed., *Private Science: Biotechnology and the Rise of the Molecular Sciences,* 65–79. Philadelphia: University of Pennsylvania Press.

———. 2001. "Principles, Property Rights and Profits: Historical Reflections on University/Industry Tensions." *Accountability in Research* (8): 293–307.

———. 2002. "Of Mice and Money: The Story of the World's First Animal Patent." *Daedalus* (131): 78.

Khurana, Rakesh. 2007. *From Higher Aims to Hired Hands.* Princeton: Princeton University Press.

King, Christopher. 2007. "Of Nations and Top Citations." *Science Watch,* May/June. Available at http://www.sciencewatch.com/may-june2007/index.html.

King, David. 2004. "The Scientific Impact of Nations: What Different Countries Get for Their Research Spending." *Nature* (430): 311–316.

Kirp, David. 2003. *Shakespeare, Einstein and the Bottom Line.* Cambridge, Mass.: Harvard University Press.

Kitch, Edmund. 1977. "The Nature and Function of the Patent System." *Journal of Law and Economics* (20:2): 265–290.

Kitcher, Philip. 1993. *The Advancement of Science.* New York: Oxford University Press.

———. 2001. *Science, Truth and Democracy.* New York: Oxford University Press.

Kjaergard, L.; and Als-Nielsen, B. 2002. "Association between Competing Interests and Authors' Conclusion: Epidemiological Study of Randomized Clinical Trials in the BMJ." *British Medical Journal* 325: 249–252.

Klaes, Matthias; and Sent, Esther-Mirjam. 2005. "A Conceptual History of the Emergence of Bounded Rationality." *History of Political Economy* (37): 27–60.

Klamer, Arjo; and van Dalen, Hendrik. 2005. "Is Science a Case of Wasteful Competition?" *Kyklos* (58): 395–414.

Kleinman, Daniel. 1995. *Politics on the Endless Frontier.* Durham, N.C.: Duke University Press.

———. 2003. *Impure Cultures: University Biology and the World of Commerce.* Madison: University of Wisconsin Press.

Kleinman, Daniel; and Vallas, Steven. 2006. "Contradiction in Convergence: Universities and Industry in the Biotechnology Field." In Frickel and Moore 2006, 35–62.

Kline, Ronald. 1995. "Constructing Technology as Applied Science." *Isis* (86): 194–221.

Kocherlakota, N.; and Yi, K. 1997. "Is There Endogenous Long-Run Growth?" *Journal of Money, Credit and Banking* (29): 235–262.

Koertge, Noretta, ed. 1998. *A House Built on Sand*. New York: Oxford University Press.

Kohler, Robert. 1991. *Partners in Science*. Chicago: University of Chicago Press.

Kolowich, Steve. 2009. "What Doomed Global Campus?" *Inside Higher Ed*, September 3.

Koopmans, Tjalling. 1957. *Three Essays on the State of Economic Science*. New York: McGraw-Hill.

Kourany, Janet. 2008. "Philosophy of Science: A Subject with a Great Future." *Philosophy of Science* (75): 767–778.

———. 2010. *Philosophy of Science after Feminism*. Oxford: Oxford University Press.

Kraemer, Sylvia. 2006. *Science and Technology Policy in the United States*. New Brunswick, N.J.: Rutgers University Press.

Kragh, Helge. 1999. *Quantum Generations*. Princeton: Princeton University Press.

Kreilkamp, Karl. 1971. "Hindsight and the Real World of Science Policy." *Science Studies* (1:1): 43–66.

Kreps, David. 1990. *A Course in Microeconomic Theory*. Princeton: Princeton University Press.

Krige, John. 2006. *American Hegemony and the Postwar Reconstruction of Science in Europe*. Cambridge, Mass.: MIT Press.

Krige, John; and Pestre, Dominique, eds. 1997. *Science in the Twentieth Century*. Amsterdam: Harwood.

Krimsky, Sheldon. 2003. *Science in the Private Interest*. Lanham, Md.: Rowman & Littlefield.

Kroto, Harry. 2007. "The Wrecking of British Science." *The Guardian*, May 22.

Ku, Katherine; and Henderson, James. 2007. "The MTA—Rip It Up and Start Again?" *Nature Biotechnology* (25:7): 721–723.

Kuemmerle, Walter. 1999. "Foreign Direct Investment in Industrial Research in the Pharmaceutical and Electronics Industries." *Research Policy* (28): 179–193.

Kunin, Stephen; Nagumo, Mark; Stanton, Brian; Therkorn, Linda S.; and Walsh, Stephen. 2002. "Reach-Through Claims in the Age of Biotechnology." *American University Law Journal* 51: 608–638.

Kupiec-Weglinski, J. W. 2003. "Interactions between the Pharmaceutical Industry and Academia Revisited." *Transplantation Proceedings* 35: 1238–1239.

Kusch, Martin. 2002. *Knowledge by Agreement*. Oxford: Oxford University Press.

Labi, Aisha. 2010. "Times Higher Education Releases New Rankings, but Will They Appease Skeptics?" *Chronicle of Higher Education*, September 15.

Labinger, Jay; and Collins, Harry, eds. 2001. *The One Culture?* Chicago: University of Chicago Press.

Lamoreaux, Naomi; Raff, Daniel; and Temin, Peter. 2003. "Beyond Markets and Hierarchies." *American Historical Review* (108): 404–433.

———. 2004. "Against Whig History." *Enterprise and Society* (5): 376–387.

Landau, Ralph, ed. 1973. *Regulating New Drugs.* Chicago: Center for Policy Study.

Landefeld, J.; and Mataloni, R. 2004. "Offshore Outsourcing and Multinational Companies." Bureau of Economic Analysis, Department of Commerce.

Landes, William; and Posner, Richard. 2003. *The Intellectual Structure of Intellectual Property Law.* Cambridge, Mass.: Harvard University Press.

———. 2004. *The Political Economy of Intellectual Property Law.* Washington: Brookings-AEI.

Langlois, Richard. 1985. "From Knowledge of Economics to the Economics of Knowledge." *Research in the History and Methodology of Economics* (3): 225–235.

———. 2004. "Chandler in the Larger Frame." *Enterprise and Society* (5): 355–375.

Larédo, Philippe; and Mustar, Philippe, eds. 2001. *Research and Innovation Policies in the New Global Economy: An International Comparative Analysis.* Cheltenham, U.K.: Elgar.

Larsen, Peder; and von Ins, Markus. 2010. "The Rate of Growth in Scientific Publication and the Decline in Coverage Provided by Science Citation Index." *Scientometrics* (84): 575–603.

Latour, Bruno. 2005. *Reassembling the Social.* Oxford: Oxford University Press.

Latour, Bruno; and Woolgar, Steve. 1979. *Laboratory Life.* Thousand Oaks, Calif.: Sage.

Lave, Rebecca, Randalls, Sam; and Mirowski, Philip. 2011. "STS and the Commercialization of Science: An Introduction." *Social Studies of Science.*

Lavoie, Don. 1985. *Rivalry and Central Planning.* New York: Cambridge University Press.

Lécuyer, Christophe. 1995. "MIT, Progressive Reform and Industrial Science." *Studies in the History of Physical and Biological Sciences* 26(1): 35–88.

———. 1998. "Academic Science and Technology in the Service of Industry." *American Economic Review* (88:2): 28–33.

Lederman, Doug. 2007. "Inexorable March to a Part-Time Faculty." *Inside Higher Ed,* March 28.

———. 2008a. "Call for Crackdown on Research Conflicts." *Inside Higher Ed,* January 21, http://www.insidehighered.com/news/archive/%28year%29/2008/%28month%29/1.

———. 2008b. "Unprecedented 2-Year Decline for US Science Funds." *Inside Higher Ed,* August 25.

———. 2009. "Painful Lesson on Patents." *Inside Higher Ed,* October 2.

———. 2010. "NIH Chasing Conflicts of Interest." *Inside Higher Ed,* May 21.

Lee, Kyu-Sang. 2006. "Mechanism Design Theory Embodying an Algorithm-Centered Vision." In Mirowski and Hands 2006, 283–304.

Lemke, Thomas. 2001. "The Birth of Bio-Politics: Michel Foucault's Lecture at the College de France on Neo-Liberal Governmentality." *Economy and Society* (30): 190–207.

Lemonick, Michael. 2006. "Are We Losing our Edge?" *Time Magazine,* February 5, available at http://www.time.com/time/magazine/article/0,9171,1156575,00.html.

Lemonick, Michael; and Goldstein, Andrew. 2000. "At Your Own Risk." *Time* (April 22): 46–56.

Lenoir, Timothy. 1998. "Revolution from Above: The Role of the State in Creating the German Research System." *American Economic Review* (88:2): 22–27.

Lenzer, Jeanne. 2008. "Contract Research Organizations: Truly Independent Research?" *British Medical Journal,* September 13 (337):602–606.

Leonard, Thomas C. (2004) "Making Betty Crocker Assume the Position." *Journal of the History of Economic Thought* 26: 115–122.

Leontief, Wassily. 1970. Comment on Chipman. In Raymond Vernon, ed., *The Technology Factor in International Trade.* New York: Columbia University Press.

Leslie, Stuart. 1993. *The Cold War and American Science.* New York: Columbia University Press.

Lessig, Lawrence. 2001. *The Future of Ideas.* New York: Random House.

———. 2004. *Free Culture.* http://www.free-culture.org.

Lester, David; and Connor, Jane. 2003. "New Tools for Clinical Practice." *American Pharmaceutical Outsourcing* (May/June), available at www.americanpharmaceuticaloutsourcing.com/articles.

Levin, Richard; Klevorick, Alvin; Nelson, Richard; and Winter, Sidney. 1987. "Appropriating the Returns from Industrial Research and Development." *Brookings Papers on Economic Activity* (3): 783–831.

Levin, Sharon. 2005. "The Health of the Scientific Enterprise in the US." Paper presented to Trilateral Seminar on Science Policy, Tokyo.

Lewin, Tamar. 2008. "US Universities Rush to Set Up Campuses Abroad." *New York Times,* February 10.

———. 2009. "State Colleges Also Face Cuts in Ambitions." *New York Times,* March 17.

Lexchin, Joel, Bero, Lisa; Djulbegovic, Benjamin; and Clark, Otavio. 2003. "Pharmaceutical Industry Sponsorship and Research Outcome and Quality." *British Medical Journal* 326: 1167–1177.

Lexchin, Joel; and Light, Donald. 2006. "Commercial Influence and the Contents of Medical Journals." *British Medical Journal* (332): 1444–1447.

Leydesdorff, Loet. 2004. "The University-Industry Knowledge Relationship: Analyzing Patents and the Science Base." *Journal of the American Society for Information Science and Technology* (55:11): 991–1001.

Leydesdorff, Loet; and Etzkowitz, Henry. 1998. "The Triple Helix as a Model for Innovation Studies." *Science and Public Policy* 25(3): 195–203.

———. 2003. "Can 'the Public' Be Considered a Fourth Helix in University-Industry-Government Relations?" *Science and Public Policy* 30(1): 55–61.

Li, Shuyu; Cutler, Gene; Liu, Jane; Hoey, Timothy; Chen, L.; Schultz, P.; Liao, J.; and Ling, X. 2003. "A Comparative Analysis of HGSC and Celera Human Genome Assemblies and Gene Sets." *Bioinformatics* (19:13): 1597–1605.

Lieberman, Office of Joseph. 2004. "Offshore Outsourcing and America's Competitive Edge: Losing Out in High Tech R&D and Services Sectors," available at: http://lieberman.senate.gov/assets/pdf/off_shoring.pdf (last visited September 2010.

Lieberwitz, Risa. 2007. "University Science Research Funding." In Stephan and Ehrenberg, 2007.

Light, Donald. 2009. "Global Drug Discovery: Europe Is Ahead." *Health Affairs,* Web exclusive at: http://content.healthaffairs.org/cgi/content/abstract/28/5/w969.

Lindley, David. 1993. *The End of Physics.* New York: Basic Books.

Lippmann, Walter. 1966. "The University." *New Republic,* May 28: 17–20.

Liu, Zimang. 2003. "Trends in Transforming Scholarly Communication and Their Implications." *Information Processing and Management* (39:6): 889–898.

Loeppky, Rodney. 2005a. "History, Technology and the Capitalist State: The Comparative Political Economy of Biotechnology and Genomics." *Review of International Political Economy* (12): 264–286.

———. 2005b. "Biomania and US Foreign Policy." *Millenium* (34): 85–113.

———. 2005c. *Encoding Capital: The Political Economy of the Human Genome Project.* London: Routledge.

Lohmann, Susanne. 2004. "Can't the University Be More Like a Business?" *Economics of Governance* (5): 9–27.

Lohr, Steve. 2009a. "The Corporate Lab as Ringmaster." *New York Times,* August 16.

———. 2009b. "Patent Auctions Offer Protections to Investors." *New York Times,* September 21.

Lomborg, Bjorn. 2001. *The Skeptical Environmentalist.* Cambridge: Cambridge University Press.

Lowen, Rebecca. 1997. *Creating the Cold War University.* Berkeley: University of California Press.

Lurie, Peter; and Zieve, Allison. 2006. "Sometimes the Silence Can Be like the Thunder: Access to Pharmaceutical Data at the FDA." *Law and Contemporary Problems* (69): 85–97.

Lyall, Katherine; and Sell, Katherine. 2006. *Change,* January: 6–13.

Lynch, Kathleen. 2006. "Neoliberalism and Marketisation: The Implications for Higher Education." *European Educational Research Journal* (5): 1–17.

Lynskey, Michael. 2006. "Transformative Technology and Institutional Transformation: Co-Evolution of Biotechnology Venture Firms and Institutional Framework in Japan." *Research Policy* (35): 1389–1422.

Macdonald, Stuart. 2004. "When Means Become Ends: Considering the Impact of Patent Strategy on Innovation." *Information Economics and Policy* (16): 135–158.

Machlup, Fritz. 1962. *The Production and Distribution of Knowledge in the US.* Princeton: Princeton University Press.

———. 1980. *Knowledge: Its Creation, Distribution and Economic Significance.* 2 vols. Princeton: Princeton University Press

———. 1984. *Knowledge, vol.*3. Princeton: Princeton University Press.

Machlup, F.; and Machlup, Fritz, eds. 1983. *The Study of Information.* New York: Wiley.

Macho-Stadler, Ines; and Perez-Castrillo, David. 2001. *An Introduction to the Economics of Information.* 2nd ed. Oxford: Oxford University Press.

Macilwain, Colin. 2010. "What Is Science Really Worth?" *Nature,* June 17 (465): 682–684.

MacKenzie, Donald. 1990. *Inventing Accuracy.* Cambridge, Mass.: MIT Press.

———. 1996. *Knowing Machines.* Cambridge, Mass.: MIT Press.

———. 2006. *An Engine, Not a Camera.* Cambridge, Mass.: MIT Press.

Magnus, David; Caplan, Arthur; and McGee, Glenn, eds. 2002. *Who Owns Life?* Amherst, MA: Prometheus.

Maher, Brendan. 2002. "Test Tubes with Tails." *The Scientist,* February 4 (16:3): 22.

Mäki, Uskali. 1992. "Social Conditioning in Economics." In Neil de Marchi, ed., *Post-Popperian Methodology of Economics.* Boston: Kluwer.

———. 1999. "Science as a Free Market: A Reflexivity Test." *Perspectives on Science* (7): 486–509.

———. 2005. "Economic Epistemology: Hopes and Horrors." *Episteme* (1): 211–220.

———. 2008. "Philosophy of Economics." In Martin Curd and Stathis Psillos, eds., *Routledge Companion to the Philosophy of Science.* New York: Routledge.

Makowski, Louis; and Ostroy, Joseph. 2001. "Perfect Competition and the Creativity of the Market." *Journal of Economic Literature* (39): 479–535.

Malerba, Fancesco; and Brusoni, Stephano, eds. 2007. *Perspectives on Innovation.* New York: Cambridge University Press.

Malerba, Francisco; and Orsenigo, Luigi. 2000. "Knowledge, Innovative Activities and Industry Evolution." *Industrial and Corporate Change* (9): 289–314.

Malone, Cheryl; and Elichirigoity, Fernando. 2003. "Information as Commodity and Economic Sector: Its Emergence in the Discourse of Industrial Classification." *Journal of the American Society for Information Science and Technology* (54:6): 512–520.

Mansfield, Edwin. 1965. "Rates of Return from Industrial Research and Development." *American Economic Review* (55): 310–322.

———. 1998. "Academic Research and Industrial Innovation." *Research Policy* (26): 773–786.

March, James; and Simon, Herbert. 1958. *Organizations.* New York: Wiley.

Marginson, Simon. 2007. "The Public/Private Divide in Higher Education: A Global Revision." *Higher Education* (53): 307–333.

Marginson, Simon; and Rhoades, Gary. 2002. "Beyond Nation States, Markets, and Systems of Higher Education." *Higher Education* (43): 281–309.

Marginson, Simon; and Van der Wende, M. 2007. "Globalization and Higher Education." OECD Education Working Paper 8. Paris: OECD.

Markoff, John. 2008. "Two Views of Innovation, Colliding in Washington." *New York Times,* January 13.

Marks, Harry. 1997. *The Progress of Experiment.* New York: Cambridge University Press.

Marron, David. 1984. "Otto Eckstein and the Founding of Data Resources Inc." *Review of Economics and Statistics* (66): 537–542.

Marshall, Elliot. 1997a. "The Mouse That Prompted a Roar." *Science* (277): 24–25.
———. 1997b. "Materials Transfer: Need a Reagent? Just Sign Here." *Science* (278): 212–213.
———. 2002. "DuPont Ups the Ante on Use of Harvard's Oncomouse." *Science*, (May 17 (296): 1212–1213.
Martin, Ben. 2003. "The Changing Social Contract for Science." In Geuna et al. 2003.
Martin, Ben; and Nightingale, Paul. eds. 2000. *The Political Economy of Science, Technology and Innovation*. Cheltenham: Elgar.
Maternowski, Kate. 2009. "Conflict with My Interests." *Inside Higher Ed*, July 10.
Matkin, Gary. 1990. *Technology Transfer and the University*. New York: Macmillan.
Maurer, Stephen. 2002. "Promoting and Disseminating Knowledge: The Public/Private Interface." Paper presented to NRC Symposium on Information in the Public Domain. Washington, September.
May, Christopher. 2002. *The Information Society: A Skeptical View*. Cambridge: Polity.
———. 2006. "The World Intellectual Property Organization." *New Political Economy* (11): 435–445.
May, Christopher; and Sell, Susan. 2006. *Intellectual Property Rights*. Boulder, CO: Lynne Rienner.
Mayer, Anna. 2004. "Setting Up a Discipline: British History of Science and the End of Ideology, 1931–48." *Studies in the History and Philosophy of Science* (35): 41–72.
Mazzucato, Mariana; and Dosi, Giovanni, eds. 2006. *Knowledge Accumulation and Industry Evolution*. Cambridge: Cambridge University Press.
McAfee, Kathleen. 2003. "Neoliberalism on a Molecular Scale." *Geoforum* (34): 203–219.
McClellan, Chris. 1996. "The Economic Consequences of Bruno Latour." *Social Epistemology* (10): 193–208.
McClintick, David. 2006. "How Harvard Lost Russia." *Institutional Investor* (40:1).
McCloskey, Donald. 1994. *Knowledge and Persuasion in Economics*. New York: Cambridge University Press.
McCormack, Richard. 2008. "Budget Crunch Forces Bureau of Economic Analysis to Cut Back Data on Foreign Direct Investment and Multinational Activity in U.S. and Abroad." *Manufacturing and Technology News*, June 15 (15:11). Available at http://www.manufacturingnews.com/news/08/0616/fdi.html (last visited September 2010).
———. 2009. *Manufacturing a Better Future for America*. Washington: Alliance for American Manufacturing.
McGarity, Thomas; and Wagner, Wendy. 2008. *Bending Science*. Cambridge, Mass.: Harvard University Press.
McGoey, Linsey. 2007. "On the Will to Ignorance in a Bureaucracy." *Economy and Society* (36:2): 212–235.
McGucken, William. 1984. *Scientists, Society and the State*. Columbus: Ohio State University Press.

McKelvey, Maureen; and Holmen, Magnus, eds. 2009. *Learning to Compete in European Universities.* Cheltenham, U.K.: Elgar.

McKelvey, Maureen; Richne, Annika; and Hellman, Jens, eds. 2004. *The Economic Dynamics of Modern Biotechnology.* Cheltenham, U.K.: Elgar.

McMeekin, Andrew; and Harvey, Mark. 2002. "The Formation of Bioinformatic Knowledge Markets: An Economics of Knowledge Approach." *Revue d'Economie Industrielle* (101): 47–64.

McNeil, Donald. 2006. "Tough-Talking Journal Editor Faces Accusations of Leniency." *New York Times,* August 1, p. D1.

McSherry, Corynne. 2001. *Who Owns Academic Work?* Cambridge, Mass.: Harvard University Press.

Medema, Steven. 2011. "The Life and Times of the Coase Theorem." *Journal of the History of Economic Thought,* forthcoming.

Mehrling, Perry. 2005. *Fischer Black and the Revolutionary Idea of Finance.* New York: Wiley.

Meier, Barry. 2004. "Group Is Said to Seek Full Drug-Trial Disclosure." *New York Times,* June 14, p. A1.

Mervis, Jeffrey. 2007. "US Output Flattens, NSF Wonders Why." *Science,* August 3 (317): 582.

Merz, J.; Kriss, D.; Leonard, D.; and Cho, Mildred. 2002. "Diagnostic Testing Fails the Test." *Nature* (415): 577–579.

Metlay, Grischa. 2006. "Reconsidering Renormalization." *Social Studies of Science* (36): 565–597.

Michaels, David. 2005. "Doubt Is Their Product." *Scientific American,* June, 96–101.

———. 2006. "Manufacturing Uncertainty: Protecting Public Health in the Age of Contested Science." *Annals of the New York Academy of Science* (1076): 149–162.

———. 2008. *Doubt Is Their Product.* New York: Oxford University Press.

Michaels, David; and Monfortin, Celeste. 2007. "How Litigation Shapes the Scientific Literature." *Journal of Law and Policy* (15): 1137–1169.

Milgrom, Paul; and Stokey, Nancy. 1982. "Information, Trade and Common Knowledge." *Journal of Economic Theory* (26): 17–27.

Miller, Boaz. 2007. "What Trust in Science?" *Spontaneous Generations* (1:1): 132–135.

Miller, Henry. 2007. "Biotech's Defining Moments." *Trends in Biotechnology* (25): 56–59.

Miller, Karen. 1999. *The Voice of Business: Hill & Knowlton and Postwar Public Relations.* Chapel Hill: University of North Carolina Press.

Mills, Ann; and Tereskerz, Patti. 2007. "Junk Patents in Biotechnology: An Illusion or Real Threat to Innovation?" *Biotechnology Law Report* (236): 226–230.

Milne, C.; and Paquette, C. 2004. "Meeting the Challenge of the Evolving R&D Paradigm: What Role for the CRO?" *American Pharmaceutical Outsourcing.* Available at www.americanpharmaceuticaloutsourcing.com.

Minkel, J. R. 2003. "The Economics of Science: An Interview with Terence Kealey." *Scientific American,* March.

Mirowski, Philip. 1989. *More Heat Than Light*. New York: Cambridge University Press.

———. 1992. "What Were von Neumann and Morgenstern Trying to Accomplish?" In E. R. Weintraub, ed. *Toward a History of Game Theory*. Durham, N.C.: Duke University Press.

———, ed. 1994. *Edgeworth on Chance, Economic Hazard and Probability*. Lanham, Md.: Rowman & Littlefield.

———. 2002. *Machine Dreams*. New York: Cambridge University Press.

———. 2004a. *The Effortless Economy of Science?* Durham, N.C.: Duke University Press.

———. 2004b. "The Scientific Dimensions of Society and Their Distant Echoes in American Philosophy of Science." *Studies in the History and Philosophy of Science A*, June (35): 283–326.

———. 2006. "Twelve Theses on the History of Demand Theory in America." In Mirowski and Hands 2006.

———. 2007a. "Markets Come to Bits: Markomata and the Future of Computational Evolutionary Economics." *Journal of Economic Behavior and Organization*, April (63): 209–242.

———. 2007b. "Naturalizing the Market on the Road to Revisionism." *Journal of Institutionalist Economics* (3:3): 351–372.

———. 2007c. "Johnny's in the Basement, Mixin' up the Medicine." *Social Studies of Science* (37): 311–327.

———. 2008. "Why There Is (as Yet) No Such Thing as an Economics of Knowledge." In Harold Kincaid and Don Ross, eds., *The Oxford Handbook of the Philosophy of Economics*. Oxford: Oxford University Press.

———. 2009. "Some Economists Rush to Rescue Science from Politics, Only to Discover in Their Haste, They Went to the Wrong Address." In Jeroen van Bouwel, ed., *The Social Sciences and Democracy*. London: Macmillan.

———. 2010. "Inherent Vice." *Journal of Institutionalist Economics* (6): 1–37.

———. Forthcoming. "On the Origins (at Chicago) of Some Neoliberal Species of Evolutionary Economics." In Rob Van Horn, Philip Mirowski and Thomas Stapleton, eds., *Building Chicago Economics*. New York: Cambridge University Press.

Mirowski, Philip; and Hands, D. Wade, eds. 2006. *Agreement on Demand*. Durham, N.C.: Duke University Press.

Mirowski, Philip; and Nik-Khah, Edward. 2007. "Markets Made Flesh." In Donald MacKenzie et al, eds., *Do Economists Make Markets?* Princeton: Princeton University Press.

Mirowski, Philip; and Plehwe, Dieter, eds. 2009. *The Road from Mont Pèlerin: The Making of the Neoliberal Thought Collective*. Cambridge, Mass.: Harvard University Press.

Mirowski, Philip; and Sent, Esther-Mirjam, eds. 2002. *Science Bought and Sold*. Chicago: University of Chicago Press.

———. 2007. "The Commercialization of Science and the Response of STS." In Ed Hackett et al, eds., *The New Handbook of Science and Technology Studies*. Cambridge, Mass.: MIT Press.

Mirowski, Philip; and van Horn, Robert. 2005. "The Contract Research Organization and the Commercialization of Science." *Social Studies of Science* (35): 503–548.

———. 2009. "The Road to a World Made Safe for Corporations: The Rise of the Chicago School." In Mirowski and Plehwe 2009.

Misa, Thomas. 2004. "Beyond Linear Models." In Grandin et al., 2004.

Miyake, Shingo. 2004. "Universities Get Taste of Business World." *Nikkei Weekly*, July 26.

Miyoshi, Masao. 2000. "Ivory Tower in Escrow." *Boundary 2* 27: 7–50.

Mody, Cyrus. 2009. "Instruments of Commerce and Knowledge: Probe Microscopy." In Richard Freeman and Daniel Goroff, eds. *Science and Engineering Careers in the United States: An Analysis of Markets and Employment*. Chicago: University of Chicago Press.

Moffat, Barton; and Elliott, Carl. 2007. "Ghost Marketing: Pharma Companies and Ghostwritten Journal Articles." *Perspectives in Biology and Medicine* (50): 18–31.

Moggridge, D. E. 2008. *Harry Johnson: A Life in Economics*. New York: Cambridge University Press.

Mokyr, Joel. 2002. *The Gifts of Athena*. Princeton: Princeton University Press.

Monaghan, Peter. 2007. "American Psychological Association Panel Urges Curbs on Corporate Influence in the Discipline." *Chronicle of Higher Education*, December 10.

Monastersky, Richard. 2002. "Atomic Lies." *Chronicle of Higher Education*, August 16.

Monbiot, George. 2003. "Guard Dogs of Perception: Corporate Takeover of Science." *Science and Engineering Ethics* 9: 49–57.

Mooney, Chris. 2005. *The Republican War on Science*. New York: Basic.

———. 2008. "The Manufacture of Uncertainty." *American Prospect*, March 28.

Mooney, Chris; and Kirshenbaum, Sheril. 2009. *Unscientific America*. New York: Basic.

Moore, Kelley. 2008. *Disrupting Science*. Princeton: Princeton University Press.

Moore, Solomon. 2009. "Science Found Wanting in Nation's Crime Labs." *New York Times*, February 5.

Morey, Ann. 2004. "Globalization and the Emergence of For-Profit Higher Education." *Higher Education* 48: 131–150.

Morgan, Steve; Barer, Morris; and Evans, Roberts. 2000. "Health Economics Meets the Fourth Tempter: Drug Dependency and Scientific Discourse." *Health Economics* 9: 659–667.

Morin, Alexander. 1993. *Science Policy and Politics*. Englewood Cliffs, N.J.: Prentice-Hall.

Morley, Ann. 2004. "Globalization and the Emergence of For-Profit Higher Education." *Higher Education* (48): 131–150.

Mowatt, Graham; Shirran, Liz; Grimshaw, Jeremy M., Rennie, Drummond; and Flanagin, Annette, et al. 2002. "Prevalence of Honorary and Ghost Authorship in Cochrane Reviews." *Journal of the American Medical Association* (287): 2769–2771.

Mowery, David. 1983. "The Relationship between Intrafirm and Contractual Forms of Industrial Research in American Manufacturing, 1900–40." *Explorations in Economic History* (20): 351–374.

———. 1990. "The Development of Industrial Research in United States Manufacturing." *American Economic Review, Papers and Proceedings* (80): 345–349.

———. 2005. "The Bayh-Dole Act and High-Tech Entrepreneurship." In Gary Libecap, ed., *University Entrepreneurship and Technology Transfer.* Amsterdam: Elsevier.

Mowery, David; Nelson, Richard; Sampat, Bhaven; and Ziedonis, Arvids. 2004. *Ivory Tower and Industrial Innovation.* Stanford: Stanford University Press.

Mowery, David; and Rosenberg, Nathan. 1989. *Technology and the Pursuit of Economic Growth.* New York: Cambridge University Press.

———. 1998. *Paths of Innovation.* New York: Cambridge University Press.

Mowery, David; and Ziedonis, Arvids. 2002. "Academic Patent Quality and Quantity before and after the Bayh-Dole Act in the US." *Research Policy* (31): 399–418.

———. 2007. "Academic Patents and Materials Transfer Agreements: Substitutes or Complements?" *Journal of Technology Transfer* (32:3): 157–172.

Murmann, Johann. 2004. *Knowledge and Comparative Advantage.* Cambridge: Cambridge University Press.

Murray, Fiona. 2006. "The Oncomouse That Roared: Resistance and Accommodation to Patenting in Academic Science." MIT Sloan School Working Paper, available at web.mit.edu/fmurray/www/papers/THE%20**ONCOMOUSE**%20THAT%20ROARED_ FINAL.pdf.

Murray, Fiona; and Stern, Scott. 2007. "Do Formal Intellectual Property Rights Hinder the Free Flow of Scientific Knowledge?" *Journal of Economic Behavior and Organization* (63): 648–687.

Nace, Ted. 2003. *Gangs of America.* San Francisco: Berrett-Koehler.

Nagoaka, Sadao. 2006. Presentation to the CSIC/OECD Conference.

Napoli, Philip. 1999. "The Marketplace of Ideas Metaphor in Communications Regulation." *Journal of Communications* (49): 151–169.

Narula, Rajneesh. 2003. *Globalization and Technology.* Cambridge: Polity Press.

Nasto, Barbara. 2008. "Chasing Biotech across Europe." *Nature Biotechnology* (26:3): 283–288.

National Academy of Sciences (NAS). 2005. *Rising above the Gathering Storm.* Washington: National Academy Press. Available at www.nap.edu/catalogue.

National Bureau of Economic Research (NBER). 1962. *The Rate and Direction of Inventive Activity.* Princeton: Princeton University Press.

National Science Board (NSB). 2004. *Science and Engineering Indicators 2002.* Arlington, Va.: National Science Foundation. Available at www.nsf.gov/sbe/srs/seind02.

———. 2008a. *Science and Engineering Indicators 2008.* Arlington, Va.: National Science Foundation.

———. 2008b. *Research and Development: Essential Foundation for U.S. Competitiveness in a Global Economy.* Arlington, Va.: National Science Foundation.

———. 2008c. *International Science and Engineering Partnerships. A Priority for US Foreign Policy.* Arlington, Va.: National Science Foundation.

National Science Foundation. 1973. *Interpretations of Science and Technology in the Innovative Process.* Columbus, Ohio: Battelle Institute.

National Science Foundation, Division of Science Resources Statistics (NSF SRS). 2007. SRS Publication Trends Study. NSF07–330. Arlington, Va.: NSF.

Nedeva, Maria; and Boden, Rebecca. 2006. "Changing Science: The Advent of Neo-liberalism." *Prometheus* (24:3): 269–281.

Nelson, Philip. 1970. "Information and Consumer Behavior." *Journal of Political Economy* (78): 311–329.

Nelson, Richard. 1959. "The Simple Economics of Basic Research." *Journal of Political Economy* (67):297–306. Reprinted in Mirowski and Sent 2002.

———. 2001. "Observations on the Post Bayh-Dole Rise of Patenting at American Universities." *Journal of Technology Transfer* 26: 13–19.

———. 2004. "The Market Economy and the Scientific Commons." *Research Policy* (33): 455–471.

———. 2006. "Reflections on 'The Simple Economics of Basic Scientific Research': Looking Back and Looking Forward." *Industrial and Corporate Change* (15): 903–917.

Nelson, Richard; and Winter, Sidney. 1982. *An Evolutionary Theory of Economic Change.* Cambridge, Mass.: Harvard University Press.

Nelson, Richard; and Wright, Gavin. 1992. "The Rise and Fall of American Technological Leadership." *Journal of Economic Literature* (30): 1931–1964.

Newfield, Christopher. 2003. *Ivy and Industry.* Durham, N.C.: Duke University Press.

———. 2008. *Unmaking the Public University.* Cambridge, Mass.: Harvard University Press.

Newman, Nathan. 2002. "Big Pharma, Bad Science." *Nation,* July 25, available at www.thenation.com/doc.mhtml%3Fi=20020805&s=newman20020725 (last visited March 18, 2005).

Nightingale, Paul. 1998. "A Cognitive Theory of Innovation." *Research Policy* (27): 689–709.

———. 2000. "Economies of Scale in Pharmaceutical Experimentation." *Industrial and Corporate Change* (9): 315–359.

———. 2003. "If Nelson and Winter Were Half Right about Tacit Knowledge, Which Half?" *Industrial and Corporate Change* (12): 149–183.

———. 2004. "Technological Capacities, Invisible Infrastructure and the Un-social Construction of Predictability." *Research Policy,* (33): 1259–1284.

———. 2008. "Meta-Paradigm Change and the Theory of the Firm." *Industrial and Corporate Change* (17): 533–583.

Nightingale, Paul; and Martin, Paul. 2004. "The Myth of the Biotech Revolution." *Trends in Biotechnology* (22): 564–569.

Nik-Khah, Edward. 2008. "George Stigler, the GSB, and the Pillars of the Chicago School." Roanoke Working Paper.

———. 2009. "Getting Hooked on Drugs." Roanoke Working Paper.

Ninov, Victor; Gregorich, K.; Loveland, W.; Ghiorso, A.; Hoffman, D. C.; Lee, D. M.; Nitsche, H.; Swiatecki, W. J.; Kirbach, U. W.; Laue, C. A.; Adams, J. L.; Patin, J. B.; Shaughnessy, D. A.; Strellis, D. A.; and Wilk, P. A. 1999. "Observation of Superheavy Nuclei Produced . . ." *Physical Review Letters* (83:6): August 9.

Noble, David. 1979. *America by Design*. New York: Oxford University Press.

Noll, Roger, ed. 1998. *Challenges to Research Universities*. Washington: Brookings.

Noruzi, Alireza. 2005. "Google Scholar: The Next Generation of Citation Indexes." *Libri* (55): 170–180.

Nowotny, Helga; Pestre, Dominique; Schmidt-Assmann, E.; Schultze-Fielitz, H.; and Trute, H., eds. 2005. *The Public Nature of Science under Assault*. Berlin: Springer.

Nowotny, Helga; Scott, Peter; and Gibbons, Michael. 2001. *Re-Thinking Science: Knowledge and the Public in an Age of Uncertainty*. Cambridge: Polity Press.

———. 2003. "Mode 2 Revisited." *Minerva* 41: 175–194.

O'Connor, Alice. 2008. "The Privatized City." *Journal of Urban History* (34: 2): 333–353.

OECD. 2006a. Summary Report on Conference on Research Use of Patented Inventions. Madrid: OECD. Available at www.oecd.org/sti/ipr.

———. 2006b. *OECD Biotechnology Statistics 2006*. Paris: OECD.

———. 2008. *The Internationalization of Business R&D*. Paris: OECD.

Office of the U.S. Inspector General (OIG). 2009. OIG Final Report: "How Grantees Manage Financial Conflicts of Interest in Research Funded by the National Institutes of Health."OEI-03-07-00700, http://oig.hhs.gov

———. 2010. "Challenges to the FDA's Ability to Monitor and Inspect Foreign Clinical Trials." OEI-01-08-00510.

Okie, Susan. 2001. "Missing Data on Celebrex: Full Study Altered Picture of Drug." *Washington Post*, August 5, p. A11.

Olivieri, Nancy. 2003. "Patients Health or Company Profits?" *Science and Engineering Ethics* 9: 29–41.

Olson, Maynard. 2002. "The Human Genome Project: A Player's Perspective." *Journal of Molecular Biology* (319): 931–942.

O'Neill, John. 1996. "Who Won the Socialist Calculation Controversy?" *History of Political Thought* (17): 431–442.

———. 2003. "Unified Science as Political Philosophy: Positivism, Pluralism, Liberalism." *Studies in the History and Philosophy of Science* (34): 575–596.

———. 2006. "Knowledge, Planning, and Markets." *Economics and Philosophy* (22): 55–78.

Oreskes, Naomi; and Conway, Eric. 2008. "Challenging Knowledge: How Climate Science Became a Casualty in the Cold War." In Proctor and Scheibinger 2008.

———. 2010. *Merchants of Doubt*. New York: Bloomsbury Press.

Oreskes, Naomi; Conway, Erik; and Shindell, Matthew. 2008. "From Chicken Little to Doctor Pangloss." *Historical Studies in the Natural Sciences* (38): 109–152.

Orsenigo, Luigi. 1989. *The Emergence of Biotechnology*. London: Pinter.

Orsi, Fabienne; and Coriat, Benjamin. 2005. "Are Strong Patents Beneficial to Innovative Activities: Lessons from the Genetic Testing for Breast Cancer?" *Industrial and Corporate Change* (14): 1205–1221.

Ostrom, Elinor; and Hess, Charlotte, eds. 2007. *Understanding Knowledge as a Commons: From Theory to Practice.* Cambridge, Mass.: MIT Press.

Owens, Larry. 1994. "The Counterproductive Management of Science in WWII." *Business History Review* (68): 515–576.

Owen-Smith, Jason. 2006. "Commercial Imbroglios: Proprietary Science and the Contemporary University." In Frickel and Moore 2006, 63–90.

Owen-Smith, Jason; and Powell, Walter. 2003. "The Expanding Role of Patenting in the Life Sciences." *Research Policy* 32(9): 1695–1711.

Owen-Smith, Jason; Riccaboni, M.; Pammoli, F.; and Powell, Walter. 2002. "A Comparison of US and European University-Industry Relationships in the Life Sciences." *Management Science* (48): 24–43.

Parexel. 2003. *Pharmaceutical R&D Statistical Sourcebook.* Waltham, MA: Parexel International.

Park, Man-Seop. 2007. "Homogeniety Masquerading as Variety." *Cambridge Journal of Economics* (31): 379–392.

Park, Paula. 2004. "Buffalo Case Highlights MTAs." *The Scientist,* August 9.

Parker, Richard. 2005. *John Kenneth Galbraith.* New York: Farrar Straus.

Parry, Bronwyn. 2007. "Cornering the Futures Market in Bio-Epistemology." *BioSocieties* (2): 386–389.

Pear, Robert. 2003. "Congress Weighs Drug Comparisons." *New York Times,* August 24.

———. 2008. "Patent Law Battle a Boon to Lobbyists." *New York Times,* April 30.

Pels, Dick. 2005. "Mixing Metaphors: Politics or Economics of Knowledge?" In Nico Stehr and V. Meja, eds. *Society and Knowledge.* New Brunswick, N.J.: Transaction.

Perraton, Jonathan. 2006. "Heavy Constraints on a Weightless World? Resources and the New Economy." *American Journal of Economics and Sociology* (65): 642–691.

Pestre, Dominique. 2003a. "Regimes of Knowledge Production in Society: Towards a More Political and Social Reading." *Minerva* 41: 245–261.

———. 2003b. *Science, Argent et Politique: un essai d'interprétation.* Paris: Institut Nationale de la Recherche Agronomique.

———. 2004. "Thirty Years of Science Studies." *History and Technology* (20): 351–369.

———. 2005. "The Technosciences between Markets, Social Worries and the Political." In Nowotny et al. 2005.

———. 2007. "The Historical Heritage of the 19th and 20th Centuries: Techno-Science, Markets and Regulations in a Long-Term Perspective." *History and Technology* (23): 407–420.

Pestre, Dominique; and Krige, John. 1992. "Some Thoughts on the Early History of CERN." In Peter Galison and Bruce Hevly, eds., *Big Science: The Growth of Large-Scale Research,* 78–99. Stanford: Stanford University Press.

Peters, John Durham. 2004. "Marketplace of Ideas: History of a Concept." In Andrew Calabrese and Colin Sparks, eds., *Towards a Political Economy of Culture,* 65–82. Lanham, Md.: Rowman & Littlefield.

Petryna, Adriana. 2007. "Clinical Trials Offshored: On Private Sector Science and Public Health." *BioSocieties* (2): 21–40.

———. 2009. *When Experiments Travel.* Princeton: Princeton University Press.

Petryna, Adriana; Lakoff, Andrew; and Kleinman, Alan, eds. 2006. *Global Pharmaceuticals: Ethics, Markets, Practices.* Durham, N.C.: Duke University Press.

Philips, Louis. 1988. *The Economics of Imperfect Information.* Cambridge: Cambridge University Press.

Pichaud, B. S. 2002. "Outsourcing in the Pharmaceutical Manufacturing Process: An Examination of CRO Experience." *Technovation* 22: 81–90.

Pichler, Frank; and Turner, Susan. 2007. "The Power and Pitfalls of Outsourcing." *Nature Biotechnology* (25): 1093–1096.

Pickering, Andrew. 1995. *The Mangle of Practice.* Chicago: University of Chicago Press.

———. 2005. "Decentering Sociology: Synthetic Dyes and Social Theory." *Perspectives on Science* (13): 352–405.

Pinch, Trevor; and Bijker, Weebe. 1987. "The Social Construction of Facts and Artifacts." In W. Bijker, T. Hughes, and T. Pinch, eds., *The Social Construction of Technological Systems.* Cambridge, Mass.: MIT Press.

Pisano, Gary. 2006a. "Can Science Be a Business?" *Harvard Business Review* (84:10): 114–125.

———. 2006b. *Science Business.* Cambridge, Mass.: Harvard Business School Press.

Plehwe, Dieter; and Walpen, Bernard. 2005. "Between Network and Complex Organization: The Making of Neoliberal Knowledge and Hegemony." In Dieter Plehwe, Bernard Walpen, and Gisela Nuenhoffer, eds., *Neoliberal Hegemony: A Global Critique.* London: Routledge.

Polanyi, Michael. 1962. "The Republic of Science: Its Political and Economic Theory." *Minerva* (1): 54–75.

Pollack, Andrew. 2002. "Despite Billions for Discoveries, Pipeline of Drugs Is Far from Full." *New York Times,* 19 April 19.

———. 2003. "Three Universities Join Researcher to Develop Drugs." *New York Times,* July 31, p. C1.

———. 2004. "Is Biotechnology Losing its Nerve?" *New York Times,* February 27, Section 3, 1.

———. 2008. "For Biotech, a Tax Break Spells Hope." *New York Times,* December 10.

Pomfret, John; and Nelson, Deborah. 2000. "In Rural China, a Genetic Mother Lode." *Washington Post,* December 20.

Porat, Mark. 1977. *The Information Economy.* Washington: Department of Commerce.

Porter, Theodore. 2009. "Measurement and Meritocracy: An Intellectual History of IQ." *Modern Intellectual History* (6): 637–644.

Posner, Richard A. 2002. "The Law and Economics of Intellectual Property." *Daedalus* (Spring): 5–12.

———. 2005. "Bad News." *New York Times Book Review,* July 31, Section 7, pp. 1, 8–11.

Powell, Walter; and Owen-Smith, Jason. 1998. "Universities and the Market for IP in the Life Sciences." *Journal of Policy Analysis and Management* (17: 2): 253–277.

Powell, Walter; Owen-Smith, Jason; and Colyvas, Jeanette. 2007. "Innovation and Emulation: Lessons from American Universities in Selling Private Rights to Public Knowledge." *Minerva* (45): 121–142.

Powell, Walter; and Snellman, Kaisa. 2004. "The Knowledge Economy." *Annual Review of Sociology* (30): 199–220.

Power, Michael. 2003. "Evaluating the Audit Explosion." *Law and Society* (25): 185–202.

———. 2007. *Organized Uncertainty.* Oxford: Oxford University Press.

Press, Eyal; and Washburn, Jennifer. 2000. "The Kept University." *Atlantic Monthly,* March: 39–54.

Pressman, Lori; Burgess, Richard; Cook-Deegan, Robert; McCormack, Stephen; Nami-Wolk, Io; Soucy, Melissa; and Walters, LeRoy. 2006. "The Licensing of DNA Patents by US Academic Institutions: An Empirical Survey." *Nature Biotechnology* (24): 31–39.

Price, Derek de Solla. 1963. *Little Science, Big Science.* New York: Columbia University Press.

———. 1976. *Science since Babylon.* New Haven: Yale University Press.

Proctor, Robert; and Scheibinger, Londa, eds. 2008. *Agnotology: The Making and Unmaking of Ignorance.* Stanford: Stanford University Press.

Public Citizen Health Research Group. 2001. "Rx R&D Myths: The Case Against the Drug Industry's R&D 'Scare Card.'" Available at http://www.citizen.org/hrg (last visited 18 March 2005).

Quackenbush, John. 2001. "The Power of Public Access." *Nature Genetics* (29): 4–6.

Quéré, Michel. 2003. "Knowledge Dynamics and the Pervasive Character of Biotechnology in the Pharmaceutical Industry." *Industry and Innovation* (10): 255–273.

Rader, Karen. 2004. *Making Mice.* Princeton: Princeton University Press.

Rai, Arti. 2004. "The Increasingly Proprietary Nature of Publicly Funded Research." In Donald Stein, ed., *Buying In or Selling Out?* New Brunswick, N.J.: Rutgers University Press.

Rai, Arti; and Eisenberg, Rebecca. 2003. "Bayh-Dole Reform and the Progress of Biomedicine." *Law and Contemporary Problems* (56): 289–314.

Rajan, Kaushik Sunder. 2006. *Biocapital.* Durham, N.C.: Duke University Press.

Raloff, Janet. 2008. "Judging Science: Courts May Be Too Skeptical." *Science News,* January 19 (173:3).

Rammert, Werner. 1997. "New Rules of Sociological Method: Rethinking Technology Studies." *British Journal of Sociology* (48): 171–191.

Rasmussen, Nicolas. 2002. "Of Small Men, Big Science, and Bigger Business: The Second World War and Biomedical Research in the US." *Minerva* 40: 115–146.

———. 2004. "The Moral Economy of the Drug Company–Medical Scientist Collaboration in Interwar America." *Social Studies of Science* 34(2): 161–186.

———. 2005. "The Commercial Clinical Trial in Interwar America: Three Types of Physician Collaborator." *Bulletin for the History of Medicine* (79): 50–80.

Rath, Matthias. 2006. "How Profits and Research Mix at Stanford." *San José Mercury News,* July 9.

Rauchway, Eric. 2007. "The State Game." *The New Republic Online,* September.

Ravetz, Jerome. 1971. *Scientific Knowledge and Its Social Problems.* Oxford: Oxford University Press.

Readings, Bill. 1997. *The University in Ruins.* Cambridge, Mass.: Harvard University Press.

Reddy, Prasada. 2000. *The Globalization of Corporate R&D.* London: Routledge.

Reder, Melvin. 1982. "Chicago Economics: Permanence and Change." *Journal of Economic Literature* (20): 1–38.

Reich, Leonard. 1985. *The Making of American Industrial Research.* Cambridge: Cambridge University Press.

Reimers, Niels. 1998. "Stanford's Office of Technology Licensing and the Cohen/Boyer Cloning Patents." Oral History, Bancroft Library, Berkeley, Calif.

Reingold, Nathan, ed. 1979. *The Sciences in the American Context.* Washington, D.C.: Smithsonian Institution Press.

———. 1995. "Choosing the Future." *Historical Studies in the Physical and Biological Sciences* (25:2): 301–327.

Reisch, George. 2005. *How the Cold War Transformed the Philosophy of Science.* New York: Cambridge University Press.

Rennie, Drummond; Flanagin, Annette; and Yank, Veronica. 2000. "The Contributions of Authors." *Journal of the American Medical Association* 284: 89–91.

Resnik, David. 2007. *The Price of Truth: How Money Affects the Norms of Science.* New York: Oxford University Press.

Rettig, Richard. 2000. "Drug Research and Development: The Industrialization of Clinical Research." *Health Affairs* 19: 129–146.

Revill, Jo. 2005. "How the Drugs Giant and a Lone Academic Went to War." *The Observer,* December 4.

Revkin, Andrew. 2006. "Climate Expert Says NASA Tried to Silence Him." *New York Times,* January 29, p. A1.

Rhoads, Richard; and Torres, Carlos, eds. 2006. *The University, the State and the Market.* Stanford: Stanford University Press.

Rich, Frank. 2008. "The Brightest Are Not Always the Best." *New York Times,* December 7.

Robbins, Lionel. 1932. *An Essay on the Nature and Significance of Economic Science.* London: Macmillan.

Robbins-Roth, C. 2000. *From Alchemy to IPO: The Business of Biotechnology.* Cambridge, Mass.: Perseus.

Robinson, Joan. 1953/1954. "The Production Function and the Theory of Capital." *Review of Economic Studies* (21): 81–106.

Rochon, P.; Gurwitz, J. H.; Simms, R. W.; Fortin, P.R.; Felson, D. T., et al. 1994. "A Study of Manufacturer-Supported Trials of Non-Steroidal Anti-Inflammatory Drugs in the Treatment of Arthritis." *Archives of Internal Medicine* 154: 157–163.

Rodrigues, Maria, ed. 2002. *The New Knowledge Economy in Europe: A Strategy for International Competitiveness.* Cheltenham, U.K.: Elgar.

———. 2003. *European Policies for a Knowledge Economy.* Cheltenham, U.K.: Elgar.

———, ed. 2009. *Europe, Globalization and the Lisbon Agenda.* Cheltenham, U.K.: Elgar.

Rodriguez, Victor. 2005. "Material Transfer Agreements: Open Science vs. Proprietary Claims." *Nature Biotechnology* (23:4): 489–491.

———. 2008. "Governance of Material Transfer Agreements." *Technology in Society* (30): 122–128.

Rodriguez, Victor; Janssens, F.; Debackere, K.; and DeMoor, B. 2007. "Do Material Transfer Agreements Affect the Choice of Research Projects?" *Scientometrics* (71): 239–269.

Roman, Joanne. 2002. "U.S. Medical Research in the Developing World: Ignoring Nuremberg." *Cornell Journal of Law and Public Policy* 11: 441–460.

Romer, Paul. 1986. "Increasing Returns and Long Run Economic Growth." *Journal of Political Economy* (94): 1002–1037.

———. 1990. "Endogenous Technical Change." *Journal of Political Economy* (98): S71–S102.

———. 1994. "The Origins of Endogenous Growth." *Journal of Economic Perspectives* (8): 3–22.

Rose, Nikolas. 1999. *Powers of Freedom.* Cambridge: Cambridge University Press.

Rosenberg, Nathan. 1976. *Perspectives on Technology.* New York: Cambridge University Press.

———. 1982. *Inside the Black Box: Technology and Economics.* New York: Cambridge University Press.

———. 1994. *Exploring the Black Box: Technology, Economics and History.* New York: Cambridge University Press.

Rosenstock, Linda. 2006. "Protecting Special Interests in the Name of 'Good Science.'" *Journal of the American Medical Association* (295:20): 2407–2410.

Ross, Joseph; Hill, Kevin; Egilman, David; and Krumholz, Harlan. 2008. "Guest Authorship and Ghost Writing." *Journal of the American Medical Association* (299): 1800–1812.

Rothschild, Michael. 1973. "Models of Market Organization with Imperfect Information: A Survey." *Journal of Political Economy* (81): 1283–1308.

Rottenberg, Simon. 1981. "The Economy of Science: The Proper Role of Government in the Growth of Science." *Minerva* (19): 43–67.

Rounsley, Steven. 2003. "Sharing the Wealth: The Mechanics of Data Release from Industry." *Plant Physiology* (133): 438–440.

Rubin, Eugene. 2005. "The Complexities of Individual Financial Conflicts of Interest." *Neuropsychopharmacology* (30): 1–6.

Rubin, Samuel. 2006. "Merck KGaA v. Integra Lifesciences I, Ltd.: Greater Research Protection for Drug Manufacturers." *Duke Journal of Constitutional Law and Public Policy* (1): 79–85.

Rudy, Alan P.; Coppin, Dawn; Konefal, Jason; and Shaw, Bradley T. 2007. *Universities in the Age of Corporate Science.* New Brunswick, N.J.: Rutgers University Press.

Ryan, Michael. 1998 *Knowledge Diplomacy.* Washington, D.C.: Brookings.
Saari, Donald; and Simon, Carl. 1978. "Effective Price Mechanisms." *Econometrica* (46): 1097–1125.
Sahlins, Marshall. 2009. "The Conflicts of the Faculty." *Critical Inquiry* (35): 997–1013.
Sampat, Bhaven. 2004. "Examining Patent Examination." Available at http://www.stiy.com/MeasuringInnovation/Sampat.pdf.
———. 2005. "Genomic Patenting by Academic Researchers: Bad for Science?" University of Michigan School of Public Health Working Paper.
———. 2006. "Patenting and U.S. Academic Research in the 20th Century: The World before and after Bayh-Dole." *Research Policy* (35): 772–789.
Sampat, Bhaven; Mowery, David; and Ziedonis, Arvids. 2003. "Changes in University Patent Quality after the Bayh-Dole Act: A Re-Examination." *International Journal of Industrial Organization* (21): 1371–1390.
Sampat, Bhaven; and Ziedonis, Arvids. 2004. "Patent Citations and the Economic Value of Patents." In Henk Moed et al., eds., *Handbook of quantitative Science and Technology Research,* 277–289. Dordrecht: Kluwer.
Samuelson, Larry. 2004. "Modeling Knowledge in Economic Analysis." *Journal of Economic Literature* (42): 367–403.
Samuelson, Paul. 1954. "The Pure Theory of Public Expenditure." *Review of Economics and Statistics* (36): 387–389.
———. 1955. "A Diagrammatic Exposition of the Theory of Public Expenditure." *Review of Economics and Statistics* (37): 350–356.
———. 2004. "An Interview with Paul Samuelson." *Macroeconomic Dynamics* (8): 519–542.
Sapsalis, Evan; and Navon, Ran. 2006. "Academic versus Industry Patenting." *Research Policy* (35): 1631–1645.
Sarkar, Husain. 2007. *Group Rationality in Scientific research.* New York: Cambridge University Press.
Saul, Stephanie. 2008. "Merck Wrote Drug Studies for Doctors." *New York Times,* April 16.
Saul, Stephanie; and Pollack, Andrew. 2008. "Big Pharma Companies Hunger for Biotech Drugs." *New York Times,* August 1.
Saunders, George. 1997. *CivilWarLand in Bad Decline.* New York: Riverhead.
Saunders, Tom. 2003. "Renting Space on the Shoulders of Giants." *Yale Law Journal* (113): 261.
Schatzberg, Eric. 2006. "*Technik* Comes to America: Changing Meanings of Technology before 1930." *Technology and Culture* (47): 486–512.
Schenckman, Rick. 2008. *Just How Stupid Are We?* New York: Basic Books.
Scherer, F. M. 2002. "The Economics of Human Gene Patents." *Academic Medicine* 77: 1348–1367.
Scherer, F. M.; and Hartoff, Dieter. 2000. "Technology Policy for a World of Skew-Distribution Outcomes." *Research Policy* (29): 559.
Scherer, F. M.; Hartoff, Dieter; and Kukies, Jorg. 2000. "Uncertainty and the Size Distribution of Rewards from Innovation." *Journal of Evolutionary Economics* (10): 175–200.

Schiller, Dan. 1988. "How to Think about Information." In V. Mosco and J. Wasko, eds., *The Political Economy of Information.* Madison: University of Wisconsin Press.

———. 2007. *How to Think about Information.* Urbana: University of Illinois Press.

Schneider, David. 2008. "A Market for Basic Science?" *American Scientist,* May-June.

Schneider, Louis. 1962. "The Role of the Category of Ignorance in Sociological Theory." *American Sociological Review* (27): 492–508.

Schofer, Evan. 2004. "Cross-National Differences in the Expansion of Science, 1970–1990." *Social Forces* (83): 215–248.

Schofer, Evan; Ramirez, Francisco; and Meyer, John. 2000. "The Effects of Science on National Economic Development, 1970 to 1990." *American Sociological Review* (65): 866–887.

Schrock, Andrew. 2007. "Opening Up the Patent Process." *Technology Review,* September.

Schulman, Bruce; and Zelizer, Julian, eds. 2008. *Rightward Bound.* Cambridge, Mass.: Harvard University Press.

Schulman, Kevin, Seils, D. M.; Timbie, J. W.; Sugarman, J.; Dame, L. A., et al. 2002. "A National Survey of Provisions in Clinical-Trial Agreements between Medical Schools and Industry Sponsors." *New England Journal of Medicine* 347: 1335–1341.

Schultz, Leslie. 2007. "What Does Society Get for the Billions Spent on Research?" Available at http://www.the-scientist.com/fragments/libraries/pdf/ros.pdf.

Schultz, Theodore W. 1979. "Distortions of Economic Research." *Minerva* (17): 460–468.

———. 1980. "The Productivity of Research: The Politics and Economics of Research." *Minerva* (18): 644–651.

———. 1987. "Are University Scholars and Scientists Free Agents?" *Minerva* (25): 349–356.

Schumpeter, Joseph A. 1976. *Capitalism Socialism and Democracy,* London: George Allen & Unwin.

Scotchmer, Suzanne. 2004. *Innovation and Incentives.* Cambridge, Mass.: MIT Press.

Scranton, Philip. 2006. "Technology, Science and American Innovation." *Business History* (48): 311–331.

Selingo, Jeffrey. 2003. "The Disappearing State in Public Higher Education." *Chronicle of Higher Education,* February 28.

Sell, Susan. 2003. *Private Power, Public Law.* New York: Cambridge University Press.

———. 2006. *Books, Drugs and Seeds.* Available at http://*media*.ffii.org/Tacd060320/proceedings/Susan%20**Sell**.doc (last visited September 2010).

Sent, Esther-Mirjam. 1998. *The Evolving Rationality of Rational Expectations.* New York: Cambridge University Press.

———. 1999. "Economics of Science: Survey and Suggestions." *Journal of Economic Methodology* 6: 95–124.

———. 2001. "Sent Simulating Simon Simulating Scientists." *Studies in the History and Philosophy of Science A* (32): 479–500.

———. 2002. "How (Not) to Influence People: The Contrary Tale of John Muth." *History of Political Economy* (34): 291–319.

———. 2005. "Behavioral Economics: How Psychology Found Its Way Back into Economics." *History of Political Economy* 36(4): 735–760.

———. 2006. "The Tricks of the (No-) Trade (Theorem)." In Mirowski and Hands 2006, 305–321.

Shah, Sonia. 2003. "Globalization of Clinical Research by the Pharmaceutical Industry." *International Journal of Health Services* 33: 29–36.

———. 2006. *The Body Hunters.* New York: New Press.

Shannon, Claude. 1948. "The Mathematical Theory of Communication." *Bell System Technical Journal* (27): 379–423, 623–656.

Shapin, Steven. 2003. "Ivory Trade." *London Review of Books,* September 11, pp. 15–19.

———. 2004. "Who Is the Industrial Scientist?" In Grandin et al. (2004).

———. 2008a. "I'm a Surfer." *London Review of Books,* March 20, pp. 5–8.

———. 2008b. *The Scientific Life: A Moral History of a Late Victorian Vocation.* Chicago: University of Chicago Press.

Shapiro, Carl. 2004. "Patent System Reform: Economic Analysis and Critique." *Berkeley Technology Law Journal,*

Shapiro, Carl; and Varian, Hal. 1999. *Information Rules.* Cambridge, Mass.: Harvard Business School Press.

Sharif, Naubahar. 2006. "Emergence and Development of the National Innovation Systems Concept." *Research Policy* (35): 745–766.

Sharma, B. L. (2003) "Indian Drug Institute Uses Computers to Reduce Animal Testing." *Associated Press Worldstream* (July 25). Available at www.siliconvalley.com/mld/ siliconvalley/6389346.htm (last visited March 18, 2005).

She Xinwei; Jiang, Zhaoshi; Clark, Royden A.; Liu, Ge; Cheng, Ze; Tuzun, Eray; Church, Deanna M.; Sutton, Granger; Halpern, Aaron L.; and Eichler, Evan E. 2004. "Shotgun Sequence Assembly and Recent Segmental Duplications within the Human Genome." *Nature* 431, 927–930.

Shelton, Robert. 2008. "Relations between National Research Investment and Publication Output: Application to an American Paradox." *Scientometrics* (74): 191–205.

Shelton, Robert; Foland, Patricia; and Gorelskyy, Roman. 2009. "Do New SCI Journals Have a Different National Bias?" *Scientometrics* (79): 351–363.

Shelton, Robert; and Holdridge, Geoffrey. 2004. "The US-EU Race for Leadership of Science and Technology." *Scientometrics* (60): 353–363.

Sherman, Brad. 1996. "Governing Science: Patents and Public Sector Research." In Michael Power, ed. *Accounting and Science.* New York: Cambridge University Press.

Sherwin, Chalmers; and Isenson, Raymond. 1967. "Project Hindsight." *Science,* June 23: 1571–1577.

Shi, Yanfei. 2001. *The Economics of Scientific Knowledge.* Cheltenham, U.K.: Elgar.

Shinn, Terry. 2002. "The Triple Helix and the New Production of Knowledge." *Social Studies of Science* (32:4): 599–614.

———. 2003. "Industry, Research and Education." In Mary Jo Nye, ed., *Cambridge History of Science,* vol. 5. New York: Cambridge University Press.

Shleifer, Andrei. 2009. "The Age of Milton Friedman." *Journal of Economic Literature* (47): 123–135.

Shreeve, James. 2004. *The Genome War.* New York: Knopf.

Shuchman, Miriam. 2007. "Commercializing Clinical Trials—Risks and Benefits of the CRO Boom." *New England Journal of Medicine,* October 4 (357): 1365–1368.

Shulman, Seth. 2006. *Undermining Science in the Bush Administration.* Berkeley: University of California Press.

Singer, Natasha. 2009a. "Trial Puts Spotlight on Merck." *New York Times,* May 14.

———. 2009b. "Medical Papers by Ghostwriters Pushed Therapy." *New York Times,* August 5.

———. 2009c. "Seeking a Shorter Path to New Drugs." *New York Times,* November 15.

Sismondo, Sergio. 2007. "Ghost Management: How Much of the Medical Literature Is Shaped Behind the Scenes by the Pharmaceutical Industry?" *PLoS Med* 4(9): e286.

———. 2008. "Pharmaceutical Company Funding and Its Consequences: A Qualitative Systematic Review." *Contemporary Clinical Trials* (29): 109–113.

———. 2009. "Ghosts in the Machine." *Social Studies of Science* (39): 171–198.

Skloot, Rebecca. 2010. *The Immortal life of Henrietta Lacks.* New York: Crown.

Slaughter, Sheila; Archerd, Cynthia; and Campbell, Teresa. 2004. "Boundaries and Quandaries: How Professors Negotiate Market Relations." *Review of Higher Education* (28:1): 129–165.

Slaughter, Sheila; Feldman, Maryann; and Thomas, Scott. 2009. "US Research Universities' Institutional Conflict of Interest Policies." *Journal of Empirical Research on Human Research Ethics,* 3–20.

Slaughter, Sheila; and Rhoades, Gary. 2002. "The Emergence of a Competitiveness R&D Policy Coalition and the Commercialization of Academic Science." In Mirowski and Sent 2002.

———. 2004. *Academic Capitalism and the New Economy.* Baltimore: Johns Hopkins University Press.

Smith, Helen; and K. Ho. 2006. "Measuring the Performance of Oxford University." *Research Policy* (35): 1554–1568.

Smith, John Kenly. 1990. "The Scientific Tradition in American Industrial Research." *Technology and Culture* (31): 121–131.

Smith, Richard. 2003. "Medical Journals and Pharmaceutical Companies: Uneasy Bedfellows." *British Medical Journal* (326): 1202–1205.

Smith, Vardaman. 1998. "Friedman, Liberalism, and the Meaning of Negative Freedom." *Economics and Philosophy* (14): 75–94.

Smolanczuk, R. 1999. "Production Mechanism of Superheavy Nuclei in Cold Fusion Reactions." *Physical Review C* (59): 2634.

Smolin, Lee. 2006. *The Trouble with Physics.* New York: Houghton Mifflin.

Snowdon, Brian; and Vane, Howard. 2005. *Modern Macroeconomics: Its Origins, Development and Current State.* Cheltenham, U.K.: Elgar.

Sokal, Alan. 2008. *Beyond the Hoax.* Oxford: Oxford University Press.

Sokal, Alan; and Bricmont, Jean. 1999. *Fashionable Nonsense*. London: Picador.

Solow, Robert. 1956. "A Contribution to the Theory of Economic Growth." *Quarterly Journal of Economics* (70): 65–94.

———. 1957. "Technical Change and the Aggregate Production Function." *Review of Economics and Statistics* (39): 312–320.

———. 1994. "Perspectives on Growth Theory." *Journal of Economic Perspectives* (8): 45–54.

———. 2007. "Heavy Thinker." *New Republic*, July.

Sommer, John, ed. 1995. *The Academy in Crisis*. New Brunswick, N.J.: Transaction Press.

Spence, A. M. 1974. "An Economist's View of Information." In C. Canadia, A. Luke and J. Harris, eds., *Annual Review of Information Science and Technology*, 57–78.

———. 2002. "Signaling in Retrospect and the Information Structure of Markets." *American Economic Review* (92): 434–459.

Starr, Paul. 2004. *The Creation of the Media*. New York: Basic Books.

Steedman, Ian. 2001. "On Measuring Knowledge in the New (Endogenous) Growth Theory." http://growthconf.ec.unipi.it/papers/Steedman1.pdf.

Steen, Kathryn. 2001. "Patents, Patriotism, and Skilled in the Art: U.S.A. vs. the Chemical Foundation." *Isis* 92: 91–122.

Stein, Donald. 2004. "Introduction." In *Buying In or Selling Out?* New Brunswick, N.J.: Rutgers University Press.

Steinzor, Rena; Wagner, Wendy; and Shudtz, Matthew. 2008. *Saving Science from Politics*. Available at www.progressivereform.org.

Stelfox, H.; Chua, G.; O'Rourke, K.; and Detsky, A. S. 1998. "Conflict of Interest in the Debate over Calcium-Channel Antagonists." *New England Journal of Medicine* 338: 101–106.

Stephan, Paula. 1996. "The Economics of Science." *Journal of Economic Literature* (34): 1199–1235.

Stephan, Paula; and Ehrenberg, Ronald, eds. 2007. *Science and the University*. Madison: University of Wisconsin Press.

Stephens, Joseph. 2000. "As Drug Testing Spreads, Profits and Lives Hang in Balance" (first of six-part series). *Washington Post*, December 17.

Stigler, George. 1961. "The Economics of Information." *Journal of Political Economy* (69): 213–225.

———. 1963. *The Intellectual and the Marketplace*. New York: Free Press.

———. 1982. *The Economist as Preacher*. Chicago: University of Chicago Press.

———. 1985. *Memoirs of an Unregulated Economist*. New York: Basic.

Stigler, George; and Becker, Gary. 1977. "De Gustibus non Est Dispudandum." *American Economic Review* (67): 76–90.

Stiglitz, Joseph. 1985. "Information and Economic Analysis." *Economic Journal—Conference Papers* (95): 21–41.

———. 1993. "Reflections on Economics." In Arnold Heertje, ed., *Makers of Modern Economics*, vol. 1. Hemel Hempstead: Harvester Wheatsheaf.

———. 1999. "Knowledge in the Modern Economy." In Romesh Vaitilingham, ed., *The Economics of the Knowledge Driven Economy*, 37–57. London: Department of Trade and Industry.

———. 2000. "The Contributions of the Theory of Information to 20th Century Economics." *Quarterly Journal of Economics* (140): 1441–1478.

———. 2002. "Information and the Change in Paradigm in Economics." *American Economic Review* (92): 460–501.

———. 2003. "Information and the Change in Paradigm in Economics." In R. Arnott, B. Greenwald, R. Kanbur and B. Nalebuff, eds., *Economics in an Imperfect World,* 569–639. Cambridge, Mass.: MIT Press.

———. 2008. "Falling Down." *New Republic,* September 10.

Stokes, Donald. 1997. *Pasteur's Quadrant.* Washington: Brookings Press.

Stone, T. Howard. 2003. "The Invisible Vulnerable: Economically and Educationally Disadvantaged Subjects of Clinical Research." *Currents in Contemporary Ethics* 31(1): 149–153.

Story, Louise. 2009. "A Rich Education for Summers." *New York Times,* April 6.

Streitz, Wendy; and Bennett, Alan. 2003. "Material Transfer Agreements: A University Perspective." *Plant Physiology* 133: 10–13.

Streitz, Wendy D.; de Bear, Isabelle; Calmettes, Caroline S.; and Reinhart, Fred. 2003. "Material Transfer Agreements: A Win-Win for Academia and Industry." Available at www.autm.net.

Stripling, Jack. 2009. "Cruel Irony." *Inside Higher Ed,* August 14.

Sunstein, Cass. 2006. *Infotopia: How Many Minds Produce Knowledge.* New York: Oxford University Press.

Surowiecki, James 2004. *The Wisdom of Crowds: Why the Many Are Smarter Than the Few and How Collective Wisdom Shapes Business, Economies, Societies and Nations.* Boston: Little, Brown.

Swann, John. 1988. *Academic Scientists and the Pharmaceutical Industry.* Baltimore: Johns Hopkins University Press.

Tassey, George. 2005. "Underinvestment in Public Good Technologies." *Journal of Technology Transfer* (30): 89–113.

Teske, Paul; and Johnson, Renee. 1994 "Moving towards an American Industrial Technology Policy." *Policy Studies Journal* 22: 296–311.

Testa, James. 2003. "The ISI Journal Selection Process." *Serials Review* (29): 210–212.

Thacker, Paul D. 2005. "In Search of Mr. Junk Science and His Influence." *SEJournal* 15 (2): 4, 14.

———. 2007. "Investigative Reporting Can Produce a Higher Obligation." *SEJournal* (17): 4, 24.

Thomsen, Esteban. 1992. *Prices and Knowledge.* New York: Routledge.

Thore, S. 1996. "Economies of Scale in the US Computer Industry." *Journal of Evolutionary Economics* (6): 199–216.

Thorpe, Charles. 2007. "Political Theory in Science and Technology Studies." In Hackett et al. 2007.

Thursby, Jerry; Jensen, Richard; and Thursby, Marie. 2001. "Objectives, Characteristics and Outcomes of University Licensing." *Journal of Technology Transfer* (26): 59–72.

Thursby, Jerry; and Thursby, Marie. 2003. "University Licensing and the Bayh-Dole Act." *Science* 301: 1052.

———. 2006a. *Here or There? A Survey of Factors in Multinational R&D Location.* Washington: National Academies Press.

———. 2006b. "Where Is the New Science in Corporate R&D?" *Science* (314): 1457–1458.

———. 2007a. "Are There Real Effects of Licensing on Academic Research?" *Journal of Economic Behavior and Organization* (63): 577–598.

———. 2007b. "Knowledge Creation and the Diffusion of Public Science with Intellectual Property Rights." In *Intellectual Property Rights and Technical Change,* Amsterdam: Elsevier.

Tijssen, Robert. 2004. "Is the Commercialization of Scientific Research Affecting the Production of Public Knowledge?" *Research Policy* 33: 709–733.

Tilney, N. L. 2003. "The Commercialization of Transplantation." *Transplantation Proceedings* 35: 1235–1237.

Tobey, Ronald. 1971. *The American Ideology of National Science, 1919–30.* Pittsburgh, Pa.: University of Pittsburgh Press.

Trembath, Jon. 2008. "KSR International vs. Teleflex" *Colorado Lawyer,* 37: 35–45.

Tremblay, Jean-François. 2008. "China's Pharma Leaps into Discovery." *Chemical and Engineering News,* February 4 (86:5): 11–15.

Tufts Center for the Study of Drug Development. 2006. "CROs Usage Associated with Faster Drug Development Speed at Comparable Quality." January/February Tufts CSDD Impact Report, Volume 8, No. 1.

Turner, Stephen. 2003. "The Third Science War." *Social Studies of Science* (33): 581–611.

Tyfield, David. 2006. "Neoliberalism and the Knowledge Economy." ESRC paper, University of Sussex.

———. 2009. "Privatizing Chinese Science: National Development vs. Neoliberal Financialization." Working Paper.

Uebel, Thomas. 2000. "Some Scientism, Some Historicism, Some Critics." In M. Stone and J. Wolff, eds., *The Proper Ambition of Science,* 151–173. London: Routledge.

Union of Concerned Scientists. 2004. "Scientific Integrity in Policy-Making." Available at www.ucsusa.org/scientific_integrity/interference.

———. 2007. *Smoke, Mirrors and Hot Air.* Available at http://www.ucsusa.org/assets/documents/global_warming/exxon_report.pdf.

———. 2008. *Federal Science and the Public Good.* Available at: http://www.ucsusa.org/assets/documents/scientific_integrity/Federal-Science-and-the-Public-Good-12-08-Update.pdf.

U.S. Federal Trade Commission. 2003. *To Promote Innovation: The Proper Balance between Competition and Patent Law.* Washington: USGPO.

United States Government Accountability Office (USGAO). 2004. *International Trade: Current Government Data Provide Limited Insight into Offshoring of Services.* GAO-04–932. Washington: USGPO.

———. 2007. *Project Bioshield.* GAO-08–88.

———. 2010. *For-Profit Colleges: Undercover Testing Finds Colleges Encouraged Fraud.* GAO-10–948T.

United States House of Representatives. Committee on Oversight and Government Reform. 2007. *Political Interference with Climate Change Science under the Bush Administration.* December.

Vaitilingham, Romesh, ed. 1999. *The Economics of the Knowledge Driven Economy.* London: Department of Trade and Industry.

Vallas, Steven; and Kleinman, Daniel. 2007. "Contradiction, Convergence and the Knowledge Economy: the confluence of academic and commercial biotechnology," *Socio-Economic Review* (2): 1–29.

Vanberg, Viktor. 2004. "The Rationality Postulate in Economics: Its Ambiguity, Its Deficiency and Its Evolutionary Alternative." *Journal of Economic Methodology,* II, S: 1–29.

———. 2008. "The Science as Market Analogy: A Constitutional Economics Perspective." Walter Eucken Institut discussion paper 08/1.

Van Dalen, Hendrik; and Klamer, Arjo. 2005. "Is Science a Case of Wasteful Competition?" *Kyklos* (58): 395–414.

VandeWall, Holly. 2007. "To the Victor Go the Stories: The Allison Commission." Unpublished Notre Dame discussion paper.

Van Horn, Rob; and Klaes, Matthias. Forthcoming. "Chicago and Intellectual Property: Barbarians ante portas." In Rob van Horn, Philip Mirowski, and Tom Stapleford, eds., *Building Chicago Economics.* New York: Cambridge University Press.

Van Noorden, Richard. 2010. "A Profusion of Measures." *Nature,* June 17 (465): 864–866.

Van Overtveldt, Johann. 2007. *The Chicago School.* Chicago: Agate.

Varian, Hal. 2002. "A New Economy with No New Economics." *New York Times,* Web version, January 17.

Veblen, Thorstein. 1918. *The Higher Learning in America.* New York: Heubsch.

Venter, Craig, et al. 2001. "The Sequence of the Human Genome." *Science* (291): 1304–1351.

Verspagen, Bart. 2006. "University Research, Intellectual Property Rights and European Innovation Systems." *Journal of Economic Surveys* (20): 607–632.

Verspagen, Bart; and Claudia Werker. 2003. "The Invisible College of the Economics of Innovation and Technical Change." Eindhoven Center Working Paper 03.21.

Vincent-Lancrin, Stephan. 2006. "What Is Changing in Academic Research?" *European Journal of Education* (41).

Vogeli, Christine; Yucel, Recal; Bendavid, Eran; Jones, Lisa; Anderson, Melissa; Louis, Karen; and Campbell, Eric. 2006. "Data Withholding and the Next Generation of Scientists: Results of a National Survey." *Academic Medicine* (81:2): 128–136.

Von Hoffman, Nicholas. 2006. "Trump U." *The Nation,* July 21, www.thenation.com/doc/20060731/trump_u.

Wagner, Wendy; and Steinzor, Rena, eds. 2006. *Rescuing Science from Politics.* New York: Oxford University Press.

Waldrop, Mitchell. 2008. "Science 2.0—Is Open Access Science the Future?" *Scientific American,* April.

Wallace, David Foster. 2004. *Oblivion.* New York: Back Bay Books.

Wallace-Wells, Benjamin. 2008. "Surfing the Universe." *New Yorker*, July 21.
Walpen, Bernard. 2004. *Die Offenen Feinde und Ihre Gesellschaft*. Hamburg: VSA.
Walsh, John; Arora, A.; and Cohen, Wesley. 2003a. "Effects of Research Tool Patents and Licensing on Biomedical Innovation." In W. Cohen and S. Merrill, eds., *Patents in the Knowledge-Based Economy*, 285–340. Washington: National Academies Press.
———. 2003b. "Working through the Patent Problem." *Science* (309): 2002–2003.
Walsh, John; Cho, Charlene; and Cohen, Wesley. 2005a. "View from the Bench: Patents and Material Transfers." *Science* (309): 2002–3.
———. 2005b. Supporting online materials for Walsh, Cho and Cohen, 2005a. Available at www.sciencemag.org/cgi/content/full/309/5743/2002/DC1.
———. 2005c. "Patents, Material Transfers and Access to Research Inputs in Biomedical Research." Final report to the National Academy of Sciences' Committee on Intellectual Property Rights in Genomic and Protein-Related Inventions, September 20, 2005.
———. 2007. "Where Excludability Matters: Material v. Intellectual Property in Academic Biomedical Research." Georgia Tech Working Paper #20.
Warsh, David. 2006. *Knowledge and the Wealth of Nations*. New York: Norton.
———. 2007. "A Nation Once Again?" http://www.economicprincipals.com/issues/2007.12.02/275.html.
———. 2008. "A Normal Professor." http://www.economicprincipals.com/issues/2008.06.01/320.html.
———. 2009a. "Up or Out?" http://www.economicprincipals.com/issues/2009.06.14/470.html.
———. 2009b. "Sound Familiar?" http://www.economicprincipals.com/issues/2009.11.15/797.html.
Warwick, Andrew. 2003. *Masters of Theory*. Chicago: University of Chicago Press.
Washburn, Jennifer. 2005. *University, Inc.* New York: Basic Books.
———. 2010. *Big Oil Goes to College*. Washington, D.C.: Center for American Progress.
Waterston, Robert; Lander, Eric; and Sulston, John. 2002. "On the Sequencing of the Human Genome." *Proceedings of the National Academy of Sciences* (99:6): 3712–3716.
———. 2003. "More on the Sequencing of the Human Genome." *Proceedings of the National Academy of Sciences*, March 18 (100): 3022–3024.
Watt, Richard. 2000. *Copyright and Economic Theory: Friends or Enemies?* Cheltenham, U.K.: Elgar.
Wazana, Ashley. 2000. "Physicians and the Pharmaceutical Industry." *Journal of the American Medical Association* (283): 373.
Weart, Spencer. 1979. "The Physics Business in America, 1919–40." In Nathan Reingold, ed., *The Sciences in the American Context*, 295–358. Washington, D.C.: Smithsonian Institution Press.
Weifang, Min; and Ding Xiaohao. 2005. "Financing Chinese Higher Education." *Harvard China Review*, Spring: 127–131.
Wellerstein, Alex. 2008. "Patenting the Bomb." *Isis* (99): 57–87.
Werskey, Gary. 1978. *The Visible College*. London: Allen Lane.

———. 2007. "The Radical Critique of Capitalist Science: A History in Three Movements." Unpublished manuscript.

Wessner, Charles. 2002. *Government-Industry Partnerships.* Washington: National Academies Press.

Westwick, Peter. 2003. *The National Labs: Science in an American System, 1947–74.* Cambridge: Harvard University Press.

Wible, James. 1998. *The Economics of Science.* London: Routledge.

Wilde, Oscar. 1898. *The Ballad of Reading Gaol.* London: Smithers.

Wilson, Duff; and Heath, David. 2001. "The Hutch Zealously Guards Its Secrets." *Seattle Times,* http://seattletimes.nwsource.com/uninformed_consent (last visited March 18, 2005).

Winner, Langdon. 2003. Testimony to the Committee on Science of the U.S. House of Representatives, 9 April.

Wise, George. 1980. "A New Role for Professional Scientists in Industry." *Technology and Culture* 21: 408–429.

Wise, M. Norton. 2006. "Thoughts on the Politicization of Science through Commercialization." *Social Research* (73): 1253–1272.

Woit, Peter. 2006. *Not Even Wrong: The Failure of String Theory.* New York: Basic.

Wong, Stanley. 1978. *The Foundations of Paul Samuelson's Revealed Preference Theory.* Boston: Routledge and Kegan Paul.

Woolgar, Steve. 2004. "Marketing Ideas." *Economy and Society* (33): 448–462.

Wright, Gavin. 1997. "Toward a More Historical Approach to Technological Change." *Economic Journal* (107): 1560–1566.

———. 1999. "Can a Nation Learn? American Technology as a Network Phenomenon." In Naomi Lamoreaux, M. Daniel, G. Raff, and Peter Temin, eds., *Learning by Doing in Markets, Firms and Countries*, 295–326. Chicago: University of Chicago Press.

Wright, Gavin; and Nelson, Richard. 1992. "The Rise and Fall of American Technological Leadership." *Journal of Economic Literature* (30): 1931–1964.

Wright, Susan. 1994. *Molecular Politics.* Chicago: University of Chicago Press.

Wysocki, Bernard. 2004. "Columbia's Pursuit of Patent Riches Angers Companies." *Wall Street Journal,* December 21, p. A1.

Yank, Veronica; and Rennie, Drummond. 1999. "Disclosure of Researcher Contributions." *Annals of Internal Medicine* 130: 661–670.

Yonezawa, Akiyosi. 2003. "The Impact of Globalization on Higher Education Governance in Japan." *Higher Education Research & Development* (22:2): 145–154.

Zachary, G. Pascal. 1997. *Endless Frontier.* New York: Free Press.

———. 2007. "Bell Labs Is Gone, Academia Steps In." *New York Times,* December 16.

Zamora Bonilla, Jesus. 2008. "The Economics of Scientific Knowledge." In Uskali Maki, ed., *Handbook of the Philosophy of Science: Philosophy of Economics.* Amsterdam: Elsevier.

Zappia, Carlo. 1996. "The Notion of Private Information in Modern Perspective." *European Journal of the History of Economic Thought* (3): 107–131.

Zarin, Deborah; and Tee, Tony. 2008. "Moving towards the Transparency of Clinical Trials." *Science* (319): 1340–1342.
Ziman, John. 1994. *Prometheus Bound.* Cambridge: Cambridge University Press.
Zittrain, Jonathan. 2008. *The Future of the Internet.* New Haven: Yale University Press.
Zuckerman, Diana. 2002. "Hype in Health Reporting: Checkbook Science Buys Distortion of Medical News." *Extra!* (September/October).

Acknowledgments

I would like to thank the following for taking time to correspond and otherwise speculate about issues covered in this volume: Elizabeth Popp Berman, Larry Busch, John Davis, Duncan Foley, Nicolas Guilhot, Diana Hicks, John O'Neill, Dieter Plehwe, Andres Ruis, Miriam Shuchman, David Tyfield, Wade Hands, Paul Nightingale, Steven Lukes, Rebecca Lave, Yves Gingras, and Don Howard. I also am grateful to the technology transfer officers at many universities too numerous to mention who took time out from their busy schedules to talk to an academic asking naïve and inconvenient questions. I especially want to thank Esther-Mirjam Sent, who collaborated with me on this research on the economics of science, until it was curtailed by her move to Radboud University in Holland.

I am grateful to All Souls College, Oxford University, for granting me a visiting fellowship in 2008 to make headway on this manuscript, and also to the International Center for Advanced Studies Fellowship, New York University, 2004–2005, for giving me the opportunity to start the composition of this book in Greenwich Village. There are few more lively places on earth to distract one from the travails of authorship. The experience completely reoriented my research trajectory. Unfortunately, NYU abruptly closed the Center soon thereafter, because it would nowhere garner them the revenue that opening a new campus at Abu Dhabi would do. Fostering local research through sustained scholarly interaction apparently constitutes an inefficient use of resources by the contemporary university. Welcome to the regime of globalized privatization. A stint as Fulbright lecturer at the Universidad de la República in Montevideo, Uruguay, brought home to me just to what extent the neoliberal movement to privatize university research has indeed become a global phenomenon.

Audiences at the following conclaves were generous in their criticism: a plenary session on the economics of science at the History of Science Meetings in Austin, Texas; the Gavin Wright Fest at Stanford University; INEM meetings in Madrid, Spain; the Science for Sale conference at Cornell University;

the All Souls Oxford visiting scholars series; the Smithsonian Institution; and the Harvard University political economy seminar.

The following universities and organizations were open minded enough to invite me to give public lectures on the book: the Heilbroner Lecture at New School for Social Research; the Brown and Haley endowed lectures at the University of Puget Sound; the Nicholas Mullins Lecture at Virginia Polytechnic Institute; the SSRC in New York; the University of Oslo Science Studies Unit; TINT at the University of Helsinki; and Wayne State University.

I just need to make clear this book and its title have nothing whatsoever to do with the actual business firm of SciMart, Inc., P.O. Box 959059 St. Louis, Mo. 63195–9059. (It is impossible to make this stuff up.) Or any other ScienceMart, for that matter.

Some portions of the book were assisted by ancillary funding from the following sources: the Seng Foundation, All Souls College Oxford, and the Society for Social Studies of Science.

Previous versions of portions of the chapters of this volume have appeared in the following forms:

A portion of Chapter 2 appeared in the *Journal of the History of Economic Thought* 29 (2007): 481–494.

An earlier incarnation of Chapter 3 appeared as Mirowski and Sent 2007.

A portion of Chaper 4 appeared in *Minerva* 46 (2008): 317–342.

A much earlier incarnation of Chapter 5 appeared as Mirowski and van Horn 2005.

I wish to thank my coauthors for their gracious permission to recycle this joint work.

Index